8th International Biochemistry of Exercise Conference,
Nagoya, September 24–28, 1991

Integration of Medical and Sports Sciences

Volume Editors
Y. Sato, Nagoya
J. Poortmans, Brussels
I. Hashimoto, Tokyo
Y. Oshida, Nagoya

140 figures, 1 color plate and 54 tables, 1992

KARGER

Basel · Freiburg · Paris · London · New York · New Delhi · Bangkok · Singapore · Tokyo · Sydney

Medicine and Sport Science

Published on behalf of the
International Council of Sport Science and Physical Education

Library of Congress Cataloging-in-Publication Data
International Biochemistry of Exercise Conference (8th: 1991: Nagoya-shi, Japan)
Integration of medical and sports sciences: proceedings of the 8th International Biochemistry of Exercise Conference, Nagoya, September 24–28, 1991 / volume editors, Y. Sato... [et al.].
(Medicine and sport science; vol. 37)
"Published on behalf of the International Council of Sport Science and Physical Education"
Includes bibliographical references and index.
1. Exercise – – Physiological effect – – Congresses. 2. Sports – – Physiological aspects – – Congresses. I. Satō, Yūzō, 1940– II. International Council of Sport and Physical Education. III. Title. IV. Series.
[DNLM: 1. Exercise – – physiology – – congresses. 2. Muscles – – metabolism – – congresses.]
ISBN 3–8055–5579–2 (alk. paper)

Drug Dosage
The authors and the publisher have exerted every effort to ensure that drug selection and dosage set forth in this text are in accord with current recommendations and practice at the time of publication. However, in view of ongoing research, changes in government regulations, and the constant flow of information relating to drug therapy and drug reactions, the reader is urged to check the package insert for each drug for any change in indications and dosage and for added warnings and precautions. This is particularly important when the recommended agent is a new and/or infrequently employed drug.

Printed in Switzerland on acid-free paper by Thür AG Offsetdruck, Pratteln
ISBN 3–8055–5579–2

Contents

Acknowledgements IX
Preface XI
Foreword XIII

Ebashi, S. (Okazaki): Muscle Research: A Cornerstone of Life Science. Opening Lecture 1
Poortmans, J.R. (Bruxelles): From Frog to Man. Farewell Lecture 8
Shimazu, T. (Ehime): The Hypothalamus and Neural Feed-Forward Regulation of Exercise Metabolism 20
Klarlund Pedersen, B. (Copenhagen): Exercise and Immunity – Mechanisms of Action 33
Yagi, K. (Gifu): Lipid Peroxides and Exercise 40

Molecular Mechanisms

Roy, R.R.; Hodgson, J.A.; Chalmers, G.R.; Buxton, W.; Edgerton, V.R. (Los Angeles): Responsiveness of the Cat Plantaris to Functional Overload 43
Totsuka, T.; Watanabe, K.; Uramoto, I.; Nagahama, M. (Kasugai); Yoshida, T. (Mie); Mizutani, T. (Nagoya): Muscular Dystrophic Mice: Apparently Hypertrophied Muscles and Bone-Muscle Growth Imbalance Hypothesis 52
Yamada, S. (Tokyo); Kimura, H. (Shiga); Fujimaki, A. (Tokyo); Strohman, R. (Berkeley, Calif.): Expression of Fibroblast Growth Factors in Exercise-Induced Muscle Hypertrophy with Special Reference to the Role of Muscle Satellite Cells 67
Oka, Y.; Ishihara, H.; Asano, T. (Tokyo): Expression of Glucose Transporter Isoforms in Atrophied and Enlarged Muscle 84

Ageing and Exercise

Holloszy, J.O.; Spina, R.J.; Kohrt, W.M. (St. Louis, Mo.): Health Benefits of Exercise in the Elderly 91
Taylor, A.W.; Noble, E.G.; Cunningham, D.A.; Paterson, D.H.; Rechnitzer, P. (London, Ont.): Ageing, Skeletal Muscle Contractile Properties and Enzyme Activities with Exercise 109

Higuchi, M. (Tokyo); Tamai, T. (Fukui); Kobayashi, S. (Tokyo); Nakai, T. (Fukui): Plasma Lipoprotein and Apolipoprotein Profiles in Aged Japanese Athletes . 126
Yamanouchi, K.; Chikada, K.; Kato, K. (Aichi); Oshida, Y.; Sato, Y. (Nagoya): Effect of Habitual Physical Activity on Glucose Tolerance and Peripheral Insulin Action in the Elderly 137

Muscle Atrophy

Booth, F.W. (Houston, Tex.); Linderman, J.K. (Moffett Field, Calif.); Kirby, C.R. (Houston, Tex.): Molecular Mechanisms of Muscle Disuse Atrophy (and Strategies of Prevention) 142
Yoshioka, T.; Takekura, H.; Yamashita, K. (Kanagawa): Effect of Endurance Training on Disuse Muscle Atrophy Induced by Body Suspension in Rats. A Structural and Biochemical Study 150
Ianuzzo, C.D.; Li, B.; Hamilton, N.; D'Costa, L.; Ianuzzo, S.E.; Barrozo, C.A.M.; Salerno, T.A.; Laughlin, M.H. (Toronto, Ont./Columbia, Mo.): Association of Cardiac Ventricular Myosin Isoforms with Hemodynamic Factors 162
Atomi, Y. (Tokyo): Decreased αB-Crystallin in Soleus Muscle Atrophy and Role of αB-Crystallin in Muscle 171

Exercise and Metabolic Disorders

Sato, Y.; Oshida, Y.; Ohsawa, I.; Sato, J. (Nagoya); Yamanouchi, K. (Aichi): Biochemical Determination of Training Effects Using Insulin Clamp and Microdialysis Techniques 193
Goodyear, L.J. (Boston, Mass.); Hirshman, M.F.; Horton, E.S. (Burlington, Vt.): The Glucose Transport System in Skeletal Muscle: Effects of Exercise and Insulin 201
Kawamori, R.; Kubota, M.; Ikeda, M.; Matsuhisa, M.; Kubota, M.; Morishima, T.; Kamada, T. (Osaka): Acute Metabolic Effects of Exercise on Glucose Fluxes in Splanchnic and Peripheral Tissues in Diabetics, Determined with an Innovative Approach 216
Galbo, H.; Linstow, M.v.; Dela, F.; Kjaer, M.; Mikines, K. (Copenhagen): Exercise and Diabetes 227
Yamashita, H.; Sato, N.; Yamamoto, M. (Tokorozawa); Gasa, S. (Sapporo); Nagasawa, J.; Sato, Y. (Nagoya); Habara, Y. (Okazaki); Ishikawa, M. (Asahikawa); Segawa, M.; Ohno, H. (Tokorozawa): Is There an Intimate Interplay between Temperature Acclimation and Exogenous Insulin? With Special Reference to the Participation of Brown Adipose Tissue 237

Regulation of Substrate Utilization in Muscle during Exercise

Wahren, J.; Katz, A. (Stockholm): Glucose Metabolism in Exercising Man 243
Richter, E.A. (Copenhagen): Interaction of Fuels in Muscle Metabolism during Exercise 252
Henriksson, J.; Hickner, R.C.; Rosdahl, H.; Fuchi, T.; Oshida, Y.; Jorfeldt, L. (Stockholm): Current Trends in Microdialysis, with Focus on the in vivo Study of Skeletal Muscle Glucose Metabolism 262

Gulve, E.A. (St. Louis, Mo.): Effects of Acute and Chronic Exercise on Insulin-Stimulated Glucose Transport Activity in Skeletal Muscle 273
Parry-Billings, M.; Newsholme, E.A. (Oxford): The Overtraining Syndrome: Some Biochemical Aspects 281

Regulation of Muscle Protein Expression with Exercise

Ullmer, S.; Mader, A. (Köln): A Mathematical Model of Regulation of Protein Synthesis by Activation Feedback: Some Reflections on Its Possibilities and Limits in Describing Muscle Mass Adaptations with Exercise 288
Essig, D.A. (Chicago, Ill.): Exercise Induction of 5′-Aminolevulinate Synthase: A Mitochondrial Enzyme in the Heme Biosynthetic Pathway 299
Wada, M.; Kikuchi, K. (Hiroshima-ken); Katsuta, S. (Ibaraki-ken): Changes in Myosin Heavy-Chain and Light-Chain Isoforms following Sustained Exercise . . 309
Ohira, Y.; Wakatsuki, T.; Inoue, N. (Kanoya City); Nakamura, K.; Asakura, T.; Ikeda, K.; Tomiyoshi, T.; Nakajoh, M. (Kagoshima City): Non-Exercise-Related Stimulation of Mitochondrial Protein Synthesis in Creatine-Depleted Rats 318
Kagawa, Y.; Ohta, S. (Tochigi-ken): Mitochondrial Myopathy and Transcriptional Control 324

Dietary Necessities and Exercise
Electrolytes, Proteins, Carbohydrates, Vitamins

Bülow, J.; Simonsen, L.; Madsen, J. (Copenhagen): Effects of Exercise and Glucose Ingestion on Adipose Tissue Metabolism. A Microdialysis Study 329
Chen, J.D.; Wang, J.F.; Wang, S.W.; Li, K.J.; Chen, Z.M. (Beijing): Recommended Dietary Allowances for Chinese Athletes. Suggestions and Illustrations . . . 336
Shimomura, Y.; Kotsuka, H.; Saitoh, S.; Suzuki, M. (Ibaraki): Effects of Exercise and Nutrition on Branched-Chain Amino Acid Metabolism. Activation of Branched-Chain Alpha-Keto Acid Dehydrogenase Complex by Exercise and Effect of High-Fat Diet Intake on the Activation of the Enzyme Complex . . 342
Terblanche, S.E. (KwaDlangezwa); Gohil, K.; Lang, J.K.; Packer, L. (Berkeley, Calif.): Sustained Exercise Endurance Capacity after Depletion of Liver Glycogen Levels 349
Coyle, E.F. (Austin, Tex.): Carbohydrate Supplementation during Exercise 356

Cardiac, Respiratory and Neurohumoral Responses during Exercise

Yoshida, T.; Watari, H. (Aichi): Noninvasive and Continuous Determination of Energy Metabolism during Muscular Contraction and Recovery 364
Itoh, H. (Tokyo): Oxygen Uptake: Work Rate Relationship in Patients with Heart Disease 374
Makiguchi, K. (Ibaraki): Characteristics of Cardiorespiratory Responses during Exercise in Patients with Chronic Airflow Obstruction 381

Yoshioka, T.; Hashizume, T.; Fuji, T.; Okano, Y.; Nakanishi, N.; Haze, K.; Shimomura, K. (Osaka): Cardiopulmonary Responses and Blood Flow Distribution during Exercise in Patients with Left Ventricular Dysfunction 396

Biochemical Effects of Exercise on the Prevention of Coronary Artery Diseases and Health Promotion

Krotkiewski, M. (Gothenburg): Possible Relationship between Muscle Morphology and Capillarisation and the Risk Factor for Development of Cardiovascular Diseases 405

Hashimoto, I.; Shimomitsu, T.; Katsumura, T.; Iwane, H. (Tokyo): Exercise, Diet and Prostaglandins 416

Hamaoka, T. (Tokyo); Albani, C.; Chance, B. (Philadelphia, Pa.); Iwane, H. (Tokyo): A New Method for the Evaluation of Muscle Aerobic Capacity in Relation to Physical Activity Measured by Near-Infrared Spectroscopy 421

Tamai, T. (Fukui); Higuchi, M. (Tokyo); Oida, K.; Nakai, T., Miyabo, S. (Fukui); Kobayashi, S. (Tokyo): Effect of Exercise on Plasma Lipoprotein Metabolism 430

Subject Index 439

Acknowledgements

Sincere gratitude is expressed to the following sponsors:

Japanese Ministry of Education, Science and Culture
Japanese Ministry of Health and Welfare
Aichi Prefecture
Nagoya City
Japan Medical Association
Aichi Medical Association
Nagoya Medical Association
Japan Diabetic Society
Japanese Society of Physical Fitness and Sports Medicine
Japanese Society of Physical Education
Nagoya University Foundation
Daiko Foundation
Aichi Health Promotion Foundation
Uehara Memorial Foundation
Meiji Life Foundation of Health and Welfare
Suzuken Memorial Foundation
Nagoya Convention Bureau

The organizing committee also wishes to acknowledge the contributions of the following sponsor companies and hospitals:

Chubu Electric Power Co., Ltd.
Combi Co., Ltd.
Sugiyama Kogyo Co., Ltd.
Toyota Motor Co., Ltd.
Tsumura & Co.
Nagoya Railroad Co., Ltd.
Boehringer-Mannheim Toho Co., Ltd.
Ihara Electric Industries Co., Ltd.
Novo Nordisk Pharma Co., Ltd.
Ohtsuka Pharmaceutical Co., Ltd.
Shinoda Naika Clinic
The Tokai Bank, Ltd.
Toho Gas Co., Ltd.
Dainippon Pharmaceutical Co., Ltd.
Hoechst Japan Co., Ltd.
Sankyo Pharmaceutical Co., Ltd.
Takeda Chemical Industries Co., Ltd.
Toenec Co., Ltd.
Yagami Co., Ltd.
Yamanouchi Pharmaceutical Co., Ltd.
All Japan Barley Processors Association
Daiichi Pharmaceutical Co., Ltd.
Eisai Pharmaceutical Co., Ltd.
Fujisawa Pharmaceutical Co., Ltd.
Honzo Pharmaceutical Co., Ltd.
Kawamoto Pump Mfg. Co., Ltd.
Kobayashi Pharmaceutical Co., Ltd.
Kowa Pharmaceutical Co., Ltd.
Matsuzakaya Co., Ltd.
Miles-Sankyo Co., Ltd.

Mizuno Co., Ltd.
Mochida Pharmaceutical Co., Ltd.
NGK Insulators Co., Ltd.
Sanwa Pharmaceutical Co., Ltd.
Shionogi Pharmaceutical Co., Ltd.
Tanabe Pharmaceutical Co., Ltd.
Toyota Gosei Co., Ltd.
Sumitomo Pharmaceutical
Iwatani International Co., Ltd.
Tokyu Shachibus Co., Ltd.
Kitamura Hospital
Aihoku Hospital
Asahi Rosai Hospital
Goodman Co., Ltd.
Mr. Shigeru Hayashi
Imamura Naika Clinic
Kainan Hospital
Kumasaka Naika Clinic
Meiji Seika Co., Ltd.
Noritake Co., Ltd.
Inazawa City Hospital
Iwaki Pharmaceutical Co., Ltd.
Prof. Hideki Ohno, National Defense Medical College

The conference could not have been successful without the medical, scientific, technical, clerical and secretarial assistance of the following:

Miss Yoshiko Hattori
Mrs. Yasuko Uno
Dr. Isao Ohsawa
Dr. Juichi Sato
Research Center of Health, Physical Fitness and Sports, Nagoya University
Dr. Kiwami Chikada
Dr. Takashi Shinozaki
Dr. Katsunori Ito
Dr. Toshihiko Nishikawa
Dr. Shoji Shimizu
First Department of Internal Medicine, Aichi Medical University

Preface

This volume represents the proceedings of the eighth in a series of international meetings pertaining to the biochemistry of exercise, organized by the Research Group on Biochemistry of Exercise of the International Council of Sport Science and Physical Education, which is sponsored by UNESCO. The objectives of the research group are to promote a better understanding of the basic biologic and biochemical mechanisms in physical exercise and to furnish information of practical value to clinicians, scientists and educators who have much interest in the role of exercise in health maintenance.

All of the previous seven meetings were held either in Europe or North America. The 1991 conference was held in Nagoya, Japan, September 24 to 28. It is an honor that we were given the privilege of hosting this meeting for the first time in the Far-East. The main theme of the conference 'The Integration of Medical and Sports Sciences' stands for the spirit and expectations of all the members of the organizing committee. The scientific programme consisted of six invited lectures, one award lecture, nine symposia, oral and poster sessions. These proceedings include the lectures and symposia. Although the content is only a contemporary reflection, we hope this information will stimulate and be useful in your work in the field of biochemistry of exercise.

On behalf of the Organizing Committee of the International Biochemistry of Exercise Conference (IBEC), 1991.

Yuzo Sato
Chairman of IBEC
Nagoya, Japan

Foreword

It is a great pleasure for the Research Group and for me to have this Conference held in Japan. I would like to thank, particularly, Prof. Sato and his local team for the excellent work they have accomplished to ensure a high-level meeting.

The previous 7 Symposia were held in Western Countries with increasing interest due to the development of analytical tools. In 1979, 3 Japanese colleagues attended the Brussels Symposium. Since then, the number has increased steadily and there are obvious reasons to believe that we, 'the gaijin' or the honorable foreigners, will be in the minority this time. Three years ago, the Research Group on Biochemistry of Exercise responded to the proposal to hold a Conference in the Far-East. Japan emerged as a good choice particularly in view of the growing number of Japanese papers appearing in outstanding scientific journals.

Japan's island position, far removed from the crossroads of Asia, created a particular cultural situation which started in 660 BC with Jimmu Tenno, the legendary founder of the actual dynasty. Since then, adaptations, modifications and improvements of local societies were combined with influences from China and Korea.

Thirty years ago, Albert Szent-Gyorgyi presented a paper 'On Scientific Creativity'. He noted that history is dominated by two types of minds. The history of human suffering is dotted with names of men of willpower and ambition, who make war, build realms which then collapse, leaving behind only destruction and misery. Already two thousand years ago, Julius Caesar said, at least in Shakespeare's words:

> Men at some time are masters of their fates:
> The fault, dear Brutus, is not in our stars,
> But in ourselves, that we are underlings, ...

The history of human progress is the story of a relatively small number of creative people, creative in art, sciences, or any other human endeavor. Nowadays, modern Japanese contributions to the biological sciences are outstanding. Sport and exercise could lead people towards brotherhood instead of hatred, sharing instead of selfishness, mental health instead of madness. The general topic of this Conference, 'The Integration of Medical and Sports Sciences', goes along the same lines in order to improve our knowledge of the benefits and limitations of exercise. On behalf of the Research Group, I wish us all a stimulating conference on both the scientific and social aspects.

I would like to say a few words about our late Friend, Phil Gollnick who passed away this summer 1991. His unexpected death came as an appalling shock to us all. I met him for the first time in 1966, in Madison (USA), where I attended my first Annual Meeting of the American College of Sports Medicine. The topic of his talk was 'Energy Production and Lactic Acid Formation'. I was already impressed by his knowledge, by the forcefulness of his arguments, with a deep sense of righteousness towards other people. He became involved in exercise biochemistry very early in his life, publishing a paper in 1961 on adenosine triphosphatase activity in skeletal and heart muscle of rats, together with George Hearn. With Bengt Saltin and Charles Tipton, he attended the First Symposium on Biochemistry of Exercise that we organized in Brussels in 1968. Being a hard worker, probably too hard a worker, he built his own Research Laboratory at the School of Physical Education of Washington State University. Together with several scientists from Japan, he published numerous papers and reviews related to the physiological chemistry of exercise. Each of us owes intellectual and emotional debts to colleagues, teachers, and students. Phil Gollnick was among those who helped shape our thinking. This is why the Proceedings of the Nagoya Conference are dedicated to him to pay respect to the memory of an unforgettable friend and scientist.

Jacques R. Poortmans
Chairman of the Research Group

Sato Y, Poortmans J, Hashimoto I, Oshida Y (eds): Integration of Medical and Sports Sciences. Med Sport Sci. Basel, Karger, 1992, vol 37, pp 1–7

Muscle Research: A Cornerstone of Life Science

Opening Lecture

Setsuro Ebashi

Okazaki National Research Institutes, Okazaki, Japan

Muscle contraction was perhaps the first scientific subject addressed by the human mind in the long history of biological and medical sciences. It was quite reasonable that Galenus (129–199), the person who first introduced the 'experiment' into studies of Nature and perhaps the first to deserve the title 'scientist' (it was generally believed by ancient people that true knowledge could be acquired only through the meditation of wise men), had a keen interest in muscle contraction.

We can enumerate a number of biological and medical fields that were pioneered or explored by muscle research. In the following, some examples are briefly described.

Bioelectricity

Modern physiology was initiated by the unexpected discovery of bioelectricity late in the 18th century. For Galvani [1], the discoverer of the phenomenon, bioelectricity was inseparably linked to muscle contraction. His followers, however, disregarded muscle and focused their attention only on nervous tissues. Thereafter, until the middle of the 20th century, many distinguished physiologists dedicated their entire scientific career to studies on the action potential of nerves, perhaps in the tacit belief that the secret of our very mind itself might be hidden in this phenomenon.

At the beginning of the 20th century, muscle research again led to important discoveries in electrophysiology. Bernstein [2] showed that the resting potential of muscle was dependent on the K^+ gradient between the outside and inside of the cell. Overton [3] found that Na^+ was essential for the action potential. Unfortunately, he could not confirm this with nerve fibers and the establishment of the Na^+ concept was postponed almost half a century until the famous work of Hodgkin and Huxley [4], which clearly and elegantly eliminated the above-mentioned mysticism ascribed to nerve electricity.

Chemical Transmission

One of the most important principles of neurophysiology is that whithin the nerve fiber the signal is propagated electrically and that the junctional transfer of the signal is mediated by chemical substances (except in a few cases of some invertebrates where the signal is transmitted electrically); however, in the brain, there is a unique structure, called the gap junction, which enables a nerve cell to influence adjacent cells electrically, thus modifying their electrical responses.

In tissues under the control of the autonomic nervous system, the chemical theory was accepted without serious objections, but in skeletal muscle a considerable number of physiologists did not admit the chemical nature of transmission even after Dale was awarded the Nobel prize in 1936 for his contribution to the chemical concept of this neuromuscular transmission. This opposition lasted until the superb work of Fatt and Katz in 1951 [5], which finally eliminated the possibility of electrical transmission at this junction.

Now we know that the mode of signal transmission from the nerve to its target cell is essentially the same throughout the body irrespective of the kind of cell. The mechanism of the neuromuscular junction in skeletal muscle is always the model for research on other junctions including non-muscle tissues.

Glycogenolysis and ATP

Although fermentation, or glycolysis, was first found in yeast, its intricate mechanism was clarified by studies of vertebrate fast skeletal muscle. This is quite reasonable because glycolysis, or glycogenolysis, is the main source of energy for contraction of this muscle and is subtly linked with excitation through Ca^{2+} (see section on Muscle as the Origin of the Ca^{2+} Era).

It was an anecdote of history that Meyerhof at one time claimed that lactic acid, an intermediate of glycolysis, must be involved in the fundamental mechanism of muscle contraction. Despite this digression, however, the effort to inquire into the glycolytic pathway eventually resulted in the discovery of ATP. This important finding was made by Lohmann [6] in Meyerhof's laboratory, and simultaneously by Fiske and SubbaRow [7] at Harvard. (Fiske, who first discovered creatine phosphate and determined its chemical structure, was a biochemical genius but has not been granted the appreciation he deserves in the history of biochemistry.)

ATP is, of course, the common currency of all metabolic systems, but its true value was realized only when its role in muscle contraction was revealed [8] (cf. next section).

Actomyosin-ATP System as Chemomechanical Transducer

'Molecular biology' is now almost a synonym for 'molecular genetics', but this term was that originally used to explain biological phenomena at the molecular level. If we adhere to this definition, an important feature of molecular biology should be one which reproduces a biological function from the molecules involved in that function. In this sense the first brilliant success in molecular biology was achieved by Szent-Györgyi in 1942 [9]; he demonstrated the contraction in vitro, i.e. the ATP-induced contraction of the actomyosin thread. Prior to this, actomyosin had been shown by Engelhardt and Liubimova [10] to have an intrinsic ATPase activity. These two important findings together gave us the impression that the essentials of muscle molecular biology had been revealed. However, no one had yet inquired penetratingly into the relationship between the contraction-inducing action of ATP and the concomitant ATP breakdown.

Further progress in this area, especially the proposal and establishment of the sliding concept [cf. ref. 11 and 12], appeared to have almost solved the molecular mechanism, leaving only the fine details to researchers of the next generation. It had been the consensus of almost all muscle scientists that sliding was carried out by rotational movement of a myosin head from one actin to the next, and that 1 ATP was broken each time when myosin came in contact with an actin.

Yanagida and his colleagues [13, 14], however, recently challanged this accepted idea. Under low load conditions, a myosin head is able to dislocate over a distance of more than 10 actin molecules at the expense of a single ATP molecule. In additon, no rotational movement of the myosin head can be found during the sliding of the myosin filament along the actin filament. Now we realize that what we have been dealing with is not an ordinary chemical reaction of materials dissolved in water, but the reaction between a gel-like system and a small soluble molecule accompanied by conversion of energy forms, i.e. chemomechanical transduction. Perhaps we need a new resolution of this difficult question; once clarified, the solution can be applied to other forms of energy conversion. In these ways then, muscle research has opened yet another front in biological sciences.

Muscle as the Origin of the Ca^{2+} Era

Keen attention is paid today to Ca^{2+} and most intracellular processes have been shown to be more or less related to Ca^{2+}-regulated reactions. This might make it appear that Ca^{2+} is a classical biological factor, but, in fact, it is a newcomer as a modulator of intracellular processes [cf. ref. 15]. Ca^{2+} action on myoplasm was first noticed by Heilbrunn around 1940 [16] but the recognition of its intracellular role at the molecular level was made around 1960 [17, 18; cf. ref. 15]. It is from the accomplishments made in 1959–1962 that the rich activity of Ca^{2+} research today originates [cf. ref. 15].

Ca^{2+} as a Common Intracellular Regulator

In 1967 [19; cf. ref. 14], Ca^{2+} was shown to attack a member of the glycogenolytic system, phosphorylase *b* kinase, and thus the process of excitation-metabolism coupling was clearly elucidated. This fact, together with the discovery of calmodulin (see next paragraph), has allowed the emancipation of Ca^{2+} from the contractile system and made it a common regulator of intracellular processes in all kinds of cells.

Ca^{2+}-Binding Protein

Troponin was isolated from native tropomyosin [20] and was shown to be the first Ca^{2+}-binding protein of vital physiological significance [21; cf. ref. 15]. This has not only confirmed the role of Ca^{2+} in contraction –

the Ca^{2+} concept was not willingly accepted even by muscle scientists until this discovery – but has opened up a new genre of biological sciences: 'Ca^{2+}-binding protein' [cf. ref. 22]. This line was eminently strengthened by the discovery of calmodulin by Kakiuchi [23, 24; cf. ref. 22] and Cheung [25; cf. ref. 22]. Now new Ca^{2+}-binding proteins are being reported next to next.

Excitation-Contraction Coupling

The facts that the fragmented sarcoplasmic reticulum takes up Ca^{2+} through an ATP-dependent reaction and that this uptake is strong enough to remove essential Ca^{2+} from the actomyosin system [18, 26; cf. ref. 16] have enabled us to understand the whole picture of how electrical potential can induce contraction [19, 27; cf. ref. 16].

Recently, dramatic progress has been achived in this area due especially to the efforts of Numa [28; cf. ref. 29]. Briefly, the transmission of an electrical signal from the T tubule to the sarcoplasmic reticulum was found to be made by the interaction of two proteins, dihydropyridine-binding protein located in the T tubule and the ryanodine-binding protein associated with the cisterns of the sarcoplasmic reticulum. The former is the voltage sensor, but acts as a Ca^{2+} channel under certain conditions; in other words, this protein has a dual function. The latter is a gigantic protein, comprising the Ca^{2+} release channel and the foot region of the sarcoplasmic reticulum. The relaxation is induced by Ca^{2+} uptake through the Ca^{2+}-ATPase channel of the sarcoplasmic reticulum. It is claimed that parvalbumin, another Ca^{2+}-binding protein, is involved in the relaxation process of twitch or short tetanus [30].

Cell Differentiation

At the beginning of the age of molecular genetics, muscle was considered to be almost a dead material because of its slow protein turnover rate. Remarkable recent progress in molecular genetics, however, has enabled scientists to investigate into the mechanism of expression of the proteins specific to individual tissues and, consequently, provoked a renewed interest in muscle.

Differentiation is, in a sense, a kind of simplification. Skeletal muscle is perhaps the most differentiated cell and therefore expresses only a relatively small number of specific proteins. It is quite understandable that studies on specific gene expression in muscle, i.e. the interaction of myogenic factors and specific genes, are now very popular [cf. ref. 31]. The findings achieved by muscle research will aid in providing a useful model for studies of other cells.

Conclusion

This brief overview has confirmed the statement in the introduction that muscle research was a key factor in the pioneering of many fields of biological sciences. Since skeletal muscle is highly differentiated and therefore greatly simplified, the essentials of biological problems are clearly visible in this tissue. In the future also, muscle research will continue to be a model for studies of other tissues.

References

1. Galvani L: De virbus electricitatis in motu musculari commentarius. De Bononiensi Scientiarum et Artium Instituto atque Academia Commentarii 1791;7:363–418. English version by E. Licht. Cambridge, Green, 1953.
2. Bernstein J: Untersuchungen zur Thermodynamik der bioelektrischen Ströme. Pflügers Arch 1902;92:521–567.
3. Overton E: Beiträge zur allgemeinen Muskel- und Nervenphysiologie. II. Ueber die Unentbehrlichkeit von Natrium- (oder Lithium-) Ionen für den Contractionsakt des Muskels. Pflügers Arch 1902;92:346–386.
4. Hodgkin A, Huxley AF: A quantitative description of membrane current and its application to conduction and excitation in nerve. J Physiol 1952;117:500–544.
5. Fatt P, Katz B: An analysis of the end-plate potential recorded with an intracellular electrode. J Physiol 1951;115:320–370.
6. Lohmann F: Ueber die Pyrophosphatefraktion im Muskel. Naturwissenschaften 1929;17:624–625.
7. Fiske H, SubbaRow Y: Phosphorus compounds of muscle and liver. Science 1929; 70:381–382.
8. Lipmann F: Metabolic generation and utilization of phosphate bond energy. Adv Enzymol 1941;1:99–162.
9. Szent-Györgyi A: The reversibility of the contraction of myosin threads. Stud Inst Med Chem Univ Szeged 1942;2:25–26.
10. Engelhardt WA, Liubimova MN: Myosin and adenosine triphosphatase. Nature 1939;144:2–3.
11. Huxley AF: Muscle structure and theories of contracction. Progr Biophys 1957;7: 255–318.
12. Huxley HE: The contraction of muscle. Sci Am 1958;199:67–82.

13 Yanagida T, Arata T, Oosawa F: Sliding distance of actin filament induced by a myosin cross-bridge during one ATP hydrolysis cycle. Nature 1985;316:336–369.
14 Ishijima A, Doi T, Sakurada K, Yanagida T: Sub-piconewton force fluctuations of actomyosin in vitro. Nature 1991;352:301–306.
15 Ebashi S, Endo M: Ca ion and muscle contraction. Progr Biophys Mol Biol 1968;18: 123–183.
16 Heilbrunn LV: The action of calcium on muscle protoplasm. Physiol Zool 1940;13: 88–94.
17 Weber A: On the role of calcium in the activity of adenosine 5′-triphosphate hydrolysis by actomyosin. J Biol Chem 1959;234:2764–2769.
18 Ebashi S: Calcium bindings and relaxation in the actomyosin system. J Biochem 1960;48:150–151.
19 Ozawa E, Hosoi K, Ebashi S: Reversible stimulation of muscle phosphorylase b kinase by low concentration of calcium ions. J Biochem 1967;61:531–533.
20 Ebashi S, Kodama K: A new protein factor promoting aggregation of tropomyosin. J Biochem 1965;58:188–190.
21 Ebashi S, Ebashi F, Kodama A: Troponin as the Ca^{2+}-receptive protein in the contractile system. J Biochem 1967;62:137–138.
22 Ebashi S, Ogawa Y: Troponin C and calmodulin as calcium receptors: Mode of action and sensitivity to drugs. Handb Exp Pharm 1988;83:31–56.
23 Kakiuchi S, Yamazaki R, Nakajima H: Studies on brain phosphodiesterase. 2. Bull Jpn Neurochem Soc 1969;8:17–20.
24 Kakiuchi S, Yamazaki R, Nakajima H: Properties of a heat stable phosphodiesterase activating factor isolated from brain extract. Proc Jpn Acad 1970;46:587–592.
25 Cheung WY: Cyclic 3′,5′-nucleotide phosphodiesterase: Demonstration of an activator. Biochem Biophys Res Commun 1970;38:533–538.
26 Ebashi S: Calcium binding activity of vesicular relaxing factor. J Biochem 1961;50: 236–244.
27 Ebashi S: The relaxing factor and the contraction-relaxation cycle of skeletal muscle. Progr Theor Phys [Suppl] 1961;17:35–40.
28 Numa S: A molecular view of neurotransmitter receptor and ionic channels. Harvey Lect 1989;83:121–165.
29 Ebashi S: Excitation-contraction coupling and the mechanism of muscle contraction. Annu Rev Physiol 1991;53:1–16.
30 Gillis JM, Thomason D, Lefevre L, Kretsinger RH: Parvalbumins and muscle relaxation: A computer simulation study. J Muscle Res Cell Motil 1982;3:377–398.
31 Ozawa E, Masaki T, Nabeshima Y (eds): Frontiers of Muscle Research. Excerpta Medica, Amsterdam, New York, Oxford, 1991.

Setsuro Ebashi, MD, PhD, Okazaki National Research Institutes,
Okazaki 444 (Japan)

Sato Y, Poortmans J, Hashimoto I, Oshida Y (eds): Integration of Medical and Sports Sciences. Med Sport Sci. Basel, Karger, 1992, vol 37, pp 8–19

From Frog to Man

Farewell Lecture

Jacques R. Poortmans

Chimie Physiologique, Institut Supérieur d'Education Physique et de Kinésithérapie, Université Libre de Bruxelles, Belgique

When the Japanese local organization committee asked me to give the farewell speech on some historical events of exercise biochemistry, I granted this as a privilege of old age. Indeed, one often asks a senior scientist to talk about his memories and reminisce on what he thinks are the major steps in progressing science. I do not feel old enough to introduce a pertinent survey full of sagacity and wisdom. Looking at the available literature in this context, I noticed that there are only a few short papers on the historical development of this growing field of interest named the biochemistry of exercise. Indeed, an introduction to muscle metabolism during exercise in man was given by Asmussen [1] at the 1970 Stockholm Symposium. He emphasized the prominent role of Lavoisier as the first experimenting work physiologist, the progressive recognition of glycogen degradation as the major source of anaerobic muscle contraction, and the importance of the phosphate compounds to the mechanical events. In their Keynote Address and Closing Session, Edwards [2] and Gollnick [3] gave a retrospect of the biochemical bases of fatigue in exercise performance. The relationship between exercise intensity and lactate formation, since the principle work of Fletcher and Hopkins [4], were described in elegant terms. Eventually, the biochemistry of muscular contraction in its historical development was reviewed by Needham [5] in that remarkable scholarly book entitled *'Machina carnis'*. I will not repeat a new attempt to stress the slow move from one to several substrates, nor to show the importance of refined analytical tools such as catheterization, biopsy and NMR techniques. On the contrary, I would like to turn my attention to the importance of some past, almost forgotten, experiments, to point out the ingenuity and cleverness of our predecessors. We ought to be modest and remember that even in science we sometimes repeat previous experiments with new advanced techniques. The development of exercise biochemistry, from frog to man, will be considered, having in mind what I believe are the foundations of this new discipline. Finally, I will include a glance at the present steps that could influence our future. As pointed out recently by Lowry [6], 'it is not necessary to be a genius to contribute to science'.

Forgotten Observations and Experiments

The Concept of Respiration in the Seventeenth Century

It is often mentioned that the metabolism of plants and animals started after the discovery of oxygen by the Englishman Priestly and the Swede Scheele, together with later experiments by the Frenchmen Lavoisier and Séguin in 1776. However, nearly a century before a young physician of Bath (England), John Mayow, dismantled the entire structure of Galenic physiology when he published at Oxford in 1674 his *Tractatus quinque medico-physici,* in which he developed his theory of a *nitro-aerial spirit,* present in the atmosphere and necessary to both combustion and respiration [cited in 7]. The theory of respiration in 1600, which was still essentially that of the ancient Greeks, showed no effect of the revolution in human anatomy brought about by Vesalius in the mid-sixteenth century. The famous Belgian anatomist taught that the lungs are as if they had been formed from solidified bloody foam. The decisive year, as far as respiration is concerned, was 1661 when the Italian Marcello Malpighi established the membranous nature of lung tissue and the existence of multitudes of vesicles enclosed in it. Malpighi made use of the lungs of frogs because of their greater simplicity and transparency. When he looked at the lung of a living frog, he clearly saw the blood rushing through the small arteries but he could not determine its ultimate destination. But in fact, as pointed out by Wilson [7], by 1553 Michael Servetus had discovered the pulmonary circuit of the blood, while by 1559 Realdus Columbus, probably independently, had not only discovered the pulmonary circuit but had published his discovery for all the world to read. The work of Columbus seems not to have taken hold upon any mind prior to that of Harvey. Mayow's own great contribution was to show direct demonstration that a portion of the air is consumed by an animal in respiration. He said:

'Let a small animal placed on a suitable support be enclosed in an inverted glass. Then let the inverted glass be sunk a little into the water, so that the water enclosed in the glass may stand at the same level as the water outside ... And so you will soon see the water sensibly rising into the cavity of the glass ...'

Then Mayow measured how much the volume of the air was reduced by the breathing of the animals taking the precaution that 'everything should remain in the glass the same as before'. His work was fundamental to the idea concerning utilisation in contraction of that part of the air which we now named oxygen. Unfortunately, Mayow died at the age of thirty-six.

The Early Biochemistry of Exercise by von Liebig

Two Germans named von Liebig were involved in the early work on muscle metabolism. Justus von Liebig enunciated in 1843 in his book *Animal Chemistry or Chemistry in Its Application to Physiology and Pathology* [8] that in living organisms the various forms of power are interconvertible without loss:

'If we reflect, that the slightest motion of a finger consumes force; that in consequence of the force expended, a corresponding portion of muscle diminishes in volume; it is obvious, that an equilibrium between supply and waste of matter can occur only when the portion separated or expelled in a lifeless form is restored in another part.'

Julius von Liebig came to stand as the virtual embodiment of the German science of his day and as Jacob Grimm said [cited in, 9]: 'Chemistry is a gibberish of Latin and German, but in Liebig's hands it becomes a powerful language'. However, if J. Liebig was so important, he attracted very little attention from either the German- or the English-speaking historical communities. As noted by W.H. Brock, even the anniversary of his death in 1973 attracted little attention. He published at least 769 printed papers and most of them are methodological and speculative contributions to the emerging discipline of physiological chemistry. Coming back to exercise, J. Liebig wrote:

'The sum of the mechanical effects produced is proportional to the amount of nitrogen in the urine: whether the mechanical force has been employed in voluntary or involuntary motions, whether it has been consumed by the limbs or by the heart and other viscera,'

This view, that protein breakdown was necessarily concerned in contraction, was generally held in the first half of the 19th century until 1865 when Fick and Wislicenus, two of his former students, proved that the supervisor's view was wrong! The Frenchman Claude Bernard was also dissatisfied with the expectations that Liebig could predict the behavior of the substances within organisms from his knowledge of chemical properties of elements and compounds in laboratory reactions. He could still be disappointed by today's theories.

Another von Liebig, Georg, in 1850, used muscles from frogs which, while living, had been injected with distilled water to remove blood [10]. He noticed carbon dioxide production and oxygen consumption in these muscles to be similar to that in muscles from frogs with normal circulation; he concluded that carbon dioxide was formed from part of the oxygen absorbed as a living activity of the muscle outside the body. Georg von Liebig also quoted that excised muscle can maintain irritability for several hours in an atmosphere of nitrogen or hydrogen.

The Less-Known Glycogen Story

A very early observation of lactic acid in muscle, with discussion of its origin, was made by Claude Bernard in 1855 [11]. He found no sugar in calf fetuses, but if the muscles were left in water at room temperature, the liquid became very acid owing to lactic acid production; if 30% alcohol was added, then lactic acid formation ceased and sugar could be detected. Four years later, glycogen in muscle was first described by Claude Bernard, particularly in young and growing animals.

However, Otto Nasse was the first author to show, in 1877, that muscle glycogen was reduced by activity [12]. When he compared glycogen loss with lactic acid content he found the former to be greater. Thus, he concluded that it was likely that the lactic acid was derived from the glycogen.

One had to wait until 1930 when Long and Grant [13], from McGill University demonstrated a relationship between the rate at which an animal was exercised and the ensuing reduction in muscle glycogen. They used a method of artificial exercise which although imperfect in some ways did yield a fairly constant reduction in body glycogen. On their abdomens rats were fastened by their limbs to a small board. Under local anesthesia electrodes were inserted in the upper dorsal and lower lumbar regions. The terminals were connected to an induction coil in the primary circuit of which a rotary contact was so arranged as to deliver the break shock only. By this method not only could the speed of exercise be regulated but also the intensity of the contractions. The authors said that 'the effects of the exercise were of some interest':

'The contractions affected practically all the muscles of the body ... When the animals were released after the exercise, they were rather cramped for a few minutes...When left to themselves, however, they usually curled up and apparently went to sleep, displaying a marked disinclination for further exercise ...'!!

Long and Grant demonstrated, for the first time in intact mammals, that the increasing intensity of exercise leads to progressive depletion of glycogen down to 29% of the original muscle content.

The Physiological Basis of Athletic Records

When I came across figure 14–1 of the textbook *'Biochemistry, a Functional Approach'* published in 1970 by McGilvery [14], I was convinced that this picture gave, for the first time, a clear plot of the world record performances in races, in which the rate of running is compared with the total time of running. The author gave a nice interpretation of the curve postulating that the breaks in the slope are dependent on the change in energy sources. In fact, this curve is an adapted one from Hill [15] published in 1925 in *Nature.* A year later he pointed out that:

'Some of the most consistent physiological data available are contained, not in books on physiology, not even in books on medicine, but in the world's records for running different horizontal distances' (Muscular Activity, 1926).

Hill believed that 'the relation shown in the curve may be accepted practically as a natural constant for the human race; it would require almost a superhuman effort to change one of the points by 2 percent'. We now know that he was wrong, looking at long-distance running for example, but this is an adaptation of training programmes rather than a misinterpretation of his data. During the same Presidential Address to the Physiology Section of the British Association he incautiously, but quite politely, inquired why the performance of men was superior to that of women and he concluded simply that the mechanical power exerted by women, per kg of body weight, was less than that by men. A storm broke out in the newspaper *Westminster Gazette* saying that he had insulted the whole female sex. As a consequence, angry strong women came to Hill's laboratory to be 'tested'.

Anyway, we can follow him when he concluded that 'athletes and healthy men give us the cleanest experiments'. If one took a patient from hospital and made him work till he could barely move, one could never be sure (a) that he had really driven himself to his limit; (b) that one would not kill him, and (c) what the cause of his stopping was. Hill [16] ended by saying that 'you can only observe your patient, but can experiment with your athlete; you can 'try out' on him the facts and theories which have reached with frogs'.

Inside Otto Meyerhof's Life: Muscle Glycogen Synthesis Revisited

The outstanding feature of Otto Meyerhof's work on muscle is the first really successful attempt to correlate the chemical and physical processes of cellular function [17–21]. He was a former student of Otto Warburg having his own team, the number of his collaborators at any time never exceeded five. He had only one trained personal technician and a part-time typist [22]. Meyerhof's laboratory made a decisive contribution to what is now called the Embden-Meyerhof pathway of glycolysis. It laid the groundwork which led to the discoveries of hexokinase, aldolase and other enzymes. During these investigations he maintained a continuous exchange of views and information with Hill. The collaboration between the two men came to a close personal friendship over decades making most fortunate the development of exercise physiology and biochemistry. A photograph published in 1950 in an article dedicated to Otto Meyerhof shows them driving together to Stockholm for the Physiological Congress [23]. After the rise to power of the Nazis, Otto Meyerhof, like other Jewish scientists, had to leave Germany. In 1938 he went to Paris where he was appointed Director of Research at the University of Paris. When the Nazi hordes invaded France he came to the United States where he was appointed Research Professor of Physiological Chemistry in the School of Medicine of the University of Pennsylvania. This was the end of the succession of papers on living muscle.

Meyerhof was the first to discover a feature of oxidative recovery in frog muscle in which some of the lactate formed during fatigue disappears when the muscle is allowed to recover in oxygen, but only about 20–25% of this lactate can be accounted for by oxidation to CO_2 and water [17–19]. Altogether, Meyerhof recorded 21 separate experiments on muscles fatigued in various ways and he concluded that the lactate unaccounted for by oxidation had been converted into carbohydrate, mostly in the form of glycogen. For example, Meyerhof found in isolated frog legs which had been stimulated to work for 15–30 min and then allowed to rest in oxygen for 21–23 h an average resynthesis of 123 mg of glycogen. Ten years later, Long and Grant [13] demonstrated in intact rats that glycogen resynthesis after exercise did not follow the removal of excess lactate from the body. They concluded that a greater part of lactate could be oxidized and only a small portion was deposited as muscle glycogen. In 1940, Flock and Bollman, working on rat intact hind legs, showed that when muscle glycogen was reduced to minimal levels by 3 min of intense electrical stimulation, it was repleted so slowly that its resynthesis appears to be quite independent of the presence of lactic acid [24]. Facing these discrepancies, Bendall and Taylor [25] reinvestigated the Meyerhof Quotient and the resynthesis of glycogen from lactate in frog and rabbit muscle using more refined techniques of assay. Their results vindicated Meyerhof's original hypothesis. However, it has to be emphasized that the samples were resting muscles, no exercise having been performed before the experiments.

More recent studies have shown two opposite opinions about the fate of muscle lactate and glycogen resynthesis after severe exercise. From calculations on human studies, Hermansen and Vaage [26] reported that approximately 75% of the lactate was reconverted into glycogen but they did not utilize the liver-muscle glucose pathway (the Cori cycle). On the contrary, Brooks and Fahey [27] concluded from their studies on rats that most of the lactate is oxidized to CO_2 and water and that only a small proportion of the lactate is converted into glycogen.

As in most cases, the truth regarding the conversion into glycogen is somewhere between the two opposing views. Indeed, McLane and Holloszy [28] in 1979 demonstrated in rats that 44% of the lactate which appeared after exercise is converted into muscle glycogen. Also, Astrand et al. [29] arrived to nearly the same estimation (45%) when looking at human subjects submitted to supramaximal short-term exercise.

From Animal to Human Model

Isolated muscle preparation studies have been used for more than a century to investigate physiological and biochemical adaptations to exercise. Owing to methodological reasons, most studies have been performed on amphibian muscles, in rather unphysiological conditions of low temperature, arrested blood flow, anaerobiosis, etc. Frog was the major species to look at alterations induced by a single bout of exercise. In 1968, Piiper et al. [30] developed an experimental setup in a dog gastrocnemius muscle preparation with intact blood supply. Oxygen uptake, lactate output and muscle high-energy phosphate concentrations were measured during different level of exercise. Permanently increased contractile activity is the basis of modern training. We all know that many factors are involved in the training response. Although hypertrophy of skeletal muscle occurs after the onset of heavy-resistance training in humans, endurance training does not produce hypertrophy of skeletal muscle [31]. Due to the lack of success in developing a training programme to promote muscle hypertrophy in laboratory animals, several investigators have turned to nonphysiological models, such as chronic electrical stimulation, tenotomy. However, these models do not mimick human physical activity and the conclusions that they brought in may be misleading for human adaptation.

Meanwhile, I want to report two animal models which have been used to produce enlargement of skeletal muscle. In 1976, Gonyea and Ericson [32] described a training apparatus for cat which could induce muscle hypertrophy by weight training. The system was very clever. The young adult cat had to move a bar for a specific distance with its right forelimb to receive a food reward. Weight increments are added until the cat is unable to move the bar for the required distance. This procedure had the advantage of inducing significant muscle hypertrophy, up to 34% during the weight-lifting regimen. Recently, Wong and Booth [33] reported on a rat model of weight training that produces skeletal muscle enlargement using strictly regimented training protocols similar to those used in human training programmes. A pulley bar mounted on a metal plate was welded to one end of the bar to which the right foot of an anesthetized rat was strapped with adhesive tape. A second free-moving pulley was soldered to the bar as was the foot plate, so that any excursion made by the foot plate due to plantar flexion of the foot of the rat resulted in comparable movement of the pulley, thus lifting the weight. Muscle wet weights of individual plantar flexors were up to 18% greater than the corresponding contralateral non-trained muscles.

Turning towards human models, several attempts have been made to limit dynamic exercise to an extremity or part of one. In 1985, Saltin and his team [34] developed an exercise model that enabled the study of muscle activity over the quadriceps femoris muscle while the rest of the body remained relaxed. All metabolic measurements were closely related to changes in the working muscle. Nowadays, one could imagine a combined model whereby a cyclist would move a treadmill belt to force a dog or a rat to run faster each day!

From Quantitative Estimation to Metabolic Regulation

For many years the biochemistry of exercise had to remain descriptive as far as the general pathways were concerned. Qualitative observations, together with local quantitative evaluations, were the common rule. Up to the mid-1950s, little information appeared on the cellular level adaptation to physical training. Most of the published work in this field stems from the Leningrad Physical Culture Research Institute [35]. Conflicting results were obtained by Hearn and Wainio [36, 37] in 1956 and 1957 looking at succinic dehydrogenase and aldolase activities of the heart and skeletal muscle of exercised rats. Gould and Rawlinson [38] extended the analyses to other enzymes and included the possible effects of growth and maturity upon the adaptative response.

My very personal estimation of the real start in the attempt to understand the biochemical adaptation in muscle of mammals subjected to a strenuous programme of exercise came from Holloszy [39] in 1967. Since then, Holloszy and his growing group investigated most of the principal biochemical pathways known to produce energy within heart and skeletal muscle. We owe most of our knowledge to those who, in Saint Louis and other places thereafter, took the lead of a systematic search for the adaptation of the largest tissue in the body.

A second major step emerged from the work of McGilvery [40] who, as the key-speaker of our 2nd Symposium on Biochemistry of Exercise in 1973 in Macolin, gave us a logical sequence of cause and event during the utilization of fuels by muscles. Estimated maximal flux for various substrates was given as a clear rationalization of the fuel economy and, according to the author, 'they were not invented for the occasion'!

A third cornerstone was placed by Newsholme and Start [41] who became involved in exercise biochemistry in the early 1970s. Indeed, after having read their book *Regulation in Metabolism* published in 1973, I was convinced that a new era of understanding could emerge from the existence of substrate cycles. This is why Newsholme was chosen as the academic lecturer of our third Symposia in Quebec City. The significance of substrate cycles in providing sensitivity and flexibility in metabolic control was first discussed in a review by Newsholme and Gevers in 1967. In fact, as early as 1947, Sir Hans Krebs foresaw that the search for cycles could be scientifically fruitful. As pointed out by Newsholme, one advantage of the substrate-cycling mechanism for increasing sensitivity is that the extent of this increase varies according to the rate of cycling; in other words, sensitivity is proportional to the ratio of cycling rate:flux. However, the potential for substrate cycling does not prove that they exist. Indeed, several researchers casted doubt on the reality of the flux generated by a substrate cycle. The use of dual isotopic labeling enabled the measurement of the glucose/glucose-6-phosphate and of the fructose-6-phosphate/fructose-1,6-bisphosphate cycles. Today, the potential quantitative importance of such cycles can perhaps be gleaned from reflection on the magnitude of substrate cycling in humans. There is little doubt that more information will be reported on the subject within the next 3 years.

And From Now On

As pointed out by Booth [42], the recent advent of recombinant DNA technology to study the regulation of gene expression can now be exploited by exercise biochemists to determine how contractile activity modulates the expression of muscle-specific genes. Also, molecular biologists appear more and more interested in muscle as a nice system to study the control of gene expression [43]. For example, vertebrate muscles mobilize four structural genes to build their α- and β-tropomyosines leading to multiple isoforms that could have different physiological functions during the ontogenesis.

A major application of molecular biology to exercise biochemistry has been to utilize messenger RNA hybridization probes to estimate changes in the levels of mRNAs in skeletal muscle during and off training [42]. As reported by Booth [42] in a recent review on the subject, the predominance of pretranslational, translational and posttranslational controls varies depending on the stage of the training and the type of muscle. The powerful revolution in molecular biology could provide new information about exercise mechanisms, such as the gene expression of mitochondrial proteins, the limiting steps in metabolic fluxes, the protein kinases involved in the transduction of several signals, the differential gene expression in respone to changes in contractile activity, the controlled DNA proliferation which induces muscle hypertrophy. These are all reasons to believe that, in the near future, molecular biologists and exercise biochemists should share the same interest towards the only healthy tissue, the largest in the human body, which can adapt its biology, rapidly, to new conditions. Let us hope that the next Conference on Biochemistry of Exercise will develop this concept to the benefit of all.

References

1 Asmussen E: Muscle metabolism during exercise in man. A historical survey; in Pernow B, Saltin B (eds): Muscle Metabolism during Exercise. New York, Plenum Press, 1971, pp 1–12.
2 Edwards RHT: Biochemical bases of fatigue in exercise performance: catastrophe theory of muscular fatigue, in Knuttgen HG, Vogel JA, Poortmans JR (eds): Biochemistry of Exercise. Champaign, Human Kinetics Publishers, 1983, pp 3–28.
3 Gollnick PD: Fatigue in Retrospect and Prospect: Heritage, Present Status and Future; in Knuttgen HG, Vogel JA, Poortmans JR (eds): Biochemistry of Exercise. Champaign, Human Kinetics Publishers, 1983, pp 909–921.
4 Fletcher WM, Hopkins FG: Lactic acid in amphibian muscle. J Physiol (Lond) 1906–1907;35:247–309.
5 Needham DM: Machina carnis. Cambridge, Universtiy Press, 1971, pp 1–782.
6 Lowry OH: How to succeed in research without being a genius. Ann Rev Biochem 1990;59:1–27.
7 Wilson LG: The transformation of ancient concepts of respiration in the seventeenth centruy. ISIS 1960;50:161–172.
8 Liebig J: Animal chemistry, or chemistry in its application to physiology and pathology. London, Taylor & Walton, 1843, pp 1–384.
9 Brock WH: Liebigiana: Old and new perspectives. Hist Sci 1981;19:219–218.
10 Liebig G: Über Respiration der Muskeln. Arch Anat Physiol Wiss Med 1850:393.
11 Bernard C: De la matière glycogène considérée comme condition de développement de certains tissus, chez le foetus, avant l'apparition de la fonction glycogénique du foie. C R Séances Acad Sci 1859:48:673–684.
12 Nasse O: Bemerkungen zur Physiologie der Kohlenhydrate. Arch Physiol 1877;14: 473–484.
13 Long CNH, Grant R: The recovery process after exercise in the mammal. I. Glycogen resynthesis in the fasted rat. J Biol Chem 1930;89:553–565.
14 McGilvery RW: Biochemistry. A Functional Approach. London, Saunders, 1970, p 269.
15 Hill AV: The physiological basis of athletic records. Nature 1925;116:544–548.
16 Hill AV: Muscular Activity. Baltimore, Williams & Wilkins, 1926, pp 1–115.
17 Meyerhof O: Die Energieumwandlungen im Muskel. I. Über die Beziehungen der Milchsäure zur Wärmebildung und Arbeitsleistung des Muskels in der Anaerobiose. Arch Ges Physiol 1920;181:258–283.
18 Meyerhof O: Über die Energieumwandlungen im Muskel. II. Das Schicksal der Milchsäure in der Erholungsperiode des Muskels. Arch Ges Physiol 1920;181:313–317.
19 Meyerhof O: Die Energieumwandlungen im Muskel. III. Kohlenhydrat- und Milchsäureumsatz im Froschmuskel. Arch Ges Physiol 1920;185:11–32.
20 Meyerhof O: Chemical Dynamics of Life Phenomena. London, Lippincott, 1924, pp 1–110.
21 Meyerhof O: Die chemischen Vorgänge im Muskel. Berlin, Springer, 1930, pp 1–350.
22 Krebs HA: Reminiscences and Reflexions. Oxford, Clarendon Press, 1981, p 1–289.
23 Hill AV: A challenge to biochemists. Biochim Biophys Acta 1950;4:4–11.

24 Flock EV, Bollman JL: Resynthesis of muscle glycogen after exercise. J Biol Chem 1940;136:469–478.
25 Bendall JR, Taylor AA: The Meyerhof quotient and the synthesis of glycogen from lactate in frog and rabbit muscle. Biochem J 1970;118:887–893.
26 Hermansen L, Vaage O: Lactate disappearance and glycogen synthesis in human muscle after maximal exercise. Am J Physiol 1977:E422–E429.
27 Brooks GA, Fahey TD: Exercise Physiology: Human Bioenergetics and Its Applications. New York, Macmillan, 1984, pp 1–726.
28 McLane JA, Holloszy JO: Glycogen synthesis from lactate in the three types of skeletal muscle. J Biol Chem 1979;254:6548–6553.
29 Astrand PO, Hultman E, Juhlin-Dannfelt A, Reynolds G: Disposal of lactate during and after strenuous exercise in humans. J Appl Physiol 1986;61:338–343.
30 Piiper J, di Prampero PE, Cerretelli P: Oxygen debt and high-energy phosphates in gastrocnemius muscle of the dog. Am J Physiol 1968;215:523–531.
31 Holloszy JO, Booth FW: Biochemical adaptations to endurance exercise in muscle. Ann Rev Physiol 1976;38:273–291.
32 Gonyea WJ, Ericson GC: An experimental model for the study of exercise-induced skeletal muscle hypertrophy. J Appl Physiol 1976;40:630–633.
33 Wong TS, Booth FW: Skeletal muscle enlargement with weight-lifting exercise by rats. J Appl Physiol 1988;65:950–954.
34 Anderson P, Adams RP, Sjogaard G, Thorboe A, Saltin B: Dynamic knee extension as model for study of isolated exercising muscle in humans. J Appl Physiol 1985;59: 1647–1653.
35 Yakovlev NN: Biochemistry of sport in the Soviet Union: Beginning, development, and present status. Med Sci Sports 1975;7:237–247.
36 Hearn GR, Waino WW: Succinic dehydrogenase activity of the heart and skeletal muscle of exercised rats. Am J Physiol 1956;185:348–350.
37 Hearn GR, Waino WW: Aldolase activity of the heart and skeletal muscle of exercised rats. Am J Physiol 1957;190:206–208.
38 Gould MK, Rawlinson WA: Biochemical adaptation as a response to exercise. I. Effect of swimming on the levels of lactic dehydrogenase, malic dehydrogenase and phosphorylase in muscle of 8-, 11- and 15-week-old rats. Biochem J 1959;73:41–44.
39 Holloszy JO: Biochemical adaptations in muscle. Effects of exercise on mitochondrial oxygen uptake and respiratory enzyme activity in skeletal muscle. J Biol Chem 1967;242:2278–2282.
40 McGilvery RW: The use of fuels for muscular work; in Howald H, Poortmans JR (eds): Metabolic Adaptation to Prolonged Physical Exercise. Basel, Birkhäuser, 1975, pp 12–30.
41 Newsholme EA, Start C: Regulation in Metabolism. London, Wiley, 1973, pp 1–952.
42 Booth FW: Application of molecular biology in exercise physiology. Exerc Sport Sci Rev 1989;17:1–27.
43 Gros F: Le muscle: Système modèle en biologie du développement. Méd Sci 1990;6: 624–625.

Prof. J.R. Poortmans, Chimie Physiologique-ISEPK, C.P. 168,
Université libre de Bruxelles, 28, avenue P.-Héger, B–1050 Bruxelles (Belgium)

Sato Y, Poortmans J, Hashimoto I, Oshida Y (eds): Integration of Medical and Sports Sciences. Med Sport Sci. Basel, Karger, 1992, vol 37, pp 20–32

The Hypothalamus and Neural Feed-Forward Regulation of Exercise Metabolism

Takashi Shimazu [1]

Department of Medical Biochemistry, Ehime University School of Medicine, Shigenobu, Ehime, Japan

A neural feed-forward regulation of substrate mobilization during exercise is essentially important for ensuring rapid supply of energy fuels required for muscles to work. At the onset of exercise, signals are transmitted from the higher brain areas not only to skeletal muscles to cause muscle contraction, but also to the autonomic center to initiate mass sympathetic discharge into visceral organs. The activation of the sympathetic nervous system evoked by exercise signals leads to glucose output from the liver and fat mobilization from adipose tissue through direct innervation of the tissues and by neural-hormonal mechanisms involving increased release of catabolic hormones like pancreatic glucagon and adrenomedullary catecholamines. These metabolic responses have been shwon to be controlled by the autonomic center of the hypothalamus [1–4]. The hypothalamus is known as a brain site for metabolic integration and can be divided structurally and functionally into the medial and lateral hypothalamus; the medial hypothalamus comprises several nuclei, the ventromedial hypothalamic nucleus (VMH) being a central position in that it is intimately associated with sympathetic facilitation [3, 5, 6]. The lateral hypothalamus is composed of the lateral hypothalamic nucleus (LH), which is related to parasympathetic facilitation in peripheral tissues [3, 6].

In this article, I will review our studies on the role of the VMH in feed-forward regulation of exercise metabolism, mainly focusing substrate utilization in muscles as well as fuel mobilization from stored organs.

[1] The author wishes to acknowledge the support for this study by grants from the Health Science Promotion Foundation and from the Nakatomi Foundation.

Substrate Mobilization in Response to Ventromedial Hypothalamic Stimulation and Exercise

Physical exercise appears to be a powerful physiological stimulus for inducing activation of the VMH-sympathetic nervous system, because there are several lines of evidence to show that the metabolic and neurohormonal responses to VMH stimulation and exercise are remarkably similar (table 1).

Our previous studies have proven that electrical and chemical stimulation of the VMH causes glucose output from the liver by rapid activation of glycogen phosphorylase, the enzyme catalyzing the rate-limiting step in glycogenolysis [7, 8], and by increasing the activity of phosphoenolpyruvate carboxykinase, a key gluconeogenic enzyme [9], leading to hyperglycemia with concomitant decrease in liver glycogen content [10]. The stimulatory effect of the VMH on hepatic gluconeogenesis is thus complementary to its glycogenolytic effect on the liver in producing hyperglycemia. These metabolic responses to VMH stimulation are caused by direct sympathetic innervation of the liver [11, 12] and by neural-hormonal mechanisms through mediation of glucoregulatory hormones from the adrenal (epinephrine) and pancreas (glucagon). Indeed, VMH stimulation was shown to increase sympathetic nervous activity of the liver [5], and to inducc increased secretion of epinephrine [13] and glucagon [14], whereas it rather suppressed insulin secretion [14]. Adrenal corticosterone [15] and growth hormone [16] secretion are enhanced by VMH stimulation. In addition, lipolysis in adipose tissue can be accelerated by VMH stimulation, as evidenced by increase in plasma levels of glycerol and free fatty acids [17, 18].

There is a good deal of evidence to show that during exercise autonomic neuroendocrine activity is augmented, which, in turn, accelerates mobilization of metabolic substrates mainly from the liver and adipose tissue [19, 20]. Increased circulating norepinephrine levels and direct measurements of nerve activity indicate that sympathetic activity is increased. Epinephrine and cortisol secretion from the adrenal glands increase with exercise. Growth hormone secretion may also increase with exercise. More importantly, physical exercise stimulates glucagon secretion from the pancreas, while plasma insulin decreases with exercise. These exercise-induced hormonal and neural changes lead to hepatic glucose production by stimulating glycogenolysis and gluconeogenesis. Gluconeogenesis becomes increasingly more important with more prolonged exercise as hepatic gly-

Table 1. Similarities of metabolic and neurohormonal responses to VMH stimulation and exercise

	VMH stimulation	Exercise
Sympathetic activity	↑	↑
Epinephrine	↑	↑
Cortisol	↑	↑
Glucagon	↑	↑
Insulin	± or ↓	↓
Growth hormone	↑	↑
Glycogenolysis	↑	↑
Gluconeogenesis	↑	↑
Blood glucose	↑	↑ or ±
Lipolysis	↑	↑
Glucose uptake	??	↑

cogen stores progressively decrease [20]. The exercise-induced rise in glucagon and fall in insulin are likely to be essential to the maintenance of glucose production during exercise. Epinephrine may become important in the regulation of hepatic glucose production during the latter stage of prolonged exercise or during heavy exercise, when this hormone may obtain critically high levels [20]. Lipolysis in adipose tissue is also accelerated by exercise. Both the increased sympathoadrenal activity and the fall in insulin are the major determinants promoting an increase in lipolysis during exercise. Like sympathoadrenal activity, lipolysis is directly related to the intensity and duration of exercise [19].

Finally, physical exercise is known to augment glucose uptake and utilization in working muscles, where exercise facilitates glucose transport directly through an insulin-independent pathway and also increases the sensitivity of the glucose transport process to insulin [21–24]. Judging from the aforementioned similarities between the metabolic and neurohormonal responses to VMH stimulation and to exercise, it can be predicted that glucose transport in skeletal muscle could also be augmented by VMH stimulation. This possibility has recently been examined and the results are presented in the following sections.

Hypothalamic Regulation of Tissue Glucose Uptake: In vivo Studies

To explore the role of the hypothalamus in glucose uptake in peripheral tissues, we have recently examined the effect of electrical and chemical stimulation of the VMH and LH on the rate of tissue glucose uptake in vivo [25]. The rate constant of tissue glucose uptake was assessed in anesthetized rats by means of the 2-deoxy-*D*-[^{3}H]-glucose (2-[^{3}H]-DG) method [26]. The method relies on the principle that 2-[^{3}H]-DG is transported by the glucose transport system and then trapped within the tissues after conversion to 2-[^{3}H]-DG-6-phosphate.

Among the tissues examined in control rats, the rate constant of net glucose uptake was greatest in the brain and followed in order by the heart, diaphragm, skeletal muscle, brown adipose tissue and white adipose tissue. When the VMH was stimulated electrically, the rate constant of glucose uptake significantly increased (2- to 10-fold) in the heart and brown adipose tissue, but not in the brain, diaphragm and white adpose tissue. In contrast, electrical stimulation of the LH had no appreciable effect on 2-[^{3}H]-DG uptake in any tissues. The increased rate of glucose uptake in brown adipose tissue in response to VMH stimulation was effectively suppressed by local sympathetic denervation, indicating a mediation of the sympathetic nerve in this effect.

The enhanced rate of glucose uptake in response to VMH stimulation was also observed in skeletal muscle at rest. Because skeletal muscle is the principal tissue involved in blood glucose utilization, three types of muscle have been analyzed. The extensor digitorum longus and soleus muscle are known to be mainly composed of fast-twitch glycolytic and slow-twitch oxidative fibers, respectively, and the quadriceps and gastrocnemius muscle are composed of mixed-type fibers. In these three types of skeletal muscle, the rate constant of net glucose uptake was increased significantly by electrical stimulation of the VMH (fig. 1).

Skeletal muscle has been recognized to be supplied with noradrenergic sympathetic axons, which are distributed to the muscle spindles and extrafusal muscle fibers [27]. Accordingly, if the effect of VMH stimulation on muscle glucose uptake is mediated by norepinephrine released from terminal sympathetic synapses with muscle fibers, it is expected that the effect of VMH stimulation can be potentiated by inhibiting physiological inactivation of norepinephrine locally. This possibility has been tested by using the monoamine reuptake inhibitor LY255485 (maleate form) [28]. Although subcutaneous injection of LY255485 alone had no detectable effect on

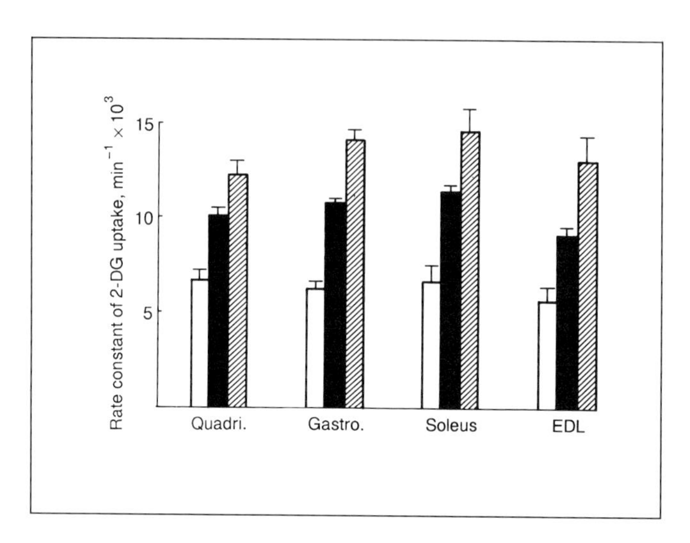

Fig. 1. Effects of electrical stimulation of the VMH and administration of LY255485 on the rate constant of 2-[^{3}H]-DG uptake in skeletal muscle of anesthetized rats. LY255485 (donation of Eli Lilly) was injected subcutaneously at the dose of 5 mg/kg, 1 h before the start of VMH stimulation. Four different hindlimb muscles, quadriceps (Quadri.), gastrocnemius (Gastro.), soleus and extensor digitorum longus (EDL) were sampled. □ = Control; ■ = VMH stimulation; ▨ = LY255485 + VMH stimulation. [From ref. 29.]

glucose uptake in skeletal muscle, this agent was found to potentiate significantly the effect of VMH stimulation on the rate of glucose uptake in all three types of skeletal muscle [29]. In fact, VMH stimulation coupled with administration of the agent resulted in an approximately 2-fold increase in the rate constant of glucose uptake in the muscles (fig. 1). This result is also suggestive of sympathetic nerve involvement in the enhancement of glucose uptake and utilization in skeletal muscle.

To confirm that neurons in the VMH are specifically concerned in neuronal regulation of glucose uptake in peripheral tissues, the effect of chemical stimulation of the VMH was also examined by microinjection of glutamate, a well-known excitatory neurotransmitter in the central nervous system. This kind of study revealed that the chemical stimulation of the VMH neurons caused significant increases in the rate of glucose uptake in the heart, brown adipose tissue and skeletal muscle as well, the effects being similar to those observed after electrical stimulation [25].

Neural Effect on Glucose Transport System in Muscle: In vitro Studies

The above results indicate that glucose uptake in resting muscles in vivo is augmented by activation of the VMH neurons. Since VMH stimulation suppresses insulin secretion from the pancreas [14] and actually plasma insulin levels did not change discernibly during electrical and chemical stimulation of the VMH in the above studies [25], our findings suggest that the effect of VMH stimulation on selective tissue glucose uptake is not mediated by insulin, but is probably mediated by the sympathetic nerve and thus may depend on an insulin-independent mechanism of glucose transport.

To corroborate this assumption, *D*-glucose transport activity has been measured in vitro, using sarcolemmal (plasma membrane) vesicles isolated from rat heart after VMH stimulation. For comparison, the *D*-glucose transport has also been measured with cardiac sarcolemma of rats treated with insulin. The initial rates of *D*- and *L*-glucose influx into the plasma membrane vesicles were assayed under equilibrium exchange conditions in the presence of the tracers *D*-[^{3}H]-glucose and *L*-[^{14}C]-glucose [30]. The specific *D*-glucose transport activity was determined by subtracting non-specific *L*-glucose uptake from the *D*-glucose value. As expected, the results showed that treatment of rats with insulin increased the rate of *D*-glucose transport in heart plasma membrane vesicles nearly 2-fold as compared with controls. VMH stimulation also caused a significantly greater rate (near-doubling) of specific *D*-glucose transport in heart plasma membrane than that observed with the control membrane [A. Takahashi and T. Shimazu, unpubl. data].

Next, to assess the role of glucose transporters in the acceleration of *D*-glucose transport after VMH stimulation or insulin-treatment, [^{3}H]-cytochalasin B binding studies were performed on the same plasma membrane preparations from rat heart [31]. Since insulin has been known to stimulate the translocation of glucose transporters from an intracellular pool (low-density microsomal membrane fraction) to the plasma membrane in perfused rat heart [32] as well as in isolated adipose cells [33, 34], the microsomal membrane was also prepared from heart muscle of control, VMH-stimulated and insulin-treated rats. With these membrane fractions, *D*-glucose-inhibitable cytochalasin B binding was measured to determine both the total number of glucose transporter molecules (R_o) and their dissociation constant (K_d) according to the method described by Cushman and Wardzala [33]. Briefly, Scatchard plots were generated from binding

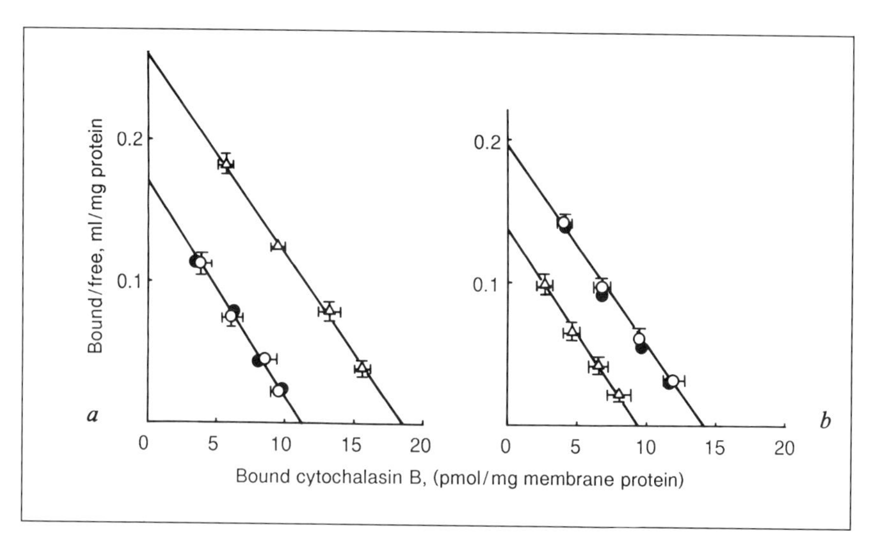

Fig. 2. Scatchard plots of specific *D*-glucose-inhibitable cytochalasin B binding to the plasma (*a*) and microsomal (*b*) membrane fractions isolated from hearts of control (○), VMH-stimulated (●) and insulin-treated (△) rats. K_d was not changed by insulin treatment or by VMH stimulation, but the number of cytochalasin B binding per milligram of membrane protein was significantly changed only after insulin treatment. [From ref. 31.]

studies in which membranes were incubated with varying concentrations of [^{3}H]-cytochalasin B in the presence or absence of *D*-glucose. The R_o and K_d were determined from a linear plot derived by subtraction along the radial axes of binding curves generated in the presence of *D*-glucose from those generated in the absence of *D*-glucose (fig. 2).

It became evident from the results shown in figure 2 that insulin injection shifted the line for plasma membrane to the right, but did not affect the slope. Thus, as expected, insulin treatment significantly increased the number of glucose transporters in the plasma membrane (18.5 ± 1.7 vs. 11.0 ± 1.0 pmol/mg protein) without altering the K_d when compared with controls. On the other hand, insulin decreased concomitantly the number of glucose transporters in the microsomal membrane (9.6 ± 1.5 vs. 14.3 ± 0.8 pmol/mg protein), but did not change the K_d value. These results demonstrate that the translocation of glucose transporters from the microsomal membrane to the plasma membrane occurs in heart muscle in response to insulin stimulation. In contrast to the effect of insulin, there were no appreciable changes in the number of cytochalasin-B-binding sites in both plasma (12.1 ± 1.1 vs. 11.0 ± 1.0 pmol/mg protein) and microsomal membranes (13.2 ± 1.0 vs. 14.3 ± 0.8 pmol/mg protein) from heart muscle of rats with VMH stumulation, as compared with controls. There were also no changes in the dissociation constants of these membrane fractions (fig. 2).

The fact that stimulation of the VMH caused no change in cytochalasin B binding indicates that VMH stimulation and insulin act by different mechanisms to facilitate glucose transport in heart muscle, and possibly in skeletal muscle. As alluded to above, insulin increases glucose transport in insulin-sensitive tissues by increasing the number of glucose transporters at the plasma membrane via recruitment from the intracellular pool. However, a similar mechanism does not seem to operate in muscle in response to VMH stimulation. Knowing that VMH stimulation enhances glucose uptake or transport without change in the number of plasma membrane glucose transporters, it is likely that the intrinsic activity of glucose transporter molecules present in the plasma membrane is increased by VMH stimulation. Similarly, an increase in the intrinsic activity of plasma membrane glucose transporters has recently been reported to occur with skeletal muscle after exercise in vivo [35] or contractile activity in situ [36], although in the latter case both increases in the number and intrinsic activity of the transporters have been noted [36, 37].

Involvement of β_3-Adrenoceptors in Skeletal Muscle Glucose Uptake

β_3-Adrenergic receptors have recently been shown to be involved in metabolic regulation and thermogenesis in brown adipocytes [38, 39]. Skeletal muscle has also been reported to have β_3-adrenoceptors, as evidenced by analysis of its mRNA expression [40]. Accordingly, to ascertain further that glucose uptake in skeletal muscle is directly facilitated by sympathetic innervation, we have investigated whether BRL35135A, a potent β_3-agonist [39], is capable of inducing increase in glucose uptake in skeletal muscle. The β_3-agonist was infused intravenously into anesthetized rats at the dose of 0.5 mg/kg/h, and tissue glucose uptake was measured by the 2-[^{3}H]-DG method 30 min after start of the infusion (fig. 3). It can be seen that the β_3-agonist caused remarkable increases in the rate constant of glucose uptake in all three types of skeletal muscle and in brown adipose tissue as well. The response of skeletal muscle to the β_3-agonist increased,

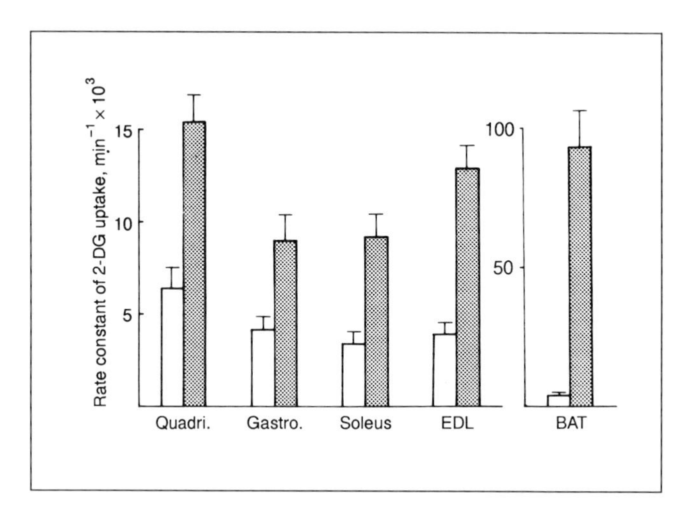

Fig. 3. Effect of infusion of BRL35135A on the rate constant of 2-[^{3}H]-DG uptake in skeletal muscle of anesthetized rats. BRL35135A (donation of SmithKline Beecham) was infused into the femoral vein of anesthetized rats at the dose of 0.5 mg/kg/h, and 30 min later 2-[^{3}H]-DG was injected for measurement of the rate of tissue glucose uptake. Four types of skeletal muscle (Quadri. = quadriceps; Gastro. = gastrocnemius; EDL = extensor digitorum longus) and the interscapular brown adipose tissue (BAT) were sampled. Results are means ± SE for 6 rats. ▩ = BRL35135A; □ = saline.

in a dose-dependent manner, with increasing dose of the agonist [H. Abe and T. Shimazu, unpubl. data]. The results show clearly that glucose uptake in skeletal muscle is directly stimulated by the sympathetic neurotransmitter, and that the β_3-adrenoceptors are involved in this mechanism.

Concluding Remarks

Our current and previous studies on hypothalamic control of energy substrate metabolism in relation to physical exercise are summarized in figure 4. During exercise, an arousal of the VMH neurons can be evoked, which then increase the sympathetic outflow to the liver, adipose tissue, skeletal muscle and some endocrine organs including the pancreas and adrenal.

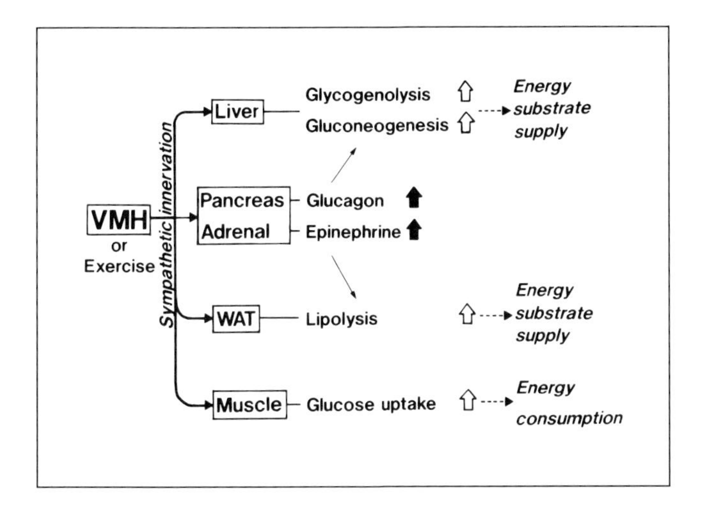

Fig. 4. Role of the VMH in a neural feed-forward regulation of substrate mobilization and utilization during exercise. WAT = white adipose tissue.

In the liver, increased sympathetic activity induces glycogen breakdown and stimulates gluconeogenesis in cooperation with glucagon released from the pancreas and epinephrine from the adrenal, ensuring supply of the energy substrate glucose to the working muscles. In white adipose tissue, the increased sympathetic outflow induces fat mobilization in concert with glucagon and adrenal catecholamines, which also contributes to energy substrate supply to muscles. In skeletal and heart muscles, a possible increase in sympathetic activity facilitates glucose uptake by a mechanism independent of insulin, but probably involving β_3-adrenoceptors, thus enabling muscle cells to take up the energy substrate during exercise in face of the decreased levels of insulin. The fact that regional blood flow to skeletal muscle increases in response to VMH stimulation [41] also favors the facilitation of glucose uptake in this tissue.

It is thus concluded that the VMH and sympathetic nervous system are importantly concerned in a neural feed-forward regulation of exercise metabolism not only by controlling substrate mobilization from the liver and adipose tissue, but also by facilitating its utilization in skeletal muscle. The notion that the VMH is deeply involved in the neural feed-forward regulation of substrate mobilization during exercise is supported by recent observations of Vissing et al. [42]. They reported that the exercise-induced increases in plasma concentrations of norepinephrine, epinephrine and glucose as well as hepatic glucose production are attenuated by local anesthesia of the VMH with marcain [42]. In addition, exercise-induced increase in plasma glucose concentration in swimming rats was abolished by infusion of the α-adrenergic blocker phentolamine into the VMH [43]. Destruction of the VMH was also reported to attenuate the increases in plasma free fatty acid and glycerol concentrations after exercise [44].

References

1 Shimazu T: Central nervous system regulation of liver and adipose tissue metabolism. Diabetologia 1981;20:343–356.
2 Shimazu T: Reciprocal innervation of the liver: Its significance in metabolic control; in Szabo AJ (ed): CNS Regulation of Carbohydrate Metabolism. Adv Metab Disord. New York, Academic Press, 1983, vol 10, pp 355–384.
3 Shimazu T: Neuronal control of intermediate metabolism; in Lightman SL, Everitt BJ (eds): Neuroendocrinology. Oxford, Blackwell Scientific, 1986, pp 304–330.
4 Shimazu T: Neuronal regulation of hepatic glucose metabolism in mammals. Diabetes Metab Rev 1987;3:185–206.
5 Saito M, Minokoshi Y, Shimazu T: Accelerated norepinephrine turnover in peripheral tissues after ventromedial hypothalamic stimulation in rats. Brain Res 1989; 481:298–303.
6 Luiten PGM, ter Horst GJ, Steffens AB: The hypothalamus, intrinsic connections and outflow pathways to the endocrine system in relation to the control of feeding and metabolism. Prog Neurobiol 1987;28:1–54.
7 Shimazu T, Matsushita H, Ishikawa K: Hypothalamic control of liver glycogen metabolism in adult and aged rats. Brain Res 1978;144:343–352.
8 Matsushita H, Shimazu T: Chemical coding of the hypothalamic neurones in metabolic control. II. Norepinephrine-sensitive neurones and glycogen breakdown in liver. Brain Res 1980;183:79–87.
9 Shimazu T, Ogasawara S: Effects of hypothalamic stimulation on gluconeogenesis and glycolysis in rat liver. Am J Physiol 1975;228:1787–1793.
10 Shimazu T, Fukuda A, Ban T: Reciprocal influences of the ventromedial and lateral hypothalamic nuclei on blood glucose level and liver glycogen content. Nature 1966; 210:1178–1179.
11 Shimazu T, Amakawa A: Regulation of glycogen metabolism in liver by the autonomic nervous system. II. Neural control of glycogenolytic enzymes. Biochim Biophys Acta 1968;165:335–348.
12 Shimazu T, Amakawa A: Regulation of glycogen metabolism in liver by the autonomic nervous system. VI. Possible mechanism of phosphorylase activation by the splanchnic nerve. Biochim Biophys Acta 1975;385:242–256.
13 Kumon A, Takahashi A, Kóri-Hara T: Epinephrine: A mediator of plasma glycerol elevation by hypothalamic stimulation. Am J Physiol 1977;233:E369–E373.

14 Shimazu T, Ishikawa K: Modulation by the hypothalamus of glucagon and insulin secretion in rabbits: Studies with electrical and chemical stimulations. Endocrinology 1981;108:605–611.
15 Shimazu T: Role of the hypothalamus in the induction of tryptophan pyrrolase activity in rabbit liver. J Biochem (Tokyo) 1964;55:163–171.
16 Bernardis LL, Frohman LA: Plasma growth hormone responses to electrical stimulation of the hypothalamus in the rat. Neuroendocrinology 1971;7:193–201.
17 Kumon A, Takahashi A, Hara T, Shimazu T: Mechanism of lipolysis induced by electrical stimulation of the hypothalamus in the rabbit. J Lipid Res 1976;17:551–558.
18 Takahashi A, Shimazu T: Hypothalamic regulation of lipid metabolism in the rat: Effect of hypothalamic stimulation on lipolysis. J Auton Nerv Syst 1981;4:195–205.
19 Galbo H: Hormonal and Metabolic Adaptations to Exercise. Stuttgart, Thieme-Stratton, 1983, pp 1–116.
20 Wasserman DH, Cherrington AD: Hepatic fuel metabolism during muscular work: Role and regulation. Am J Physiol 1991;260:E811–E824.
21 Holloszy JO, Constable SH, Young DA: Activation of glucose transport in muscle by exercise. Diabetes Metab Rev 1986;1:409–423.
22 Richter EA, Ploug T, Galbo H: Increased muscle glucose uptake after exercise: No need for insulin during exercise. Diabetes 1985;34:1041–1048.
23 Richter EA, Garetto LP, Goodman MN, Ruderman NB: Muscle glucose metabolism following exercise in the rat: Increased sensitivity to insulin. J Clin Invest 1982;69:785–793.
24 Wallberg-Henriksson H, Constable SH, Young DA, Holloszy JO: Glucose transport into rat skeletal muscle: Interaction between exercise and insulin. J Appl Physiol 1988;65:909–913.
25 Sudo M, Minokoshi Y, Shimazu T: Ventromedial hypothalamic stimulation enhances peripheral glucose uptake in anesthetized rats. Am J Physiol 1991;261:E298–E303.
26 Hom FG, Goodner CJ, Berrie MA: A [^{3}H]2-deoxyglucose method for comparing rates of glucose metabolism and insulin responses among rat tissues in vivo. Diabetes 1984;33:141–152.
27 Barker D, Saito M: Autonomic innervation of receptors and muscle fibres in cat skeletal muscle. Proc R Soc Lond 1981;B212:317–332.
28 Wong DT, Robertson DW, Bymaster FP, Krushinski JH, Reid LR: LY227942, an inhibitor of serotonin and norepinephrine uptake: Biochemical pharmacology of a potential antidepressant drug. Life Sci 1988;43:2049–2057.
29 Shimazu T, Minokoshi Y, Saito M, Sudo M: The role of the brain in the control of energy expenditure; in Oomura Y, Tarui S, Inoue S, Shimazu T (eds): Progress in Obesity Research 1990. London, Libbey, 1991, pp 67–74.
30 Sternlicht E, Barnard RJ, Grimditch GK: Mechanism of insulin action on glucose transport in rat skeletal muscle. Am J Physiol 1988;254:E633–E638.
31 Shimazu T, Sudo M, Minokoshi Y, Takahashi A: Role of the hypothalamus in insulin-independent glucose uptake in peripheral tissues. Brain Res Bull, 1991;27:501–504.

32 Zaninetti D, Greco-Perotto R, Assimacopoulos-Jeannet F, Jeanrenaud B: Effects of insulin on glucose transport and glucose transporters in rat heart. Biochem J 1988; 250:277–283.
33 Cushman SW, Wardzala LJ: Potential mechanism of insulin action on glucose transport in the isolated rat adipose cell: Apparent translocation of intracellular transport systems to the plasma membrane. J Biol Chem 1980;255:4758–4762.
34 Suzuki K, Kono T: Evidence that insulin causes translocation of glucose transport activity to the plasma membrane from an intracellular storage site. Proc Natl Acad Sci USA 1980;77:2542–2545.
35 Sternlicht E, Barnard RJ, Grimditch GK: Exercise and insulin stimulate skeletal muscle glucose transport through different mechanisms. Am J Physiol 1989;256: E227–E230.
36 Goodyear LJ, King PA, Hirshman MF, Thompson CM, Horton ED, Horton ES: Contractile activity increases plasma membrane glucose transporters in absence of insulin. Am J Physiol 1990;258:E667–E672.
37 King PA, Hirshman MF, Horton ED, Horton ES: Glucose transport in skeletal muscle membrane vesicles from control and exercised rats. Am J Physiol 1989;257: C1128–C1134.
38 Arch JRS, Ainsworth AT, Cawthorne MA, Piercy V, Sennitt MV, Thody VE, Wilson C, Wilson S: Atypical β-adrenoceptor on brown adipocytes as target for anti-obesity drugs. Nature 1984;309:163–165.
39 Cawthorne MA: Thermogenic drugs: The potential of selective β-adrenoceptor agonists as anti-obesity drugs; in Oomura Y, Tarui S, Inoue S, Shimazu T (eds): Progress in Obesity Research 1990. London, Libbey, 1991, pp 567–574.
40 Emorine LJ, Marullo S, Briend-Sutren MM, Patey G, Tate K, Delavier-Klutchko C, Strosberg AD: Molecular characterization of the human β_3-adrenergic receptor. Science 1989;245:1118–1121.
41 Iwai M, Hell NS, Shimazu T: Effect of ventromedial hypothalamic stimulation on blood flow of brown adipose tissue in rats. Pflügers Arch 1987;410:44–47.
42 Vissing J, Wallace JL, Scheurink AJW, Galbo H, Steffens AB: Ventromedial hypothalamic regulation of hormonal and metabolic responses to exercise. Am J Physiol 1989;256:R1019–R1026.
43 Scheurink AJW, Steffens AB, Benthem L: Central and peripheral adrenoceptors affect glucose, free fatty acids, and insulin in exercising rats. Am J Physiol 1988;255: R547–R556.
44 Nishizawa Y, Bray GA: Ventromedial hypothalamic lesions and the mobilization of fatty acids. J Clin Invest 1978;61:714–721.

Prof. Takashi Shimazu, Department of Medical Biochemistry,
Ehime University School of Medicine, Shigenobu, Ehime 791-02 (Japan)

Sato Y, Poortmans J, Hashimoto I, Oshida Y (eds): Integration of Medical and Sports Sciences. Med Sport Sci. Basel, Karger, 1992, vol 37, pp 33–39

Exercise and Immunity – Mechanisms of Action

Bente Klarlund Pedersen

Department of Infectious Diseases, University Hospital, Copenhagen, Denmark

In recent years, considerable interest has been directed to the effect of physical exercise on the immune system [1, 2]. There are at least two good reasons to examine the immune system in relation to exercise: (1) to explore how exercise performance affects susceptibility to infections, and (2) that exercise is a replicable and quantifiable stressor, which means that the effect of exercise on the immune system can be used as a general model of the effect of stress on the immune system.

In this paper I will focus on our own data regarding (1) the effect of acute exercise (a) on several immune parameters, (b) at different exercise levels and (c) its mechanisms of action, and (2) the effect of long-term training (a) on several immune parameters and (b) in healthy persons and in patients.

Effect of Acute Physical Exercise

Young, healthy volunteers were exercised on an ergometer bicycle at 75% of VO_2max for 1 h.

Exercise leads to a neutrophilia and a lymphocytosis during exercise. Two hours after bicycle exercise the lymphocyte concentrations decreased, while the neutrophils increased 4-fold [3]. The percentage of pan-T cells ($CD3^+$) declined during exercise, this was due to a fall in percentage of $CD4^+$ T cells, whereas the percentage of $CD8^+$ T cells did not change. The percentage of $CD20^+$ B cells did not change. The percentage as well as the absolute concentration of $CD16^+$ natural killer (NK) cells increased during exercise but was back to prevalues 2 h after exercise. Recently, it was shown that the $CD16^+$ cells isolated during exercise also expressed the CD8 marker ($CD16^+/CD8^+$) [Ullum and Pedersen, unpubl. observations]. The $CD14^+$ monocytes increased 2- to 3-fold following exercise and these cells expressed the HLA-DR receptors [3, 4].

Natural Killer Cell Activity

During exercise the NK cell activity increased and the NK cell fraction ($CD16^+$ cells) of blood mononuclear cells (BMNC) increased proportionally. When BMNC were preincubated with interferon-α (IFN-α), interleukin-2 (IL-2) or the prostaglandin inhibitor indomethacin, a significant increase in NK cell activity was registered at all times studied. During exercise the IL-2-enhanced NK cell activity increased significantly more than the IFN-α and indomethacin-enhanced NK cell activity. These results indicated that NK cells with a high IL-2 response capacity are recruited to the blood during exercise [3].

Two hours after maximal exercise, the NK cell activity dropped to a low point. The decreased NK cell activity was probably not due to fluctuations in the NK cell pool since the proportion of $CD16^+$ cells was normal. The fall in NK cell activity was probably due to inhibition by prostaglandins released by the elevated number of monocytes. In agreement with this, increased prostaglandin production was found after work and the NK cell activity of monocyte-depleted mononuclear cells did not decrease after exercise. Furthermore, indomethacin in vitro and in vivo fully restored the suppressed NK cell activity [5].

The NK cell activity increased during severe, moderate as well as light exercise (75, 50 and 25% of VO_2max, respectively), but the NK cell function was suppressed only following severe exercise, but returned to preexercise levels following moderate and light exercise. Interestingly, while an increased proportion of monocytes was demonstrated 2 h after exercise at 75% of VO_2max, the monocyte concentration did not change following exercise at 50 and 25% of VO_2max [1].

T Cell Function

Tvede et al. [4] showed that the phytohemagglutinin (PHA)-induced proliferative response decreased during exercise. To determine whether a single subpopulation of BMNC was responsible for the suppressed PHA proliferative response during work, BMNC were cultured in the presence of PHA, pulsed with [^{3}H]-thymidine, followed by FACS sorting into $CD4^+$ and $CD8^+$ subgroups. It was shown that the proliferative response per

$CD4^+$ cell did not change but that the contribution of the $CD4^+$ subgroup to proliferation declined [4]. Activation of T lymphocytes leads to production of IL-2 and to the expression of the TAC-low-affinity IL-2 receptor. During physical exercise an increased proliferative response to IL-2 was found despite unchanged IL-2 receptor expression. Reversal to preexercise value was found 2 h after exercise. The most responsive cells among the subpopulations were found to be the $CD16^+$ NK cells. The contribution of $CD16^+$ cells increased during exercise, while the $CD4^+$ cells decreased although they were expressing the highest number of low-affinity IL-2 receptors on the cell surface. These results suggest that the increased IL-2 responsiveness during physical activity reflects changes in cell composition and is not a consequence of cellular activation [Tvede et al., submitted].

B Cell Function

Using a reverse plaque-forming cell assay, Tvede et al. [6] showed that stimulation with pokeweed mitogen, recombinant IL-2 and Epstein-Barr virus (EBV) resulted in significant decreases in numbers of IgG-, IgM- and IgA-secreting blood cells during as well as 2 h after bicycle exercise with reversal to preexercise values 24 h later. Since the proportion of $CD20^+$ B cells was found unchanged, the suppression of immunoglobulin-secreting cells was not due to changes in numbers of B lymphocytes. A decline in the percentage of T cells, mainly $CD4^+$ cells, was measured only during exercise with normalization after exercise. Therefore, the B lymphocyte suppression, most pronounced 2 h after exercise, was probably not due to changes in T lymphocytes, as also indicated in the experiments using EBV-stimulated cultures since EBV acts directly on B lymphocytes. Two hours after exercise, an increased level of $CD14^+$ monocytes was observed. Purified B lymphocytes produced plaques only after EBV stimulation and in these cultures no exercise-induced suppression was found. Altogether these results indicate that the exercise-induced suppression of the plaque-forming cell response was mediated by monocytes [6].

Cytokines

Cytokines act as molecular signals between immunocompetent cells. BMNC were stimulated in vitro with lipopolysaccharide and PHA, and the following cytokines were estimated by bioassays as well as ELISA techniques. The production of IL-1α, β and IL-6 increased significantly 2 h following exercise. This increase is at least in part due to increased concentration of monocytes. It is, however, also possible that this increase is in part due to increased production of monokines, but this remains to be shown [7].

Other Immune Parameters

We were unable to show any significant effect of heavy short-term exercise on the levels of erythrocyte CR1, circulating immune complexes and the complement split products C3c and C3d [8].

Possible Mechanisms

Exercise induces increase in a number of stress hormones, including adrenaline, noradrenaline, growth hormone, β-endorphins and cortisol. Severe bicycle exercise induces also rise in body temperature to 39.5 °C. We have investigated the possibility that stress hormones or heating of the body are responsible for the exercise-induced immunomodulation.

Adrenaline. Selective administration of adrenaline to obtain plasma concentrations identical with those observed during bicycle exercise (75% of VO_2max, 1 h) showed that the modulation of BMNC subsets, NK activity and lymphocyte function was closely mimicked by administration of adrenaline; however, adrenaline caused minor increase in neutrophil concentration as compared with that induced by exercise [9].

Growth Hormone. Recently, we have obtained results showing that in vivo injection of growth hormone in humans had no effect on BMNC subset, NK activity, cytokine production or lymphocyte function, but induced a highly significant increase in neutrophil concentration [Kappel and Pedersen, unpubl. observations].

β-Endorphins. We have preliminary results showing that when healthy young men were given an epidural analgesia that blocked the afferent impulses and inhibited increase in β-endorphins and ACTH during exercise, this did not inhibit the exercise-induced increase in NK cell function or NK cell concentration. Based on these results we do not think that the β-endorphin response play an important role in exercise-induced modulation of NK cells [Klokker and Pedersen, unpubl. observations].

Cortisol. During bicycle exercise (75% of VO_2max, 1 h) only a minor increase in cortisol concentration is found (0.63 μmol/l before and 0.79

μmol/l after). It cannot easily be predicted from the literature how this minor increase in cortisol concentration can account for the exercise-induced immunomodulation, but cortisol may play a role in exercise training of longer duration.

Hyperthermia. To examine the selective effect of hyperthermia on the immune system, normal healthy young volunteers were immersed into a hot water bath whereby their rectal temperature increased to 39.5 °C. In vivo hyperthermia induced alterations that resembled the changes observed in relation to exercise. Hyperthermia induced, however, only minor increases in plasma adrenaline and noradrenaline concentrations [10, 11].

In conclusion, we found that adrenaline can account for the effect of physical exercise on NK activity, BMNC subsets and proliferative responses, while adrenaline together with growth hormone may be responsible for the increase in neutrophil concentration following exercise. It is, however, not clear whether the increase in body temperature during exercise can explain part of the immunomodulation in relation to exercise.

Effect of Long-Term Training

Can Overtraining Impair the Immune System?

There are several anecdotal reports from athletes and their coaches that hard training and in particular overtraining is associated with increased likelihood of respiratory infections including persistent colds, sore throats and flu-like illnesses. Sebastian Coe and Diane Edwards are 2 well-known examples who have suffered from toxoplasmosis, generally regarded as an opportunistic infection only affecting individuals with depressed immune system [12]. We do not know how common overtraining-related immune depression is and there is a lack of controlled studies on this.

The Immune System in Trained and Untrained?

The main problem while examining if the immune system is dependent on training status is to eliminate the effect of acute physical exercise. We examined top-trained elite cyclists, who were not overtrained, and untrained controls, who were not allowed to exercise 20 h prior to blood sampling.

The NK cell activity and concentration of CD16+ cells were found to be elevated in elite cyclists compared to sex- and age-matched controls. BMNC subpopulations, lymphocyte proliferative response and NK activity were measured under resting conditions in 29 highly trained male racing cyclists and in 15 untrained persons. The trained had elevated NK cell function and monocyte concentration and in vitro IL-1 production, but there were no significant differences in any T and B cell subsets, proliferative responses or in vitro production of IL-2 or IFN-α [13, 14].

The elite cyclists were examined during a period of high as well as low training degree, but the function of the immune system did not differ between seasons [14]. The immune system of male endurance athletes was unaffected by whether the diet in the preceding 6 weeks had been meat-rich diet or a lacto-ovo vegetarian diet [15].

When top-trained athletes due to anecdotal information are often found in a state of immune depression this is probably due to either overtraining or due to repetitive severe short-term exercise that is followed by severe immune depression lasting a few hours, during which time microorganisms might evade early immunologic recognition and establish infection. There are, however, no reasons to believe that a high training status influences the immune system in a bad manner.

Conclusion

Today we know quite a lot about how different immune parameters change in relation to physical exercise, while there is a lack of controlled studies on the clinical significance of this immunomodulation. Hopefully, future research will examine the clinical significance of exercise-induced immunomodulation.

References

1 Pedersen BK: The effect of physical exercise on the cellular immune system – Mechanisms of action: A review. Int J Sports Med 1991;12(suppl 1):23–29.
2 Fitzgerald L: Exercise and the immune system. Immunol Today 1988;9:337–339.
3 Pedersen BK, Tvede N, Hansen FR, Andersen V, Bendix T, Bendixen G, Bendtzen K, Galbo H, Haahr PM, Klarlund K, Sylvest J, Thomsen B, Halkjær-Kristensen J: Modulation of natural killer cell activity in peripheral blood by physical exercise. Scand J Immunol 1988;26:673–678.

4 Tvede N, Pedersen BK, Hansen FR, Bendix T, Christensen LD, Galbo H, Halkjær-Kirstensen J: Effect of physical exercise on blood mononuclear cell subpopulations and in vitro proliferative responses. Scand J Immunol 1989;29:383–389.
5 Pedersen BK, Tvede N, Klarlund K, Christensen LD, Hansen FR, Galbo H, Kharazmi A, Halkjær-Kristensen J: Indomethacin in vitro and in vivo abolishes post-exercise suppression of natural killer cell activity in peripheral blood. Int J Sports Med 1990;11:127–131.
6 Tvede N, Heilmann C, Halkjær-Kristensen J, Pedersen BK: Mechanisms of B lymphocyte suppression induced by acute physical exercise. J Clin Lab Immunol 1989; 30:169–173.
7 Haahr PM, Pedersen BK, Fomsgaard A, Tvede N, Diamant M, Klarlund K, Pedersen BK: Effect of physical exercise on in vitro production of IL-1, IL-6, TNFα, IL-2 and IFN-γ. Int J Sports Med 1991;12:223–227.
8 Thomsen BS, Haahr PM, Rødgaard A, Tvede N, Hansen FR, Steensberg J, Halkjær-Kristensen J, Pedersen BK: Complement receptor type one (CR1, CD35) on erythrocytes, circulating immune complexes and complement C3 split products after short-term physical exercise and training. Int J Sports Med, in press.
9 Kappel M, Tvede N, Galbo H, Haahr PM, Kjær M, Linstouw M, Klarlund K, Pedersen BK: Epinephrine can account for the effect of physical exercise on natural killer cell activity. J Appl Physiol 1991;70:2530–2534.
10 Kappel M, Stadeager C, Tvede N, Galbo H, Pedersen BK: Effect of in vivo hyperthermia on natural killer cell activity, in vitro proliferative responses and blood mononuclear cell subpopulations. Clin Exp Immunol 1991;84:175–180.
11 Kappel M, Diamant M, Hansen M, Klokker H, Pedersen BK: Effects of hyperthermia in vitro on proliferative responses of individual blood mononuclear cell subsets, interferon, lymphotoxin, tumour necrosis factor, interleukin 1, 2 and 6. Immunology 1991;73:304–308.
12 Fitzgerald L: Overtraining increases the susceptibility to infection. Int J Sports Med 1991;12(suppl 1):5–8.
13 Pedersen BK, Tvede N, Christensen LD, Klarlund K, Kragbak S, Halkjær-Kristensen J: Natural killer cell activity in peripheral blood of highly trained and untrained. Int J Sports Med 1989;10:129–131.
14 Tvede N, Steensberg J, Baslund B, Halkjær-Kristensen J, Pedersen BK: Cellular immunity in highly-trained elite racing cyclists during periods of training with high and low affinity. Scand J Med Sci Sports 1991;1:163–166.
15 Richter E, Tvede N, Kiens B, Pedersen BK: Immune parameters in male endurance athletes after a lacto-ovo vegetarian diet and a meat-rich Western diet. Med Sci Sports Exerc 1991;23:517–521.

Bente Klarlund Pedersen, Department of Infectious Diseases, M 7721,
Rigshospitalet, University Hospital, Tagensvej 20,
DK–2200 Copenhagen N (Denmark)

Sato Y, Poortmans J, Hashimoto I, Oshida Y (eds): Integration of Medical and Sports Sciences. Med Sport Sci. Basel, Karger, 1992, vol 37, pp 40–42

Lipid Peroxides and Exercise

Kunio Yagi

Institute of Applied Biochemistry, Yagi Memorial Park, Mitake, Gifu, Japan

Our basic and clinical investigations have indicated that lipid peroxide increases in the body, especially in the blood, are a direct cause of many diseases including age-related diseases and even aging itself [1, 2]. Among these diseases, cardiovascular disorders deserve special attention. The increase in lipid peroxides in the blood is induced by exogenous and/or endogenous causes. The former is mainly due to an intake of foods containing lipid peroxides, but the latter is due to different factors such as radiation, certain drugs or hazardous substances that produce free radicals when taken into the body, and metabolic disorders.

Forced intensive exercise, which is often accompanied by stress, is known to increase lipid peroxide levels in the body. Stress-induced lipid peroxide increase can be explained by our recent data on catecholamine-iron complexes [3]. Figure 1 shows that norepinephrine induces an enormous increase in microsomal lipid peroxidation in the presence of NADPH when it forms a complex with iron. The result with epinephrine is similar to that with norepinephrine.

Intensive exercise, even without stress, was found to increase the serum lipid peroxide level. Kanter et al. [4] reported that the level in humans was increased from 2.26 ± 0.39 to 3.60 ± 0.84 nmol/ml (in terms of malondialdehyde) after an 80-kilometer race. Intensive exercise would bring about local acidosis, which might induce liberation of iron from ferritin, which, in turn, would result in the formation of lipid peroxides. If local acidosis provokes injury to the biomembranes, iron-containing enzymes might be denatured and their iron as well would contribute to the formation of lipid peroxides.

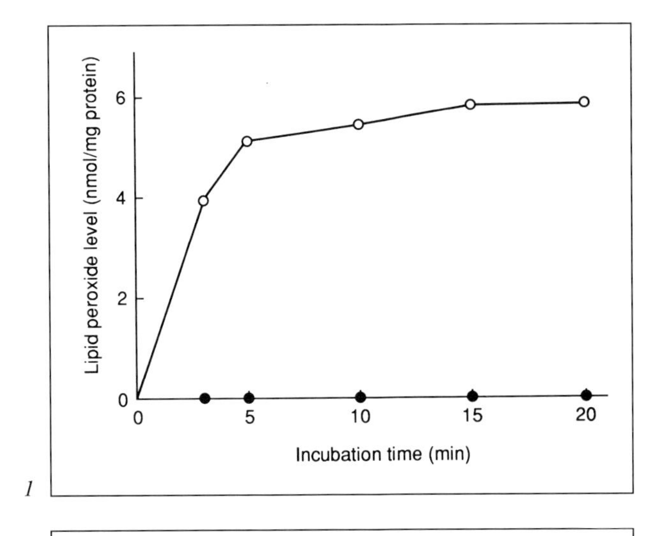

1

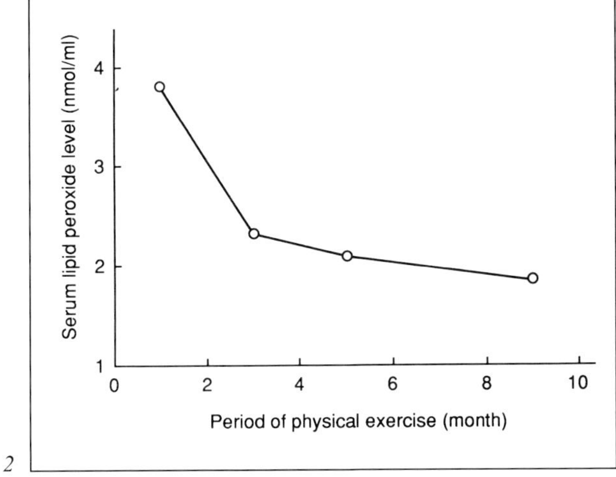

2

Fig. 1. Effect of norepinephrine-iron complex on microsomal lipid peroxidation. Reaction mixture containing rat liver microsomes (3 mg protein/ml) and norepinephrine-iron complex (norepinephrine, 10 μ*M*; $FeCl_3$, 20 μ*M*) was incubated in the presence of 0.1 m*M* of NADPH at pH 7.5 and 30 °C (○). As a control, the reaction mixture containing norepinephrine instead of norepinephrine-iron complex was also examined (●). The lipid peroxide level was measured according to Ohkawa et al. [6], and expressed in terms of malondialdehyde.

Fig. 2. Changes in serum lipid peroxide level during long-term nonexhaustive physical exercise. The serum lipid peroxide level of a man aged 60 years was followed for 9 months. The serum lipid peroxide level was measured according to Yagi [7], and expressed in terms of malondialdehyde.

On the contrary, moderate exercise was found to decrease the level of lipid peroxides in the serum. Figure 2 shows a typical result; when moderate exercise was continued for 3 months, the serum lipid peroxide level decreased by half of that before the exercise and decreased even further after 9 months of moderate exercise [5]. Since a sedentary life tends to produce obesity, in which case serum lipid peroxide levels increase, moderate exercise would prevent obesity and the situation would become better than that in the sedentary life. However, if moderate exercise has some effects other than the prevention of obesity, the mechanism for its effect of decreasing the lipid peroxide level should be studied in more detail.

References

1 Yagi K: Lipid peroxides and human diseases. Chem Phys Lipids 1987;45:337–351.

2 Yagi K: Role of lipid peroxides in aging and age-related diseases; in Ozawa T (ed): New Trends in Biological Chemistry. Tokyo, Japan Scientific Societies Press, 1991, pp 207–223.

3 Yagi K, Ishida N, Komura S, Ohishi N: Induction of lipid peroxidation in rat liver microsomes by catecholamine-iron complex. J Clin Biochem Nutr 1990;9:179–184.

4 Kanter MM, Lesmes GR, Kaminsky LA, Ham-Saeger JL, Nequin ND: Serum creatine kinase and lactate dehydrogenase changes following an eighty kilometer race: Relationship to lipid peroxidation. Eur J Appl Physiol 1988;57:60–63.

5 Igarashi S, Uemura M, Watanabe C, Kikkawa K, Aoki Y, Hashima Y, Meguro S, Yagi K: Unpublisched data.

6 Ohkawa H, Ohishi N, Yagi K: Assay for lipid peroxides in animal tissues by thiobarbituric acid reaction. Anal Biochem 1979;95:351–358.

7 Yagi K: A simple fluorometric assay for lipoperoxide in blood plasma. Biochem Med 1976;15:212–216.

Kunio Yagi, Institute of Applied Biochemistry, Yagi Memorial Park, Mitake, Gifu 505-01 (Japan)

Sato Y, Poortmans J, Hashimoto I, Oshida Y (eds): Integration of Medical and Sports Sciences. Med Sport Sci. Basel, Karger, 1992, vol 37, pp 43–51

Responsiveness of the Cat Plantaris to Functional Overload

Roland R. Roy, John A. Hodgson, Gordon R. Chalmers, William Buxton, V. Reggie Edgerton[1]

University of California at Los Angeles, Brain Research Institute, UCLA School of Medicine, Center for the Health Sciences, Los Angeles, Calif., USA

Functional overload (FO) of a skeletal muscle by removal of its synergists results in a compensatory hypertrophy and an adaptation in the mechanical and metabolic properties towards those normally observed in a 'slower' muscle. These adaptations in the rat plantaris have been characterized extensively [1–3]. It is generally assumed that the adaptations are a direct result of the mechanical overload, i.e. increased force, on the muscle [3, 4], although there are little or no data to substantiate this assumption. Gardiner et al. [5] have demonstrated that the amount of activation of the rat plantaris (as determined from EMG amplitudes) is elevated after FO in comparison to control. No in vivo data, however, are available comparing the forces produced by any muscle before and after FO in the same animal. One purpose of this study, therefore, was to determine the forces produced by the overloaded cat plantaris using implanted tendon force transducers. Another purpose was to provide some insight into the load-hypertrophy interrelationship using an animal model in which load could easily be titrated due to the sedentary nature of the cat. Some data on the use of the cat plantaris FO model have been published recently [6–8].

Methods

Four groups of adult cats were studied: (1) normal, control (NC; n = 6); (2) FO, control (FO-C; n = 3); (3) FO, treadmill-trained (FO-T; n = 6), and (4) FO, high-intensity play (FO-P; n = 6). Under aseptic conditions, the soleus and gastrocnemius were removed

[1] The authors thank I. Meadows and C. Koutures for their technical assistance. This work was supported, in part, by NIH grant NS16333.

bilaterally from each FO cat as described by Baldwin et al. [2] for rats. All cats were housed in large spacious cages. FO-T cats were exercised on a treadmill for ~ 5 min/day, 3–4 times/week, and FO-P cats were made to chase table tennis balls around a room for about 25–30 min/day, 5 days/week. FO cats were generally maintained for ~ 3 months after surgery, although some were maintained for up to 7 months.

The plantaris of some cats assigned to the FO-C and FO-T groups were implanted bilaterally with tendon force transducers [9, 10] and intramuscular EMG recording electrodes [11] ~2 weeks prior to FO surgery. The transducers were calibrated prior to implantation [9]. Most transducers were also calibrated on the tendon during an in situ terminal experiment. Force and EMG signals were recorded periodically during treadmill locomotion the week prior and up to 1 month after FO surgery. At this time the tendon transducers were removed and only EMG recordings were made. In one cat, a force transducer was reimplanted ~ 10 weeks after the initial FO surgery. Recordings were made during the week prior to the terminal experiment, i.e. in the 12th week of FO. During each recording session, each cat was tested at a variety of treadmill speeds (0.22–2.30 $m \cdot s^{-1}$) and inclines (0, 10, 20%). Force and EMG signals were recorded using an FM tape recorder (bandwidth 0–1 kHz) for later analysis.

The terminal experiment for the instrumented FO-C and FO-T cats involved the in situ calibration of the tendon transducer [see ref. 10 for details] and the testing of the isometric and isotonic mechanical properties of the plantaris [see ref. 12 for details]. Briefly, the leg was stabilized in a horizontal position. The plantaris was freed from surrounding tissues and its nerve isolated for stimulation. The plantaris tendon was cut distally and tied tightly to a noncompliant metal rod that was attached securely to a force lever. The transducer was left on the tendon so that its output could be compared directly with force values determined from the calibrated lever. Maximum isometric tension (P_0) was produced via supramaximal stimulation of the nerve at a frequency of 200 Hz. The muscle was maintained in a mineral oil bath kept at ~ 37 °C. At the end of the testing, the muscle was removed, cleaned of excess fat and connective tissue, weighed (wet weight) and prepared for histochemical analysis. The plantaris of the FO-P group was not tested physiologically. Rather, the plantaris on one side was injected with horseradish peroxidase for back-labeling of motoneurons, while the contralateral plantaris was simply removed, weighed (wet weight) and processed for histochemical analysis [6]. Raw force and EMG records were processed using the techniques described in detail previously [13].

Results

Amplitudes, but not durations, of both the plantaris force and integrated EMG associated with treadmill stepping were elevated following FO. During the first 2–3 weeks after FO, the mean force and integrated EMG were more than twice that observed in NC. For example, the mean plantaris force during locomotion at 1.84 $m \cdot s^{-1}$ was ~ 5 and 13 kg for NC and FO conditions in a large male cat (~ 5 kg), respectively (fig. 1). After ~ 4 weeks, the forces began to decline, whereas the EMG signals continued

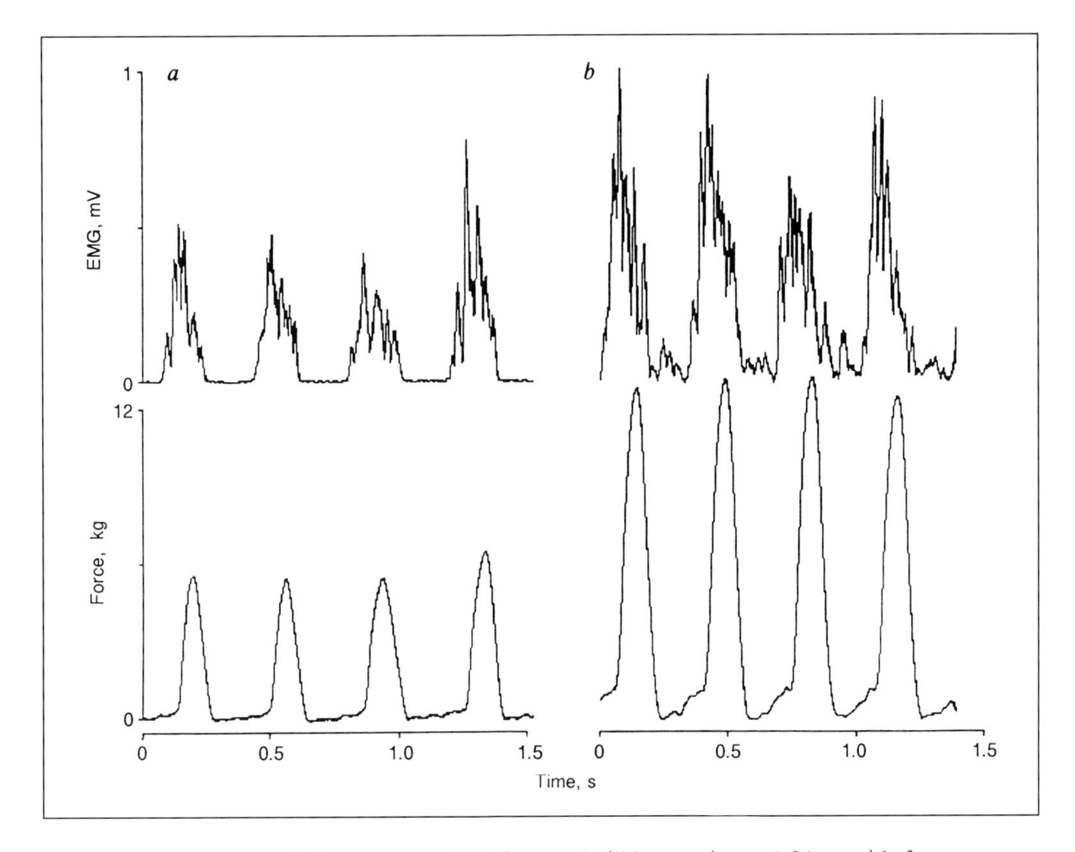

Fig. 1. Plantaris force and EMG during treadmill locomotion at 1.84 $m \cdot s^{-1}$ before FO *(a)* and 3 weeks after FO *(b)* in a male cat of ~ 5.0 kg from the FO-T group. Note that the amplitude, but not the duration, of the signals increased following overload.

to be elevated. In fact, after about 5 weeks the observed force had returned to ~ pre-FO levels. To determine the probable cause of the decrease in force, we anesthetized one cat during the 5th week of FO and investigated the tendon implant. The girth of the tendon had grown considerably and was beginning to encapsulate the transducer. Portions of the 'new' tendon were on the outside of the crosspiece of the transducer and tended to 'unload' the original tendon during contraction. One cat was investigated after 6 weeks of FO and the tendon was completely encapsulated in the 'new' tendon. In a later experiment, the transducer implanted during the FO surgery was removed after 2 weeks and then reimplanted after ~ 10 weeks

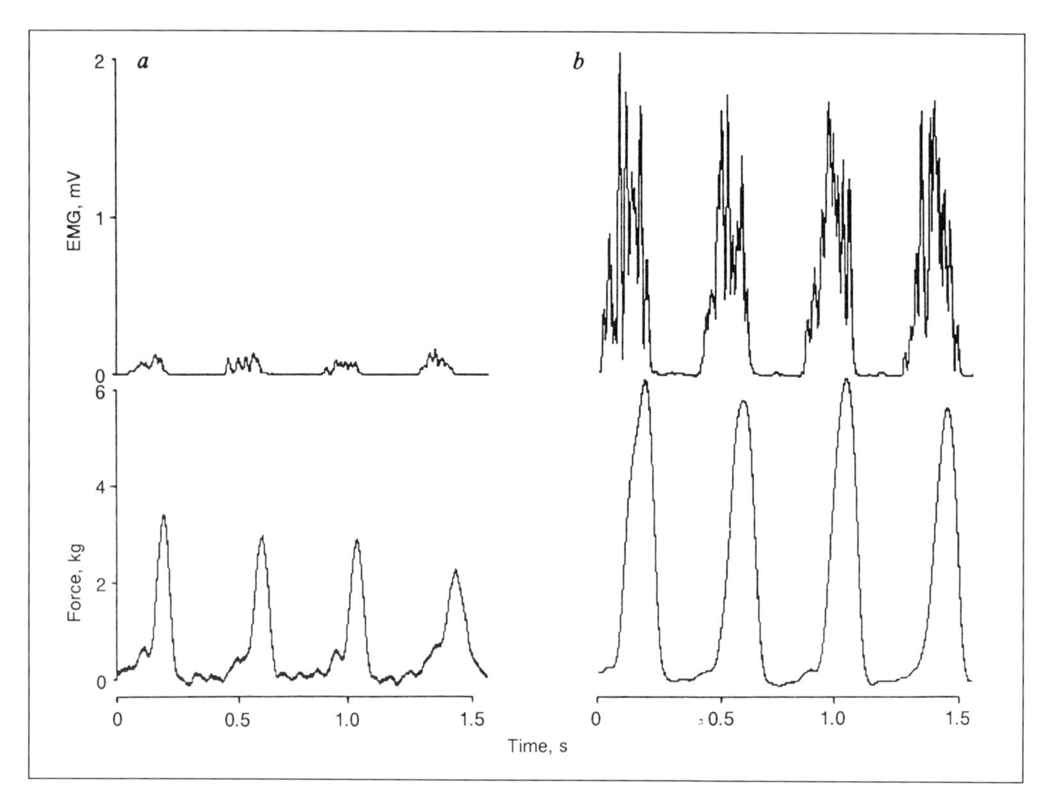

Fig. 2. Plantaris force and EMG during treadmill lomotion at 1.61 $m \cdot s^{-1}$ before FO *(a)* and 12 weeks after FO *(b)* in a female cat of ~ 3.0 kg from the FO-C group. Note that the force recordings were made from the same tendon transducer implanted twice, i.e. once prior to FO and once during the 10th week after FO.

of FO. The force and EMG patterns during locomotion were recorded at each time period. A comparison of the recordings made at a speed of 1.61 $m \cdot s^{-1}$ before FO and after 12 weeks of FO is shown in figure 2. These data show an ~ 2-fold increase in the mean amplitude of the force and an ~ 8-fold increase in the EMG activity after FO compared to NC. Note that the mean plantaris force in the NC condition was about 40% of that observed in the cat in figure 1 (~ 2 vs. 5 kg). This difference is consistent with the fact that the cat represented in figure 2 was a relatively small female cat (~ 3 kg) from the FO-C group, whereas the data in figure 1 were taken from a relatively large male cat (~ 5 kg) from the FO-T group.

After ~ 4 months of overload, mean plantaris weights in the FO-C, FO-T and FO-P groups were ~ 10 ($p > 0.05$), 65 and 130% heavier than control, respectively. In each case the mean body weight was slightly higher in NC than FO cats. P_0's paralleled the changes in muscle weight, i.e. there was about a 10 and 65% increase in the FO-C and FO-T groups. Thus, a normal specific tension (tension/unit tissue) was maintained after FO. No mechanical data were obtained from FO-P cats.

Discussion

The cat appears to be an excellent model to study the relationship between an increase in muscle loading and the associated hypertrophic response. The cat is, by nature, a very sedentary animal. Thus, after removal of the synergists of a muscle, the functionally isolated muscle will be subjected to minimum overload if the cat is permitted to remain sedentary. Careful postsurgical manipulation of activity patterns then can be used to titrate the amount and pattern of overload on the FO muscle. In the present study, a dramatic difference was observed in the hypertrophic response of the FO plantaris among cats that were allowed to be sedentary (FO-C; 10%), walked on a treadmill for a few minutes every other day (FO-T; 65%) or subjected to very intense play activity involving sprinting and jumping for some 30 min/day (FO-P; 130%).

The large difference in muscle mass between FO-C and trained FO cats contrasts with the much smaller difference in muscle mass between control and trained FO rats. Compared to sedentary rats, Riedy et al. [14] reported 67% hypertrophy in the plantaris of FO rats with normal cage activity and an increase to ~ 92% in FO rats that were treadmill-trained (up to 60 min/day at 27 $m \cdot min^{-1}$). Others have found virtually no effect of treadmill training [15] or weight lifting [Roy and Edgerton, unpubl. observations] on the hypertrophic response of overloaded rat muscles. This difference in responsiveness to increased activity between rats and cats may be related to the high activity levels of the rat, even when housed in small cages. The weight-bearing function during routine activity apparently is sufficient to induce a doubling of the rat plantaris mass within a few weeks [1].

Beginning ~ 1 week after FO surgery, plantaris EMG activity and forces produced in vivo were about twice those observed in the same muscle prior to FO. No changes in burst durations were observed. Similarly, Gardiner et al. [5] reported that EMG amplitude, but not duration, was increased in the rat FO plantaris during stepping. Kinematic data on 2 cats (and observation on all FO cats) locomoting on a treadmill showed that the FO cat walked plantigrade ('flatfooted') as opposed to digitigrade ('on the toes') during the first 3–4 weeks after surgery. A similar change in gait has been associated with FO when synergists of the plantaris were denervated [16]. As a consequence, the greater than normal lengthening excursion of the plantaris while it is activated should enhance force output as well as EMG activity [9]. These data indicate that the plantaris was heavily recruited and produced strong lengthening contractions during walking in the initial stages of FO.

By about the 4th week after FO, the cat was walking plantigrade, suggesting that compensatory mechanisms had developed to overcome the initial deficit and to reduce the amount of lengthening that occurred in the plantaris during stance. The cat that was reimplanted with a transducer after 12 weeks of FO showed a mean peak in vivo force during trotting of ~6 kg (fig. 2). The P_0 produced by this plantaris muscle during the in situ experiment performed less than a week after these recordings was ~9 kg. One might have expected this muscle to produce about 1.5 times the isometric tension in a maximum eccentric contraction, i.e. about 14 kg when fully activated. Thus, these data indicate that the plantaris in this FO-C cat was not maximally recruited during locomotion after 12 weeks of FO.

The expected overload on the FO plantaris can be estimated from the normal forces recorded during locomotion from the plantaris and triceps surae (the muscles removed during the FO surgery). The best estimate of the peak tendon force output from each of these muscles in a cat of ~3 kg body weight during trotting on a treadmill is: soleus, 2 kg [17]; plantaris, 2 kg (present data), and medial gastrocnemius, 2 kg [17]. No comparable data are available for the lateral gastrocnemius. Based on force data from both heads of the gastrocnemius during slow walking overground [18] and assuming a relatively constant relationship in the force output of the two heads during treadmill locomotion at the speeds studied, a force of ~1.5 kg is estimated for the lateral head. Thus, the total force at the Achilles tendon would be expected to be ~7.5 kg during trotting, i.e. 2.5 times body weight. The forces observed in the FO plantaris were in approximate agreement with this estimate, although the values recorded from the cat 3 months after FO are somewhat low.

The high forces produced on the plantaris tendon provide interesting insight into how the nervous system adapts to a sudden loss of its motor output potential followed by a rapid adjustment, both peripherally and centrally, following FO surgery. The average P_0 was ~8 and 16 kg for NC and FO-T cats. Thus, the peak eccentric force potential would be 12 and 24 kg, respectively (i.e. assuming that maximum eccentric contractions produce ~1.5 times P_0). The NC cats produced ~15% (2 out of 12 kg) and the FO-T cats at least 50% (13 out of 24 kg) of their eccentric potential during routine stepping. If there was a doubling of fiber size (2 times the normal force potential) but no change in the neural control system, then about 4 kg of force would be produced when the same proportion of motor units were recruited in FO cats. Thus, it appears that some compensatory events occurred in the central nervous system as well as in the muscle. Perhaps as many as twice the number of muscle fibers were activated by recruiting more motor units.

Further insight into the compensatory mechanisms associated with FO can be gained from the consideration of the adaptations at the motor unit, motoneuron and muscle fiber levels. Following 12 weeks of FO-P, the mean and range of soma sizes and oxidative capacities of the motoneurons innervating the plantaris were similar to NC [6, 7]. In contrast, decreases in soleus motoneuron oxidative capacity 60 days after FO [19] and in motoneuron size 4–8 weeks after FO [20] have been reported in rats. There is no clear explanation for these species differences. The plantaris fibers of FO as well as NC cats showed a wide range in metabolic properties and in size [6, 7]. In spite of the near-maximal recruitment of the FO muscle during movement, a mixture of fast glycolytic, fast oxidative glycolytic and slow oxidative fibers [21] persisted in FO muscles. As reported previously in cats [16] and rats [2], there was an increase in the percentage of slow fibers in the FO plantaris.

The normal heterogeneous properties of motor units after FO could reflect a persistence of variation of recruitment or a range of sensitivity to a given level of recruitment or overload. Some fibers in the FO cats were up to 3 times larger than the largest fibers, but some were much smaller than the smallest fibers observed in NC. Fiber size distributions indicated that fast fibers had a greater potential to enlarge than slow fibers. The very small fibers reacted with a neonatal myosin heavy-chain antibody, indicating the presence of newly formed fibers in FO plantaris [22]. Increases in the average tensions of all motor unit types in the cat medial gastrocnemius [23] and the rat plantaris [24] following FO substantiate that a sufficient overload can impact most of the fibers in FO muscles.

When there is a modest level of overload, a preferential increase in the P_0 of slow and fast fatigue intermediate units in the cat can occur [23].

With higher levels of chronically imposed overload, more or even most of the enhancement in fiber size can occur among the fast motor units (muscle fibers) that are presumably excited the least often [6, 7]. Thus, the stimulus:response ratio as a result of FO seems to differ substantially among different unit and fiber types. This difference in responsiveness may also contribute to the persistence of a heterogeneous population of motor units and fibers after overload. In any case, it is unlikely that all motor units are recruited similarly following FO.

References

1 Roy RR, Meadows ID, Baldwin KM, Edgerton VR: Functional significance of compensatory overloaded rat fast muscle. J Appl Physiol 1982;52:473–478.
2 Baldwin KM, Valdez V, Herrick RE, MacIntosh AM, Roy RR: Biochemical properties of overloaded fast-twitch skeletal muscle. J Appl Physiol 1982;52:467–472.
3 Roy RR, Baldwin KM, Edgerton VR: The plasticity of skeletal muscle: Effects of neuromuscular activity; in Holloszy J (ed): Exercise and Sports Sciences Reviews. Baltimore, Williams & Wilkins, 1991, vol 19, pp 269–312.
4 McDonagh MJN, Davies CTM: Adaptive response of mammalian skeletal muscle to exercise with high loads. Eur J Appl Physiol 1984;52:139–155.
5 Gardiner P, Michel R, Bowman C, Noble E: Increased EMG of rat plantaris during locomotion following surgical removal of its synergists. Brain Res 1986;380:114–121.
6 Chalmers GR, Roy RR, Edgerton VR: Motoneuron and muscle fiber succinate dehydrogenase activity in control and overloaded plantaris. J Appl Physiol 1991;71:1589–1592.
7 Chalmers GR, Roy RR, Edgerton VR: Enzymatic and morphologic response of cat plantaris fibers to functional overload. J Appl Physiol, in press.
8 Meadows ID, Roy RR, Powell PL, Edgerton VR: Contractile and fatigue properties of the compensatory overloaded cat plantaris. Physiologist 1982;25:260.
9 Gregor RJ, Roy RR, Whiting WC, Lovely RG, Hodgson JA, Edgerton VR: Mechanical output of the cat soleus during treadmill locomotion: In vivo vs in situ characteristics. J Biomech 1988;21:721–732.
10 Sherif MH, Gregor RJ, Liu LM, Roy RR, Hager CL: Correlation of myoelectric activity and muscle force during selected cat treadmill locomotion. J Biomech 1983;16:691–701.
11 Pierotti DJ, Roy RR, Gregor RJ, Edgerton VR: Electromyographic activity of cat hindlimb flexors and extensors during locomotion at varying speeds and inclines. Brain Res 1989;481:57–66.
12 Roy RR, Sacks RD, Baldwin KM, Short M, Edgerton VR: Interrelationships of contraction time, Vmax and myosin ATPase after spinal transection. J Appl Physiol 1984;56:1594–1601.
13 Lovely RG, Gregor RJ, Roy RR, Edgerton VR: Weight-bearing hindlimb stepping in treadmill-exercised adult spinal cats. Brain Res 1990;514:206–218.

14 Riedy M, Moore RL, Gollnick PD: Adaptive response of hypertrophied skeletal muscle to endurance training. J Appl Physiol 1985;59:127–131.
15 Baldwin KM, Cheadle WG, Martinez OM, Cooke DA: Effect of functional overload on enzyme levels in different types of skeletal muscle. J Appl Physiol 1977;42:312–317.
16 Wetzel MC, Gerlach RL, Stern LZ, Hannapel LK: Behavior and histochemistry of functionally isolated cat ankle extensors. Exp Neurol 1973;39:223–233.
17 Walmsley B, Hodgson JA, Burke RE: Forces produced by medial gastrocnemius and soleus muscles during locomotion in freely moving cats. J Neurophysiol 1978;41: 1203–1215.
18 Fowler EG: Relationship between Individual Muscle Moments and the Generalized Muscle Moment about the Ankle Joint in the Cat during the Stance Phase of Locomotion; PhD thesis, University of California, Los Angeles, 1990.
19 Pearson JK, Sickles DW: Enzyme activity changes in rat soleus motoneurons and muscle after synergist ablation. J Appl Physiol 1987;63:2301–2308.
20 Finkelstein DI, Lang JG, Luff AR: Functional and structural changes of rat plantaris motoneurons following compensatory hypertrophy of the muscle. Anat Rec 1991; 229:129–137.
21 Peter JB, Barnard RJ, Edgerton VR, Gillespie CA, Stempel KE: Metabolic profiles of three fiber types of skeletal muscle in guinea pigs and rabbits. Biochemistry 1972; 11:2627–2633.
22 Schiaffino S, Gorza L, Pitton G, Saggin L, Ausoni S, Sartore S, Lomo T: Embryonic and neonatal myosin heavy chain in denervated and paralyzed rat skeletal muscle. Dev Biol 1988;127:1–11.
23 Walsh JV Jr, Burke RE, Rymer WZ, Tsairis P: Effect of compensatory hypertrophy studied in individual motor units in medial gastrocnemius muscle of the cat. J Neurophysiol 1978;41:496–508.
24 Olha AE, Jasmin BJ, Michel RN, Gardiner PF: Physiological responses of rat plantaris motor units to overload induced by surgical removal of its synergists. J Neurophysiol 1988;60:2138–2151.

Roland R. Roy, PhD, University of California at Los Angeles, Brain Research Institute, UCLA School of Medicine, Center for the Health Sciences, 10833 Le Conte Avenue, Los Angeles, CA 90024-1761 (USA)

Sato Y, Poortmans J, Hashimoto I, Oshida Y (eds): Integration of Medical and Sports Sciences. Med Sport Sci. Basel, Karger, 1992, vol 37, pp 52–66

Muscular Dystrophic Mice: Apparently Hypertrophied Muscles and Bone-Muscle Growth Imbalance Hypothesis

Tsuyoshi Totsuka[a], *Kimi Watanabe*[a], *Isao Uramoto*[a], *Masato Nagahama*[b], *Takeshi Yoshida*[c], *Takaharu Mizutani*[d]

Departments of [a]Physiology and [b]Morphology, Institute for Developmental Research, Aichi Prefectural Colony for the Handicapped, Kasugai, Aichi; [c]Department of Physiology, Mie University School of Medicine, Tsu, Mie; [d]Department of Biochemistry, Faculty of Pharmaceutical Sciences, Nagoya City University, Nagoya, Japan

Progressive muscular dystrophies are widely believed to be diseases characterized by degeneration and impaired regeneration of skeletal muscle fibers. As a challenge to this 'muscle degeneration theory', we previously proposed a 'muscle defective maturation theory' and a further 'bone-muscle growth imbalance hypothesis' in order to explain our findings in hereditary muscular dystrophic mice (C57BL/6J *dy/dy*). This hypothesis implies a taboo and possible treatments for the disease, and also that the dystrophic mouse might be a good tool for research on stretch-induced hypertrophy of muscles because it appears to suffer from a defect in the mechanism of bone-growth-induced muscle growth. Stretched-contraction-induced damages followed by repair, instead of degeneration-regeneration in the muscle degeneration theory, would frequently occur locally along dystrophic fibers especially on movement. Consistent with this hypothesis, various interesting facts have already been reported by many authors. New ideas about the pathogenesis (e.g. central nuclei) are also presented.

Bone-Muscle Imbalance Hypothesis

The disease process is composed of two stages: (1) the early pathosis is a defective maturation (growth) [1–6] of muscle fibers (fig. 1), and (2) bones grow almost normally at least during the first month of life.

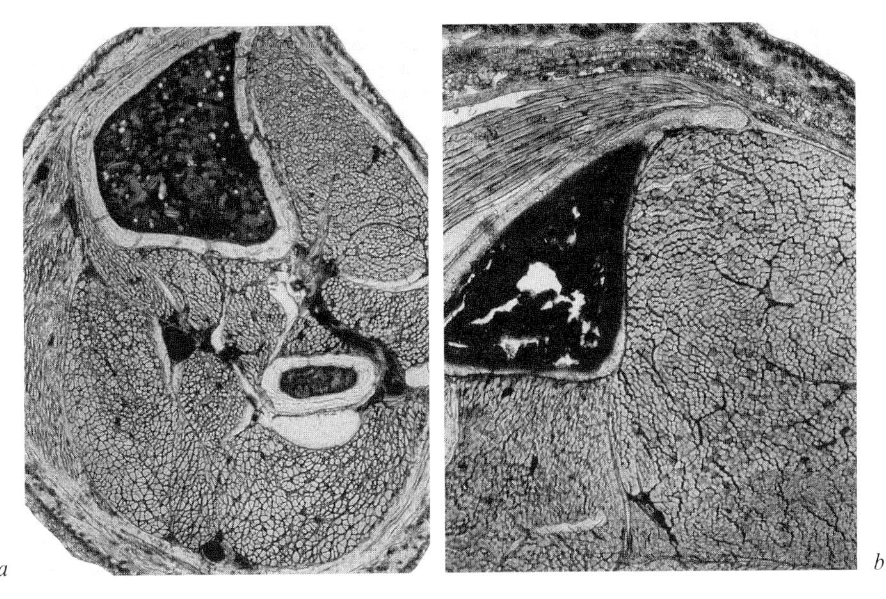

Fig. 1. Whole and partial views of cross-sections of crura from 5-month-old dystrophic (*a*) and normal (*b*) mice. The hindlegs were quickly fixed by swinging in a large volume of 70% ethanol, decalcified in 15% HNO_3, embedded in 22% gelatin and fixed in 10% formalin containing 1.5% glutaraldehyde and 1.5% paraformaldehyde [7]. They were then immersed in 10% crude egg albumin, fixed again in the formalin solution, frozen, cross-sectioned with a cryotome and stained with hematoxylin. Albumin may act as a good linker to stabilize the tissue. Specimens were mounted with 50% glycerin containing 7% gelatin and 1% phenol, and sealed with a manicure. It is obvious that the dystrophic muscles are severely hypotrophied compared to the growing bones (dystrophic and normal tibiae seem to be similar in size). × 10.

Consequently, with age, an imbalance between bones and muscles increases. This age-related aggravation in bone-muscle imbalance is responsible for the secondary pathogensis and progression of the disease [8–15].

Forelegs of Dystrophic Mice

Dystrophic mice [16] are obtained by mating heterozygotes (*dy*/+: the autosomal gene *dy* is recessive). One fourth of the offspring is dystrophic (*dy*/*dy*). Dystrophic mice first manifest the distinct symptom of dragging their hindlegs from about 2 weeks of age on. By about 3 months (adulthood), their hindlegs become immobile whereas their forelegs continue to

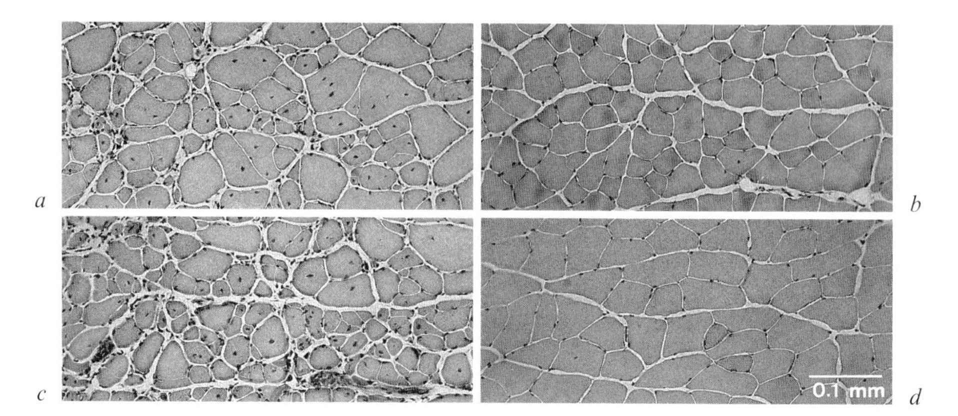

Fig. 2. Cross-sectional images of foreleg triceps (*a, b*) and hindleg rectus (*c, d*) muscles from 6-month-old dystrophic (*a, c*) and normal (*b, d*) mice. The fore- and hindlegs, bone and all tissue, were frozen in isopentane chilled with liquid nitrogen, sectioned with a disposable blade in a cryotome and stained with hematoxylin. It is noteworthy that this method enabled us to obtain good specimens with extremely less artefacts. Dystrophic fore- and hindleg muscles are similarly affected, showing (1) an abnormal increase in fiber size variation and in the frequency of central nuclei, and (2) disarrangement of muscle bundles. Note: the centronucleated fibers are not always small and have peripheral nuclei as well, suggesting an opposition to the generally and widely accepted idea that they are the regenerating and regenerated fibers. We have previously demonstrated that the size variation of dystrophic muscle fibers might be caused by a variation in their growth capability greatly shifted toward incompetence; i.e. some fibers grew like normal fibers, but a large part of them remained without growth [22, 23].

function. Accordingly, the foreleg muscles had been believed to be normal. However, we have shown [17] that: (1) the endurance of forelegs decreased rapidly with age; (2) the fore- and hindleg muscles were involved concurrently and almost similarly (fig. 2), and (3) the functional difference between them could be explained by a bone-growth difference in them. Also, some other authors have demonstrated that the foreleg muscles were affected [16, 18–21].

Hypotrophy of Motor Nerves

According to the muscle degeneration theory [see reviews in ref. 24–26], one might generally suppose that a secondary degeneration of motor nerves could occur with progression of the disease. Contrary to this

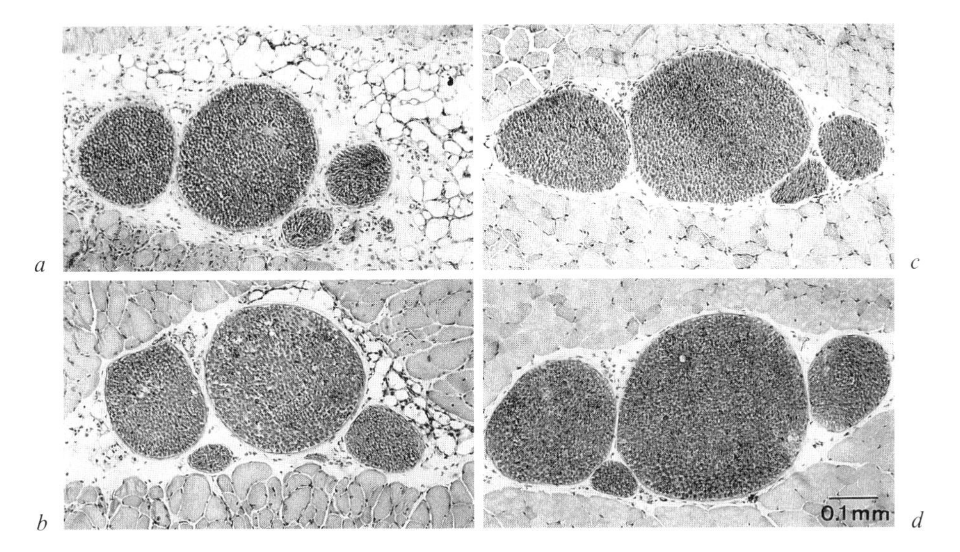

Fig. 3. Cross-sectional images of tibial nerves in thighs of dystrophic (*a, c*) and normal (*b, d*) mice of 32 (*a, b*) and 180 (*c, d*) days of age. The hindlegs were frozen, sectioned with a cryotome and stained with hematoxylin, as described in figure 2. The nerve with the largest diameter in each section is the tibial nerve.

assumption, however, the tibial nerves in thighs of dystrophic mice appeared not to become atrophied but to remain hypotrophied (fig. 3), even at advanced stages of the disease, by about 20%; i.e. the least diameters (mean ± SD in mm) of dystrophic and normal nerves in cross-sections were, respectively, 0.29 ± 0.01 (n = 3) and 0.31 ± 0.02 (n = 3) at 30–40 days, and 0.29 ± 0.01 (n = 5) and 0.35 ± 0.02 (n = 3) at 6 months (the average life span of dystrophic mice). It remains to be revealed whether the hypotrophy of tibial nerves is the cause or the result of the hypotrophied muscles in dystrophic mice.

Apparently Hypertrophied Muscles

From the bone-muscle imbalance hypothesis, one can expect various phenomena to occur or to be induced. Some of them have already been observed, as for example: (1) Small stature or such treatments that cause it are beneficial to dystrophic animals [9, 27–30] and Duchenne dystrophic

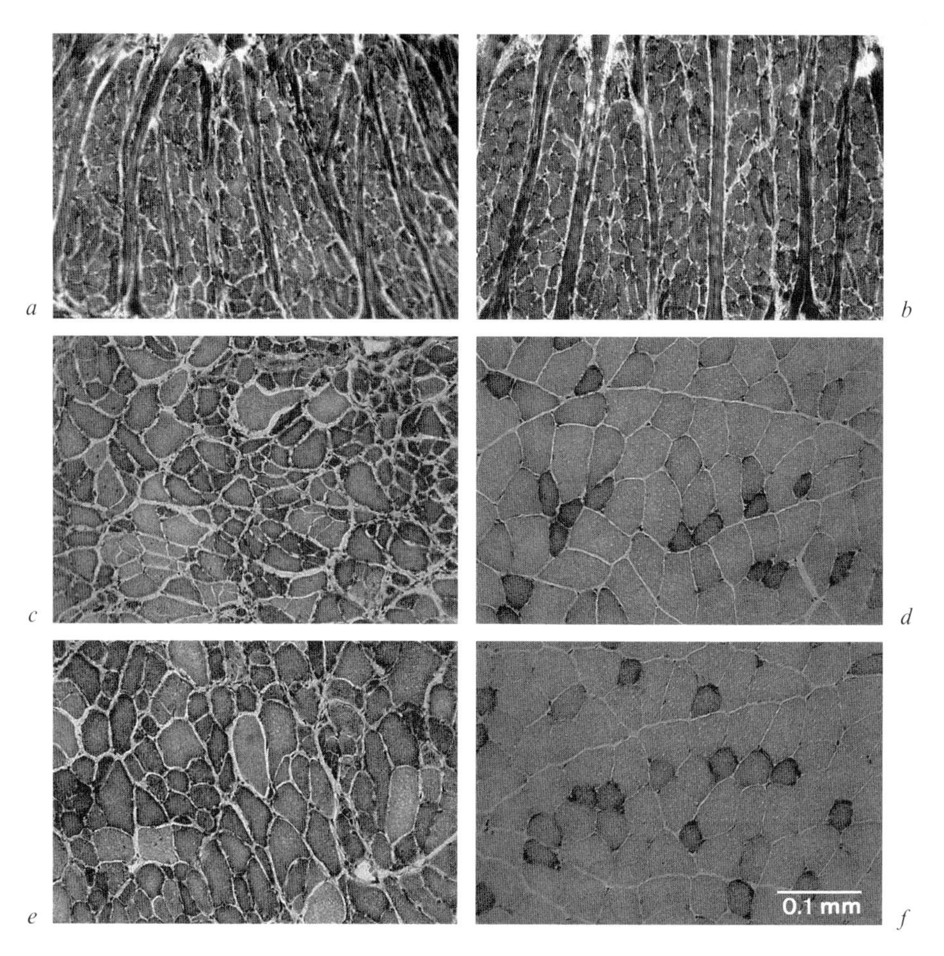

Fig. 4. Cross-sectional images of tongue (*a, b*), foreleg triceps (*c, d*) and hindleg rectus (*e, f*) muscles from 5-month-old dystrophic (*a, c, e*) and normal (*b, d, f*) mice. Tissues were excised, frozen, sectioned and stained for NADH-nitroblue tetrazolium oxidoreductase activity and with hematoxylin. The dystrophic tongue appears to be normal, and the dystrophic fore- and hindleg muscles are affected similarly (stained so densely as tongues).

patients [31–35]. (2) Growth hormone treatment is harmful [36, 37]. (3) Reduced load on muscles appears to improve the prognosis of dystrophic animals [38–41]. (4) The sarcomere of dystrophic chicken muscle at the resting state is abnormally extended [42]. (5) Passive stretch of adult

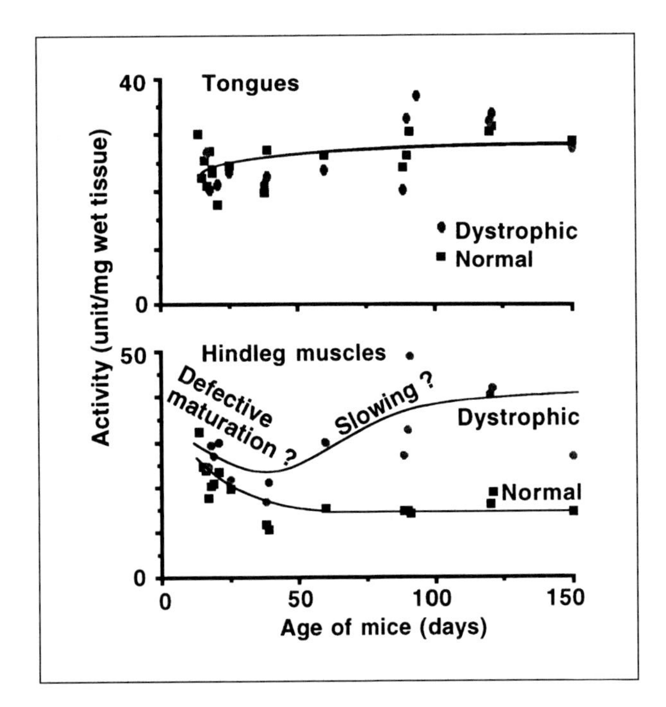

Fig. 5. NADH-ferricyanide oxidoreductase activities in tongues and hindleg rectus muscles from dystrophic and normal mice. Enzyme activity was measured in principle according to the method of Galante and Hatefi [46]. In dystrophic mice, tongues are normal against abnormal hindleg muscles. We speculate at present that: (1) the abnormally high oxidative nature, as also seen in figure 4, of dystrophic fast (white)-type muscles might be due to a defective maturation and slowing of their fibers [47–49], and (2) slowing of fibers might be a reflection of a continuous stimulus of stretching by growing bones. Interestingly, Silverman and Atwood [50] have previously reported a similar age-related change, i.e. a defective maturation and a shift toward a high oxidative state (high volume percentage of mitochondria), of gastrocnemius muscles of dy^{2J} mice.

normal chicken muscle produces a myopathy similar to muscular dystrophy [43]. (6) Thus far, our results on dystrophic mice indicate that striated muscles which are not or relatively less affected by bone growth are almost free of pathological changes [44, 45]. That is, we have previously found that the tongue (fig. 4, 5) and esophagus muscles are normal and that the costal diaphragm muscle fibers, which are distributed radially and are attached perpendicularly to the ribs, are also virtually normal. Recently, we further observed that the muscle fibers in the ears (pinnae aurium; data

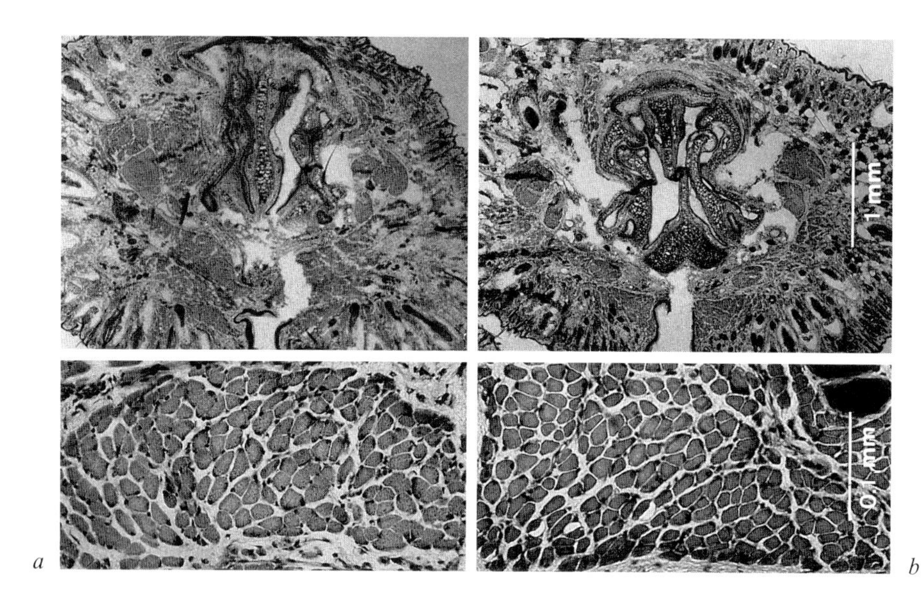

Fig. 6. Cross-sectional images of noses from 3-month-old dystrophic (*a*) and normal (*b*) mice. Noses were frozen, sectioned and stained with hematoxylin, as described in figure 2. Muscle fibers in dystrophic nose appear almost normal.

not shown) and noses (fig. 6) are normal. (7) The tongues [14], ears (table 1) and esophagi (table 2), which are independent of bones and would thus grow up normally, become apparently hypertrophied beyond the emaciated body in dystrophic mice. Hypertrophy of the tongue is known also in patients with Duchenne muscular dystrophy [51], though the reason has not been revealed.

Possible Treatments for Muscular Dystrophies

The bone-muscle imbalance hypothesis [8, 9, 11] itself implies a taboo (intensive exercise) and two possible symptomatic treatments for muscular dystrophies: (1) bone (body) growth inhibition (see above), and (2) antiinflammatory drugs such as the steroid prednisone. Subclinical minimal damages, which should not be followed by overinflammation, might frequently occur in dystrophic muscle fibers because flexor and extensor muscles which were already abnormally stretched at the resting state would be overstretched on movement [45].

Table 1. Apparent hypertrophy of ears (pinnae aurium) as compared with the emaciated body in 5-month-old dystrophic mice

Mouse group	Sex	n	Body weight g	Ear weight/ body weight mg/g	Ear area/ body weight $mm^2/g^{2/3, a}$
Normal (N)	♂	3	31.6±0.8	1.4±0.06	14.0±0.5
Normal (N)	♀	3	23.4±0.6	1.4±0.06	15.8±0.6
Dystrophic (Dy)	♀	6	9.7±1.2	2.2±0.22	25.7±2.1
Dy ♀/N	♀, %		41[b]	157[c]	163[c]
N ♀/N	♂, %		74	100	113

[a] The dimension was adjusted.
[b] Emaciated body.
[c] Hypertrophy.

Table 2. Apparent hypertrophy of esophagi beyond the emaciated body in dystrophic mice

	Body weight g	Organ weight/body weight, mg/g		
		esophagus	spleen	diaphragm
Normal	27.0±0.6	0.46±0.06	2.43±0.25	2.50±0.23
Dystrophic	10.3±0.6	0.65±0.09	2.32±0.30	3.11±0.37
Dy/N, %	38[a]	141[b]	95	124

Six pairs of 3-month-old dystrophic (Dy) and normal (N) mice were used to obtain each mean ± SD. Esophagus: thoracic part; diaphragm: costal part.
[a] Emaciated body.
[b] Hypertrophy.

Recently, as a conclusion of a conference where leading researchers on dystrophin [52, 53] got together which genetic deficiency is assumed to be the primary cause of muscular dystrophy in Duchenne patients and *mdx* mice, Mastaglia [54] recommended three forms of treatment which could be implemented right away in centers where there are already well-developed protocols for assessing therapeutic response and side effects: (1) growth hormone inhibitors; (2) prednisone [55], and (3) kinesiologic approach.

Other Dystrophic mdx *Mice*

The other dystrophic *mdx* mice [56, 57] are attractive objects though it remains to be examined to what extent the bone-muscle imbalance hypothesis is applicable to them. Violent changes very similar to inflammatory images are seen in muscles of young *mdx* mice (right and middle sections of fig. 7). These pathological changes are prominent in skeletal muscles of *mdx* mice only around the postnatal period from 30 to 60 days. It is widely believed that *mdx* mice manifest a dystrophic symptom indicative of muscle weakness [58] for this period but not afterward, and that the degeneration of their muscle fibers could be compensated for by a complete regeneration. However, figure 8 clearly shows that the endurance of forelegs decreases rapidly with age (fig. 8a) and that their muscles are severely affected and never seem to have recovered so well at 20 months of age yet (fig. 8b). Different from *dy* mice [45], the costal diaphragm muscles [59] in *mdx* mice are also affected as severely as the limb and backbone muscles, showing an abnormal increase in fiber size variation (shifted toward smaller caliber) and in the frequency of central nuclei [60]. This difference in severity of costal diaphragm muscles between *dy* and *mdx* mice might be attributable in part to some unknown cause of the difference in their body dimensions; at adulthood, *dy* mice are very lean but *mdx* mice are apparently well built, suggesting that diaphragm muscle fibers are stretched more severely in the latter than in the former. It would be very interesting to know whether anti-inflammatory drugs are effective in preventing these changes to improve the prognosis of the disease.

Pathological Meaning of Central Nuclei

In foreleg triceps muscles of *mdx* mice (fig. 9), the frequency of centronucleated fibers with ordinary diameters increases suddenly after about 30 days of age when violent pathological changes become first observable, indicating that the centronucleated fibers cannot be the regenerating fibers which at first should have an extremely small caliber. Therefore, based on

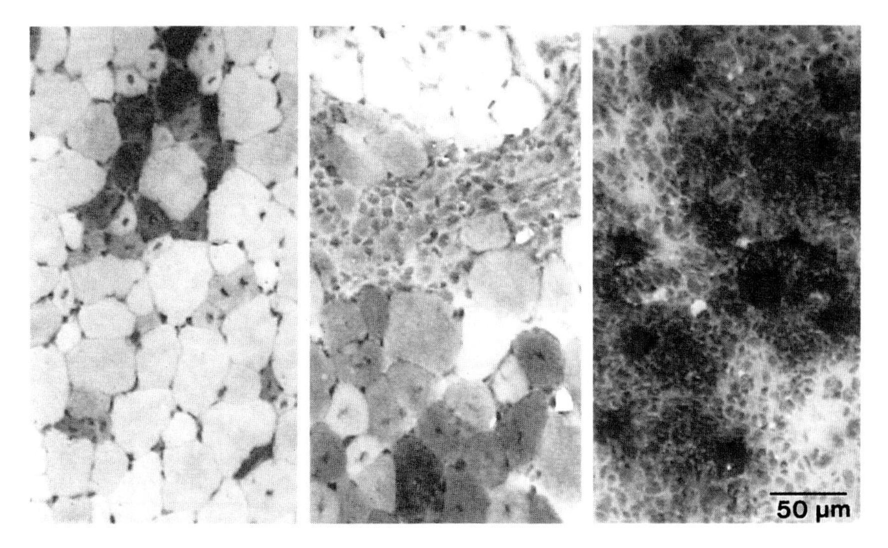

Fig. 7. Representative pathological images observed in foreleg triceps muscles of 32-day-old *mdx* mice. Muscles were excised, frozen, sectioned transversely and stained with periodic acid-Schiff and hematoxylin. It is noteworthy that centronucleated fibers are within the range of other fibers in size or stainability.

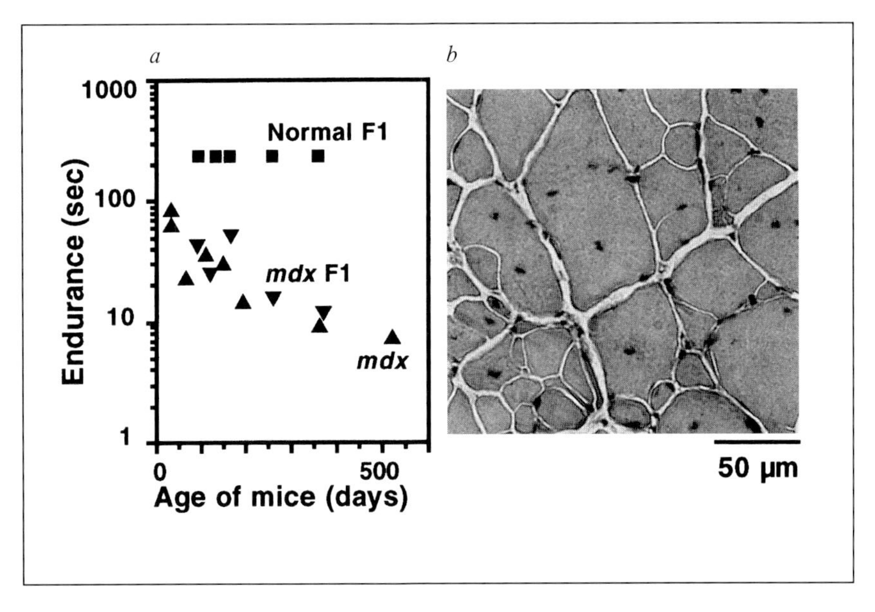

Fig. 8. Endurance of forelegs (*a*) and cross-sectional image of their muscles (*b*) in *mdx* mice. *a* Endurance of forelegs in *mdx* and *mdx* F1 mice decreases rapidly with age compared to the powerful forelegs of normal F1 mice. Endurance was measured according to our method [17]. In brief, mice were forced to hang on a nylon-covered wire (1 mm thick) which was stretched horizontally (40 cm high). The duration in seconds of hanging until the mouse fell was measured as endurance of their forelegs. Normal F1 and *mdx* F1 are female (X^{mdx}/X^{+}) and male (X^{mdx}/Y) crossbreds obtained by pairing female mice (X^{mdx}/X^{mdx}) of the C57BL/10ScSn strain with male mice (X^{+}/Y) of the DW/J strain, respectively. The plots denoted as *mdx* (▲) are results of male (X^{mdx}/Y) or female (X^{mdx}/X^{mdx}) mice of the inbred C57BL/10ScSn. Normal F1 mice would not fall beyond 4 min and their values were plotted at 240 s in the figure. *b* Cross-sectional image of the foreleg triceps muscle from a 20-month-old *mdx* mouse, stained with hematoxylin. One can see a fiber size variation, centronucleated fibers with large diameter and a fiber with a nucleus in a fusing or splitting line.

this and our other findings [61–63], we now consider three possibilities: peripheral nuclei (and/or satellite cells) move rapidly toward the inner part of the fiber to become active when fibers are (1) damaged or (2) forced to work extraordinarily. (3) On fusion of fibers, peripheral nuclei, satellite cells and fibroblasts are involved in the junctional area to become central nuclei with heterogeneity. We assume that central nuclei might play some important role in the growth and homeostasis of normal muscles where

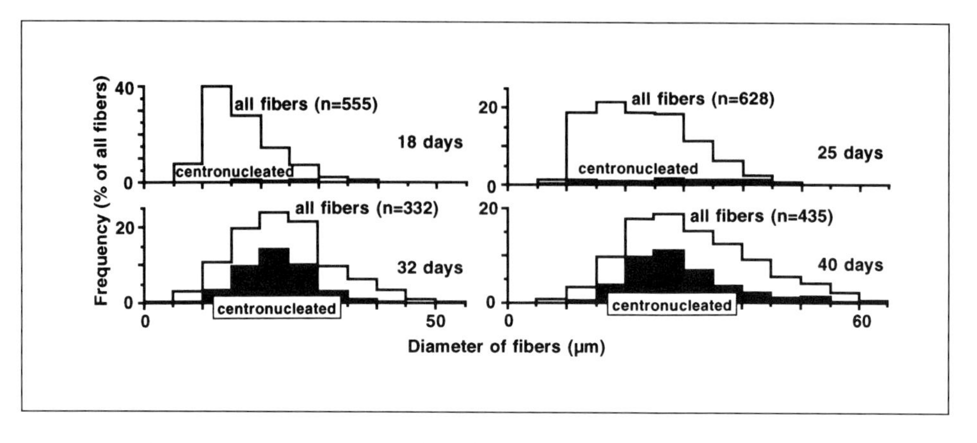

Fig. 9. Age-related changes in fiber diameter frequency distribution in foreleg triceps muscles of *mdx* mice. The frequency is the percentage of all fibers counted for each muscle. The black area in each graph represents centronucleated fibers.

centronucleated fibers are observed, though apparently not so frequently (about 1% or less of total fibers), in cross-sections. It is noteworthy that this frequency would never be so small as negligible considering that muscle fibers are very long.

References

1 Watanabe K, Totsuka T: Developmental defects of hindleg muscle cells in hereditary dystrophic mice. Med Biol 1976;93:203–206.
2 Farnbach GC, Brown MJ, Barchi RL: A maturational defect in passive membrane properties of dystrophic mouse muscle. Exp Neurol 1978;62:539–554.
3 Totsuka T, Watanabe K, Kiyono S: Maturational defects of muscle fibers in the muscular dystrophic mouse. Congen Anomal 1981;21:253–259.
4 Skau KA: The acetylcholinesterase abnormality in dystrophic mice is a reflection of a maturational defect. Brain Res 1983;276:192–194.
5 Watanabe K, Uramoto I, Totsuka T: Ontogenetic aspects of changes in muscular potentials at medial gastrocnemius muscles of dystrophic mice due to prolonged stimulation. J Neurol Sci 1984;66:59–66.
6 Totsuka T, Watanabe K, Uramoto I, Kiyono S: Comparison of fatigue resistant properties of gastrocnemius and soleus muscles between muscular dystrophic mice and other strains of mice. Congen Anomal 1984;24:163–172.
7 Ito M: Response properties and topography of vibrissa-sensitive VPM neurons in the rat. J Neurophysiol 1988;60:1181–1197.

8 Totsuka T, Watanabe K: A working hypothesis for the mechanism involved in the pathogenesis of murine muscular dystrophy and for the development of a possible symptomatic treatment for the disease. Med Biol 1979;99:1–5.
9 Totsuka T, Watanabe K, Kiyono S: Masking of a dystrophic symptom in genotypically dystrophic-dwarf mice. Proc Jap Acad 1981;57B:109–113.
10 Totsuka T, Watanabe K, Uramoto I: Dystrophy-specific alterations in muscular potentials under prolonged sciatic stimulation of 5 per sec in dystrophic, dwarf and their crossbred mice. Congen Anomal 1982;22:207–210.
11 Totsuka T, Watanabe K, Uramoto I: A bone-muscle imbalance hypothesis for the pathogenesis of murine muscular dystrophy; in Ebashi S, Ozawa E (eds): Muscular Dystrophy: Biomedical Aspects. Tokyo, Japan Scientific Societies Press, 1983, pp 29–38.
12 Totsuka T, Watanabe K, Uramoto I: Differences among dystrophic, dwarf, and their crossbred mice in the time course of changes in extracellular muscle action potentials induced by 5-Hz stimulation. Exp Neurol 1984;84:616–626.
13 Futo T, Hitaka T, Mizutani T, Okuyama H, Watanabe K, Totsuka T: Fatty acid composition of lipids in tongue and hindleg muscles of muscular dystrophic mice. J Neurol Sci 1989;91:337–344.
14 Hitaka T, Mizutani T, Watanabe K, Totsuka T: The high content of natural suppressor serine tRNA in dystrophic mouse muscle. Biochem J 1990;266:201–206.
15 Mizutani T, Hitaka T, Maruyama N, Totsuka T: The role of natural opal suppressor tRNA in incorporation of selenium into glutathione peroxidase; in Tomita H (ed): Trace Elements in Clinical Medicine. Tokyo, Springer, 1990, pp 307–314.
16 Michelson AM, Russell ES, Harman PJ: Dystrophia muscularis: A hereditary primary myopathy in the mouse. Proc Natl Acad Sci USA 1955;41:1079–1084.
17 Totsuka T, Watanabe K: Some evidence for concurrent involvement of the fore- and hindleg muscles in murine muscular dystrophy. Exp Anim 1981;30:465–470.
18 Rowe RWD, Goldspink G: Muscle fiber growth in five different muscles in both sexes of mice. II. Dystrophic mice. J Anat 1969;104:531–538.
19 Komatsu K, Tsukuda K, Hosoya J, Satoh S: Elevations of cathepsin B and cathepsin L activities in forelimb and hindlimb muscles of genetically dystrophic mice. Exp Neurol 1986;93:642–646.
20 Aoyagi T, Wada T, Kojima F, Nagai M, Harada S, Umezawa H: A multiple study on enzymatic changes in limb muscles and heart muscle of dystrophic mice. Biotech Appl Biochem 1987;9:355–361.
21 Geaves DS, Dufresne MJ, Fackrell HB, Warner AH: Age-related changes and tissue distribution of parvalbumin in normal and dystrophic mice of strain 129 ReJ. Muscle Nerve 1991;14:543–552.
22 Totsuka T, Uramoto I: A variation in growth capability of myofibers shifted to incompetence as the pathogenesis of mouse muscular dystrophy. J Physiol Soc Jap 1986;48:703–706.
23 Totsuka T: A paradoxical growth of large myofibers in growth-arrested rectus femoris muscles of muscular dystrophic mice. Congen Anomal 1986;26:157–167.
24 Kakulas BA, Adams RD: Diseases of Muscle: Pathological Foundations of Clinical Myology. Philadelphia. Harper & Row, 1985.
25 Mrak RE: Muscle Membranes in Diseases of Muscle. Florida, CRC Press, 1985.

26 Emery AEH: Duchenne Muscular Dystrophy. Oxford, Oxfrod University Press, 1987.
27 King DB, King CR, Jacaruso RB: Avian muscular dystrophy: Thyroid influence on pectoralis muscle growth and glucose-6-phosphate dehydrogenase activity. Life Sci 1981;28:577–585.
28 Karpati G, Jacob P, Carpenter S, Prescott S: Hypophysectomy mitigates skeletal muscle fiber damage in hamster dystrophy. Ann Neurol 1985;17:60–64.
29 Valentine BA, Cooper BJ, Lahunta A, O'Quinn R, Bleu JT: Canine X-linked muscular dystrophy: An animal model of Duchenne muscular dystrophy. Clinical studies. J Neurol Sci 1988;88:69–81.
30 Kurtenbach E, Moraes SS, Trocado MT, Lobo GF, Nascimento PS, Verjovski-Almeida S: Beneficial effects of anti-growth hormone antiserum in avian muscular dystrophy. FASEB J 1989;3:2189–2193.
31 Zatz M, Betti RTB, Levy JA: Benign Duchenne muscular dystrophy in a patient with growth hormone deficiency. Am J Med Genet 1981;10:301–304.
32 Zatz M, Frota-Pessoa O: Suggestion for a possible mitigating treatment of Duchenne muscular dystrophy. Am J Med Genet 1981;10:305–307.
33 Zatz M, Betti RTB, Frota-Pessoa O: Treatment of Duchenne muscular dystrophy with growth hormone inhibitors. Am J Med Genet 1986;24:549–566.
34 Zatz M, Rapaport D, Vainzof M, Rocha JML, Pavanello RCM, Colletto GMDD, Peres CA: Relation between height and clinical course in Duchenne muscular dystrophy. Am J Med Genet 1988;29:405–410.
35 Wu RH, Blethen SL, Chasalow FI, Spiro A, Buiumsohn A, Saenger P: Evidence for abnormal growth hormone (GH) in a body with Duchenne muscular dystrophy (DMD) and growth failure. Pediatr Res 1988;23:560A.
36 Chyatte SB, Rudnam D, Patterson JH, Gerron GG, O'Beirne I, Jordan A, Shavin JS: Human growth hormone and estrogens in boys with Duchenne muscular dystrophy. Arch Phys Med Rehabil 1973;54:248–253.
37 Logghe K, Wit JM, Jennekens F, Pruijs JEH: Respiratory deterioration during growth hormone therapy in a case of congenital nemaline myopathy. Eur J Pediatr 1990;150:69–71.
38 Wirtz P, Loermans H, Jonge WW: Long term functional improvement of dystrophic mouse leg muscles upon early immobilization. Br J Exp Pathol 1986;67:201–208.
39 Elder GCB: Effects of hind limb suspension on the development of dystrophic hamster muscle. Exp Neurol 1988;99:187–200.
40 Dangain J, Vrbova G: Response of dystrophic muscles to reduced load. J Neurol Sci 1988;88:277–285.
41 Lane RJM, Watmough NJ, Jaros E: Effects of tenotomy on muscle histology and energy metabolism in normal and dystrophic mice. J Neurol Sci 1989;92:307–316.
42 Ashmore CR, Mechling K, Lee YB: Sarcomere length in normal and dystrophic chick muscles. Exp Neurol 1988;101:221–227.
43 Ashmore CR, Hitchcock L, Lee YB: Passive stretch of adult chicken muscle produces a myopathy remarkably similar to hereditary muscular dystrophy. Exp Neurol 1988; 100:341–353.
44 Totsuka T: Normal diameter distribution of tongue muscle fibers in muscular dys-

trophic mice: Consistent with the bone-muscle imbalance hypothesis for the pathogenesis. Proc Jap Acad 1987;63B:131–134.

45 Totsuka T, Watanabe K, Uramoto I, Katoh-Semba R, Sano M, Nagahama M. Yoshida T, Mizutani T: Muscular dystrophic (*dy*) mice: Bone-muscle imbalance hypothesis. Proc JAAMHD 1990;6:20–30.

46 Galante YM, Hatefi Y: Resolution of complex I and isolation of NADH dehydrogenase and an iron-sulfur protein. Methods Enzymol 1978;53:15–21.

47 Parslow HG, Parry DJ: Slowing of fast-twitch muscles in the dystrophic mouse. Exp Neurol 1981;73:686–699.

48 Uramoto I, Watanabe K, Totsuka T: Changing patterns in muscular potentials at MG and SOL muscles of dystrophic and normal mice due to prolonged stimulation. J Physiol Soc Jap 1984;46:99–102.

49 Kato K, Shimizu A, Totsuka T: Developmental changes in fiber type-related proteins in soleus, rectus femoris, and heart muscles of normal and dystrophic mice. J Neurol Sci 1988;85:161–171.

50 Silverman H, Atwood H: Increase of muscle mitochondrial content with age in murine muscular dystrophy. Muscle Nerve 1982;5:640–644.

51 Nagaoka M, Minami R, Wakai S, Okabe M, Kameda K, Annaka S, Tachi N, Shinoda M: Duchenne-gata shinkousei kin-jisutorofii-shou ni okeru zetsu-hidai to koukuu keitai ni tsuite (Japanese). No To Hattatsu 1987;19:422–424.

52 Hoffman EP, Brown RH, Kunkel LM: Dystrophin: The protein product of the Duchenne muscular dystrophy. Cell 1987;51:919–928.

53 Sugita H, Arahata K, Ishiguro T, Suhara H, Tsukahara T, Ishiwa S, Eguchi C, Nonaka I, Ozawa E: Negative immunostaining of Duchenne muscular dystrophy (DMD) and *mdx* muscle surface membrane with antibody against synthetic peptide fragment predicted from DMD cDNA. Proc Jap Acad 1988;64B:37–39.

54 Mastaglia FL: Workshop II. Recommendations; in Kakulas BA, Mastaglia FL (eds): Pathogenesis and Therapy of Duchenne and Becker Muscular Dystrophy. New York, Raven Press, 1990, pp 243–254.

55 Griggs RC, Moxley RT III, Mendell JR, Fenichel GM, Brooke MH, Pestronk A, Miller JP: Prednisone in Duchenne dystrophy: A randomized, controlled trial defining the time course and dose response. Arch Neurol 1991;48:383–388.

56 Bulfield G, Siller WG, Wight PAL, Moore K: X chromosome-linked muscular dystrophy (*mdx*) in the mouse. Proc Natl Acad Sci USA 1984;81:1189–1192.

57 Tanabe Y, Esaki K, Nomura T: Skeletal muscle pathology in X chromosome-linked muscular dystrophy (*mdx*) mouse. Acta Neuropathol (Berl) 1986;69:91–95.

58 Makiejus RV, Patel VK, Krishna G, Dierdorf SF, Bonsett C: Effects of adenylosuccinate on the stretch of dystrophin lacking muscles of *mdx* mice. Biochem Arch 1991;7:95–103.

59 Stedman HH, Sweeney HL, Shrager JB, Maguire HC, Panettieri RA, Petrof B, Narusawa M, Leferovich JM, Sladky JT, Kelly AM: The *mdx* mouse diaphragm reproduces the degenerative changes of Duchenne muscular dystrophy. Nature 1991;352:536–539.

60 Totsuka T, Watanabe K, Uramoto I: Shoujou-naki kin-jisutorofii-shou(?) *mdx* mausu: Soredemo tashikani kinsen'i wa ishuku (?) shite iru (Japanese). Proc JAAMHD 1991;7:76.

61 Totsuka T: Centronucleated myofibers having also peripheral nuclei in rectus femoris muscles of muscular dystrophic mice. Congen Anomal 1987;27:51–60.
62 Totsuka T: Muscular dystrophic mice: Cross-sectional images of muscle fibers are not round. Neurosci Res 1988;5:S63.
63 Totsuka T, Uramoto I: Muscular dystrophic mice: Centronucleated fibers are otherwise within the other fibers. J Physiol Soc Jap 1988;50:533.
64 Stauber WT, Fritz VK, Clarkson PM, Riggs JE: An injury model myopathy mimicking dystrophy: Implications regarding the function of dystrophin. Med Hypoth 1991;35:358–362.

Note added in Proof

Stauber et al. [64] proposed that repetitive injury due to a more fragile myofiber accounts for the morphologic pattern seen in dystrophic muscle, and suggested that the injury occurrence may be more pronounced during periods of rapid growth.

Dr. T. Totsuka, Department of Physiology, Institute for Developmental Research, Aichi Prefectural Colony for the Handicapped, Kasugai, Aichi 480–03 (Japan)

Sato Y, Poortmans J, Hashimoto I, Oshida Y (eds): Integration of Medical and Sports Sciences. Med Sport Sci. Basel, Karger, 1992, vol 37, pp 67–83

Expression of Fibroblast Growth Factors in Exercise-Induced Muscle Hypertrophy with Special Reference to the Role of Muscle Satellite Cells

S. Yamada[a], *H. Kimura*[b], *A. Fujimaki*[c], *R. Strohman*[d]

[a] University of Tokyo;
[b] Shiga University of Medical Science, Shiga;
[c] Public Welfare Scientific Research Foundation, Tokyo, Japan;
[d] University of California, Berkeley, Calif., USA

It has been generally assumed that muscle strength is proportional to the cross-sectional area of a muscle [1]. Muscle training nowadays, therefore, mainly aims at increasing the quantity of muscle. However, the mechanisms underlying training-induced muscle enlargement remain unclear. The induction of rapid skeletal muscle growth can be achieved and evaluated by various methods using isolated muscles or in vitro preparations. These include hyperactivation of intact animals (forced exercise) [2], electrical and specific stimulation [3], cross-innervation [4], passive muscle stretch [5], isometric exercise [6] and compensatory hypertrophy by tenotomy of synergistic muscles [7].

The lattermost method using tenotomy is among the most widely used models. When a skeletal muscle acting together with other muscles is deprived of these cooperative synergists, the muscle remarkably increases in weight after a very short time [8]. This growth, which is referred to as 'compensatory hypertrophy' or 'experimental hypertrophy', can easily be produced in animals. Such an animal model has utilized in many studies on the regulation of muscle metabolism and growth. Recent interest appears to be focused on the mechanism of compensatory hypertrophy. Many experimental studies have been conducted to detect hypertrophy-associated changes, including amino acid transport, protein metabolism and muscle fiber proliferation [7, 9]. Although the absolute number of skeletal muscle fibers was once believed to remain unchanged after birth, recent studies suggest a further possibility. In addition to hypertrophy of individual muscle fibers, compensatory growth may also involve some increase in absolute number of muscle fibers, connective tissue and satellite cells [10, 12]. Within several days after tenotomy, moreover, a hypertrophied soleus muscle has been shown to increase significantly both in content of the total DNA and in number of the cellular nuclei consisting of various muscle components [7].

In recent reports concerning exercise-induced muscle growth, similar new muscle fiber formation [13, 14] and satellite cell hyperplasia [12, 13] have been proven, with one exception [14].

Our own data support the idea of new fiber formation in that embryonic isoforms of myosin heavy-chain (MHC) are expressed specifically in a population of small fibers after tenotomy. This is because embryonic myosin gene expression is known to be characteristic of newly formed fibers in adult avian and mammalian muscles [15]. Our hypothesis is that the small embryonic-type fibers seen in hypertrophied muscle induced by ecercise may represent hyperplasia and subsequent fusion of satellite cells.

Exercise-induced satellite cell proliferation seems to be regulated by a number of possible biochemical mechanisms. Firstly, it is well known that growth hormone, testosterone and insulin are essential to the normal postnatal growth of skeletal muscle. In muscle exercise, these endocrine hormones may also be responsible for promoting protein synthesis. However, when tenotomy of the gastrocnemius muscle was performed in hypophysectomized rats, both the soleus and plantaris muscles of the operated limb underwent compensatory hypertrophy in a similar manner to that observed during the normal growth [7]. Although muscle wasting is a prominent feature of the diabetic state, work-induced hypertrophy can occur in alloxan-diabetic rats [16]. The hormonal regulation, therefore, appears to participate little in muscle hypertrophy. Secondly, a recent attention has been paid to the roles of local growth factors such as fibroblast growth factor (FGF) [17–20], insulin-like growth factor (IGF-I) [21, 22] and transforming growth factor-β (TGF-β) [23, 24] in muscle growth regulation.

Here we describe the cellular localization of FGFs [acidic (aFGF) and basic (bFGF) FGF] and the biological activity of bFGF in muscle tissues. The possible role of satellite cells in exercise-induced muscle hypertrophy is discussed.

Materials and Methods

Wistar rats weighing 200–250 g were used for all animal experiments.

Tenotomy

For induction of compensatory hypertrophy in the rats, the tendon of the right gastrocnemius was resected under deep anesthesia with sodium pentobarbital. The left soleus, gastrocnemius and plantaris muscles were carefully separated from each other with forceps. After recovery from the operation, the animals were allowed to free walk. To provide additional training, food and water were placed high in the cage so that the animals had to keep standing on their hindlimbs to reach their goal. One week after such training, compensatory hypertrophy was observed in plantaris and soleus muscles of all rats examined.

Northern Blotting

Plasmids containing MHC gene sequences were a gift of Margaret Buckingham. The plasmid DNAs coded for adult (pMHC32), neonatal (pMHC16.2) and embryonic (pMHC2.2) MHC [25, 26] in mice. These DNA probes were found here to recognize also rat MHC mRNA sequences. Each DNA probe contained segments complementary to the 3′ coding and untranslated regions of the corresponding mRNA. Nick translation of 200 ng of total plasmid DNA with [^{32}P]-dATP and [^{32}P]-TTP (3,000 Ci/mmol, New England Nuclear) was performed using a nick translation kit. Preparation of total mRNA from normal and tenotomy-induced hypertrophied rat plantaris muscle was based on the procedure described by March et al. [27]. The extracted mRNAs were analyzed by Northern blotting using the above DNA probes.

Immunohistochemical Study

Fresh muscle tissues were embedded in Tissue-Tek (Miles Lab.) and frozen by immersing in liquid nitrogen. Cryostat sections of 6 μm thickness were collected and thaw-mounted on glass slides. The sections were incubated for 2 h with 20 μl of primary antiserum diluted with PBS (10 m*M* sodium phosphate buffer containing 150 m*M* NaCl, pH 7.2) at room temperature in a moist chamber, washed twice with PBS, incubated for 1 h with fluorescein-conjugated goat antimouse IgG (Sigma) diluted 1:20 with PBS and washed again in PBS.

For FGF immunohistochemistry, bFGF polyclonal antibody (F547, diluted 1:1,000) or aFGF polyclonal rabbit antiserum (diluted 1:10,000) was used as a primary antibody. The antigen product was separated according to the FGF isolation procedure of Bohlen et al. [28]. The details have previously been described by us for aFGF [29].

For MHC immunohistochemistry, antiserum specific for either embryonic, neonatal or adult myosin isoforms was used. The antibodies were a gift of Dr. D. Fischman.

Preparation of Fibroblast-Growth-Factor-Like Proteins from Muscle Tissue

To obtain a rat skeletal muscle extract enriched in FGF-like proteins, whole rat plantaris muscle was homogenized in 5 vol of 20 m*M* Tris-HCl (pH 7.4) containing 1 m*M* EDTA, 1 m*M* EGTA, 0.1 m*M* leupeptin and 0.1 m*M* PMSF. The starting muscle specimens were collected from either nontreated control muscles or hypertrophied muscles 7 days after tenotomy. The homogenate was centrifuged at 16,000 rpm for 30 min at 4 °C, and the supernatant was collected as a crude cytosolic fraction. This cytosolic fraction was used directly for Western blotting. Part of the crude fraction was applied to a heparin Sepharose column (Sigma) equilibrated with 10 m*M* Tris-HCl (pH 7.0) containing 0.6 *M* NaCl. After washing with the same buffer, the column was eluted with 1.0 *M* NaCl followed by 2.0 *M* NaCl in Tris-HCl (pH 7.0). Each eluate was dialyzed against distilled water and used as a rat heparin-binding fraction for bioassay using myogenic cell culture. On the basis of both SDS-PAGE analysis and Western blotting, it was confirmed that the 1.0-*M* NaCl fraction contained aFGF and that the 2.0-*M* NaCl fraction contained bFGF.

Western Blot Analysis

For immunoblot tests of FGFs, the rat crude cytosolic fraction was electrophoresed on SDS-polyacrylamide gel (12% polyacrylamide gel; reducing conditions) and then transferred to nitrocellulose membrane (25 m*M* Tris-glycine buffer, pH 8.3, containing 4% methanol). The membrane was incubated at room temperature for 3 h with 5% skim milk in 25 m*M* Tris-HCl (pH 7.4) containing 150 m*M* NaCl (TBS), and then incubated with either antiserum to aFGF or bFGF diluted 1:10,000 in TBS containing 1% skim milk overnight at 4 °C. The membranes were washed in TBS containing 0.1% Tween 20 and then incubated for 2 h with biotin-labeled antirabbit IgG (BRL, 1:5,000) in TBS containing 0.1% Tween 20 and 1% skim milk at room temperature. After wash, the membranes were incubated for 1 h with avidin-biotin-peroxidase complex (ABC kit, 1:1,000, Vector) at room temperature. The labeling was visualized with 0.01% 3,3′-diaminobenzidine·4HCl and 0.005% H_2O_2 in 0.1 *M* Tris-HCl buffer (pH 8.5).

Bioassay of Fibroblast Growth Factor Activity in Myogenic Culture

Neonatal chick skeletal muscles were removed from hindlimbs. Tissues were dissociated with 0.1% trypsin, and the myogenic cells remaining in suspension were collected by centrifugation and resuspended in DMEM/Ham's F-12 (50:50, Hazelton) with 20% horse serum (Gibco). After filtration through nylon meshes, myogenic cells were plated on poly-*L*-lysine-precoated dishes at a density of 3 × 105 cells/cm^2 and incubated at 37 °C in a 5% CO_2 incubator.

Collection of Macrophage-Rich Cell Clusters

For visualizing macrophages in muscle tissue, the antimacrophage antibody (Mac1) was used for immunofluorescence histochemistry. To collect macrophages, the control and hypertrophied muscles were dissected in the MEM and was incubated on a noncoated slide glass at 37 °C in a CO_2 incubator. Muscle cells attached to the noncoated slide glass, and unattached macrophages among the muscle cells were collected into the MEM. The collected cells were stained with Giemsa solution.

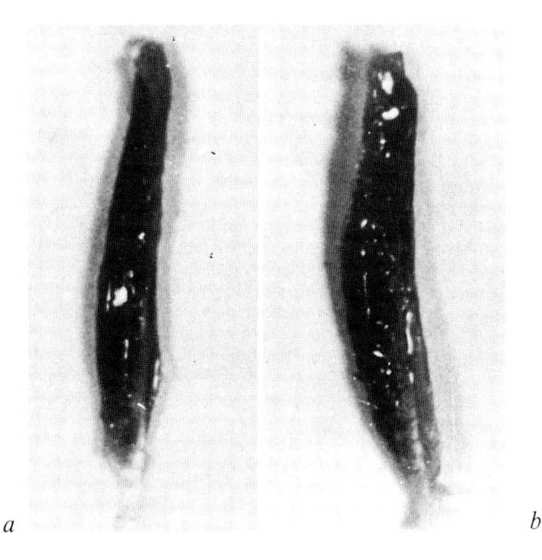

Fig. 1. Macroscopic photographs of cross-sections at distal portion of the plantaris muscle in rats. *a* Sham-operated. *b* tenotomized, examined at 7 days after surgery.

Results

Basic Morphological Study

The distal portion of the experimental muscles was larger than in the control (fig. 1). To quantitate the hypertrophy we measured the dry and wet weights of the plantaris. Both the dry and wet weights of the experimental muscle had increased significantly (fig. 2). When the 7-day hypertrophied plantaris was examined microscopically, we found an obvious increase in the diameter of most fibers together with the presence of clusters of much smaller fibers (fig. 3). These small fibers are very rarely seen in normal muscle. Most of these fibers had centrally situated nuclei, probably as a characteristic sign of regeneration.

Northern Blot Analysis for Myosin Heavy-Chain Gene Expression

As expected, the adult-specific probe pMHC32 hybridized strongly to both control and hypertrophied muscle mRNA (fig. 4a). Differential hybridization, however, occurred when control and hypertrophied muscle was reacted with the embryo-specific probe, pMHC2.2. Evidently, stronger hybridization is always observed with mRNA from hypertrophied muscle than from controls examined on days 3, 7 and 14 after surgery (fig. 4b). The weak signals seen in the controls may be due to nonspecific or cross-

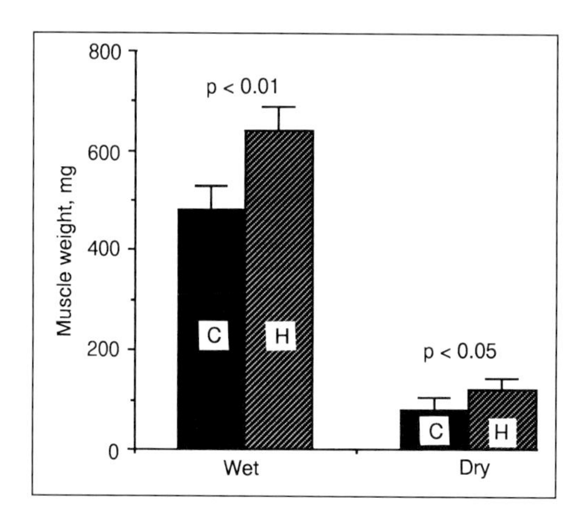

Fig. 2. Effects of tenotomy and training on the dry and wet weight of the plantaris muscles of rats. H and C show the muscle weight of the tenotomized and sham-operated limbs after training, respectively. The muscle weights are represented as means.

hybridization with the adult-type MHC. The result strongly suggests that new fiber is formed soon after the induction of hypertrophy by tenotomy.

Myosin Heavy-Chain Immunohistochemistry

Figures 5a and b demonstrate that all fibers stained positively with MF20 antibody which recognizes all three myosin isoforms (adult, neonatal and embryonal). Figure 5c shows the section stained with MF30 anti-

---

Fig. 3. Light-microscopic observation of the induction of plantaris muscle hypertrophy in rats. *a* Control. *b* Hypertrophied. The control muscle from the contralateral sham-operated leg and hypertrophied muscles were studied 7 days after surgery. Hematoxylin-eosin. × 400.

Fig. 4. Northern blot analysis of the total RNA from normal and hypertrophied rat plantaris muscles treated similarly as described in figure 1. *a* Northern blot hybridized to nick-translated adult MHC RNA-specific cDNA probe, pMHC32. Lanes A, C and E contain total RNA (20 μg) from control muscle at 3, 7 and 14 days, respectively, after sham-operation. Lanes B, D and F contain total RNA (20 μg) from hypertrophied muscle at 3, 7 and 14 days after ablation. *b* Same as *a* except that hybridization was carried out using the embryonic-specific cDNA probe, pMHC2.2. Lanes contain total RNA as described for *a*.

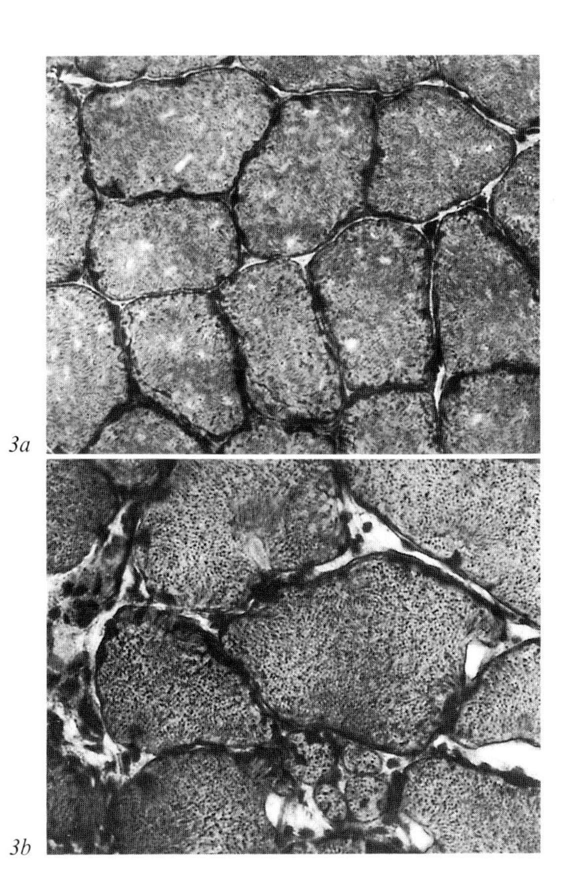
3a
3b

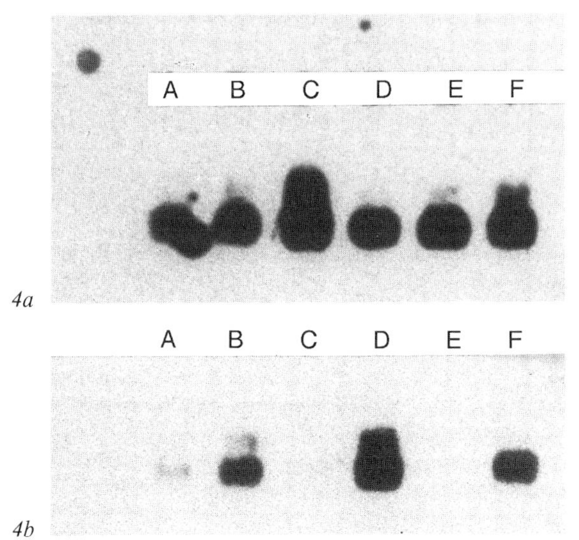
A B C D E F
4a
A B C D E F
4b

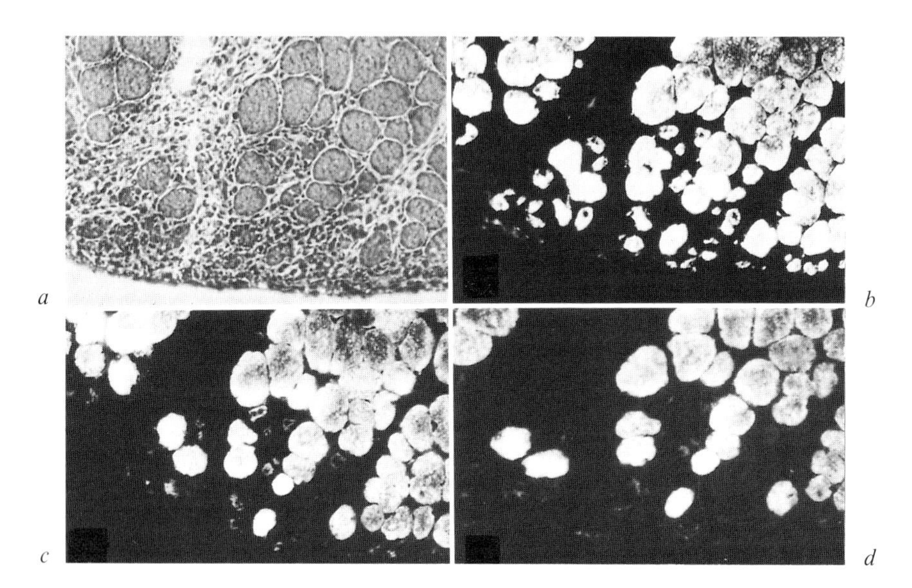

Fig. 5. Immunofluorescence analysis of MHC localization in hypertrophied rat plantaris muscle using antibodies for adult, neonatal and embryonic MHC. *a* Hematoxylin-eosin staining. *b* Stained with MF20 antibody which reacts with adult, neonatal and embryonic MHCs. *c* Stained with MF30 antibody which reacts with adult and neonatal MHCs. *d* Stained with MF-14 antibody which reacts only with adult MHC.

body, which detects both adult and neonatal, but not embryonal MHC. Here all large fibers are stained, but small fibers, distributing mostly in peripheral regions of the muscle bundle, are not stained or are only weakly positive. With the MF14 antibody (fig. 5d), which recognizes only adult MHC, again all large fibers are stained but not staining is detected in small fibers. The result clearly indicates that the small fibers are characterized by the absence of adult MHC and instead by the presence of abundant embryonic and a few neonatal MHCs. Since newly formed fibers seen in adult animals express embryonal but not adult MHC isoforms, it is reasonable to assume that this small fiber population in hypertrophied muscle has indeed formed de novo and is not derived from splitting or branching of preexisting muscle fibers.

Fibroblast Growth Factor Immunohistochemistry

bFGF-positive fibers are seen in the control (fig. 6a) and hypertrophied (fig. 6b) muscles. The positive staining is located in the endomysial

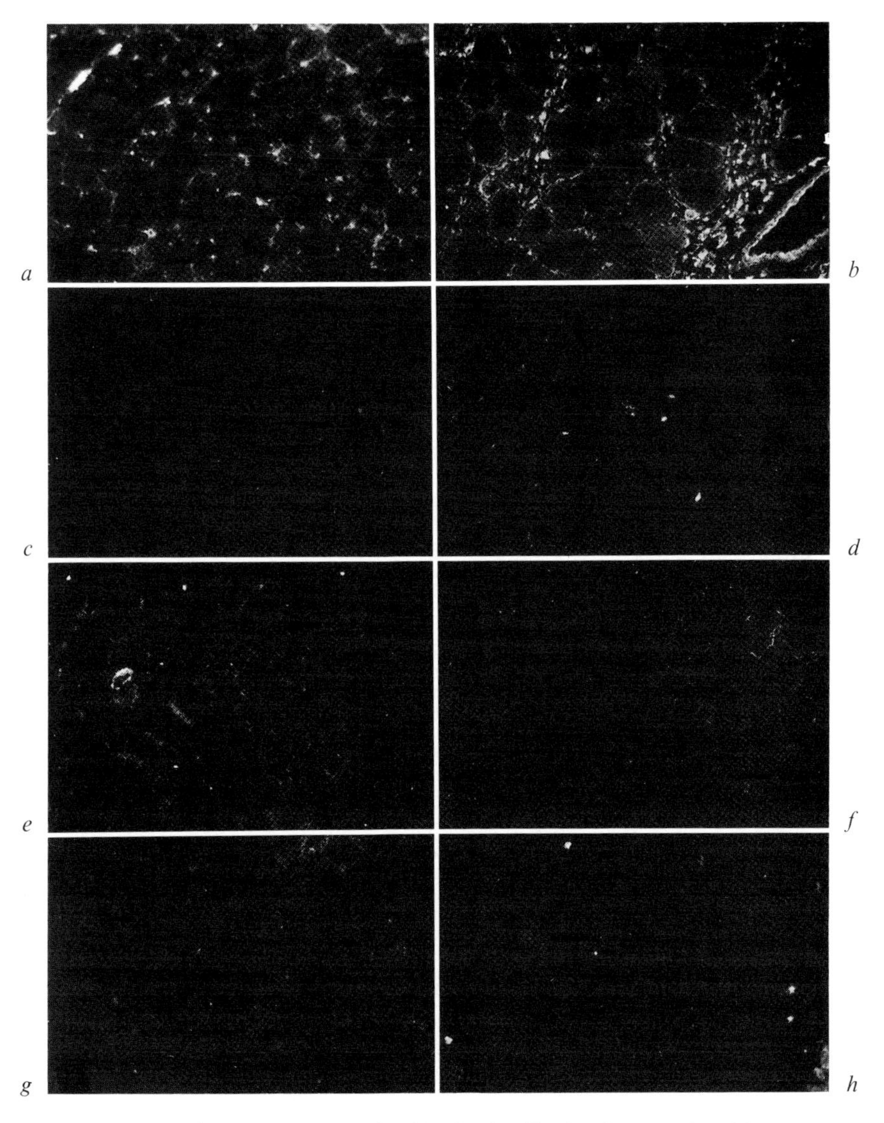

Fig. 6. Immunofluorescence analysis of bFGF localization in control and hypertrophied plantaris muscle. The sections stained with anti-bFGF represent the control (*a*) and hypertrophied muscle (*b*), respectively. *c, d* Control muscle stained with preimmune serum and with bFGF antibody preabsorbed by an excess amount of purified bFGF, respectively, as the primary antibody. *e, f* Control muscle stained with bFGF antibody after washing the sections with 1 and 2 *M* NaCl, respectively. *g, h* bFGF staining of control muscle after pretreating the sections with heparinase and collagenase, respectively.

Table 1. Effects of heparin-binding fractions extracted from control and hypertrophied muscles on muscle cell growth in culture

	Myosin synthesis cpm × 10[a]	Myosin accumulation μg[b]	Myosin DNA synthesis cpm × 10[c]
Control	4.0	4.70	72
Hypertrophy	10.0	7.00	180

Each value is expressed as per one dish. Total amount of heparin-binding protein added is 2 ng/ml. Heparin-binding material was estimated from the optical density of eluted fractions with 2 *M* NaCl from the column.
[a] cpm of [^{3}H]-methionine incorporated into MHC in single culture dish following a pulse administration on culture day 4.
[b] Micrograms of total MHC isolated from single culture dishes at culture day 4.
[c] cpm of [^{3}H]-thymidine incorporated into DNA following a pulse treatment on culture day 2. Plantaris muscle was approximately 70% hypertrophied 7 days after removal of gastrocnemius muscle.

areas. With a high magnification, bFGF immunoreactivity is distributed continuously along the surface of muscle fibers. A minor positive staining is also observed in and around small cells of both control and experimental groups (fig. 6b). Such immunopositive small cells are always scattered in gaps between myofibers. Although the light-microscopic observation does not allow to identify the cell type, these small cells seem clearly positive for bFGF in the hypertrophy model. This impression is supported by bioassay for FGF-like substances as described below (table 1). Figures 6c–h demonstrate the specificity for bFGF immunoreactivity. Immunopositive staining is no longer observed by using a preimmune serum (fig. 6c, d), using the primary antibody preabsorbed with the antigen (fig. 6e, f), in sections prewashed with 2 *M* NaCl but not with 1 *M* NaCl. In an additional specificity study (fig. 6g, h), an immunopositive structure is no longer seen in sections preincubated with either heparinase or collagenase, enzymes being capable to digest endomysial components in the sections.

By using antibody to aFGF, positive stainings are distributed in a manner essentially similar to that described for bFGF (fig. 7a). Both large muscle fibers and small cells seemed to be positive, though positive reaction products were apparently deposited in the extracellular space between

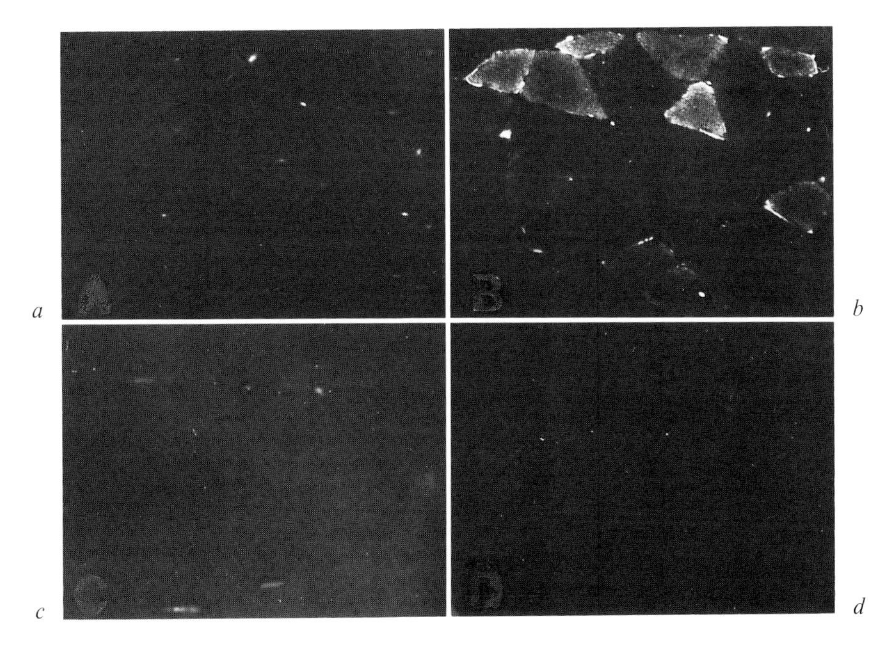

Fig. 7. Immunofluorescence analysis of aFGF localization in control and hypertrophied plantaris muscle. The sections stained with anti-aFGF represent the control (*a*) and hypertrophied muscle (*b*), respectively. *c, d* Control and hypertrophied muscle, respectively, stained with aFGF antibody after washing the sections with 1 *M* NaCl. Washing with 2 *M* NaCl did not give positive staining.

muscle cells. The positive small cells were too tiny to identify the cell type. A specificity study similar to the bFGF staining was carried out, and the validity of the method for aFGF was confirmed. One particular difference from bFGF immunoreaction was that aFGF-positive reaction was eliminated in sections prewashed with either 1 or 2 *M* NaCl (fig. 7b).

Western Blot Analysis of Fibroblast Growth Factors

On Western blots, both aFGF and bFGF antibody react with a partially purified extract of rat plantaris muscle (data not shown). The positive bands on nitrocellulose membranes transferred from SDS-PAGE roughly corresponded to the molecular weights of aFGF and bFGF. The staining intensity of these immunoreactivities was apparently higher in specimens from hypertrophied muscle than from the control, though no precise quantification was made.

Bioassay of Fibroblast-Growth-Factor-Like Activity Extracted from Muscle

By the addition of a bFGF-containing 2.0 *M* NaCl fraction eluted from the heparin affinity column, apparent promoting effects on both cell division and myosin accumulation were observed in the chick myogenic cell culture. Such activities were twice stronger with the fraction from the hypertrophied model as compared with that from the control (table 1). To support the validity of this experiment is our preliminary test where increased DNA synthesis is detectable when rat muscle extracts are added on day 1 of culture (data not shown).

Macrophage Analysis

Many macrophages were localized in the epimysium of hypertrophied muscle but not in control muscle. In order to examine possible FGF storage in macrophages, both aFGF and bFGF antibody were used for immunohistochemical study. In agreement with previous studies, bFGF immunoreactivity was present in the macrophage, but no positive staining for aFGF was observed in hypertrophied muscle.

Discussion

New Muscle Fiber Formation in the Hypertrophied Muscle

Recently, there have been direct demonstrations of new fibers associated with hypertrophy in the chicken [13] and rat [2], enhancing interest in earlier, similar and perfectly adequate data on the same subject [30] . We have also presented evidence of new fiber formation during hypertrophy. Firstly, we found an expression of embryonic MHC mRNA in hypertrophied but not in control muscle, suggesting a formation of new fibers. It has been reported that embryonic MHC largely disappears following early development [31] but reappears during new fiber formation when adult muscle regenerates [32, 33]. Secondly, we used antibodies specific for embryonal MHC to demonstrate the expression in a population of small fibers in hypertrophied but not in control adult muscle. Since these small fibers are interspersed between large myofibers and are sometimes encapsulated on the edge of large muscles, it seems very likely that the small fibers are formed newly rather than derived from splitting or branching of large fibers. The fact that cellular nuclei are increased in hypertrophied muscle also favors our assumption, because fiber splitting or branching may not involve such nuclear proliferation. Although the exact nature of the small fibers is unknown at present, it may be noteworthy that embryonal MHC is only expressed by satellite cell replication and fusion in adult muscle [34].

We have reasons, therefore, to consider possible roles of satellite cells in association with the mechanisms of hypertrophy. Satellite cells may be released from the basal lamina of large myofibers and become activated to form growing populations of new fibers. It is becoming fairly clear that, as exercise levels increase, so do the extent of muscle inflammation and the possibility of basal lamina alteration [35]. If new fibers are formed in significant numbers, then one might also expect fast/slow fiber ratios to be also significantly altered in extreme exercise. Inflammatory responses which accompany hypertrophy and regeneration include partial digestion of basement membranes by invading lymphocytes [36]. Thus, any activity that elicits an inflammatory response might provoke changes in binding of the FGF to muscle endomysium, in the binding of satellite cells to endomysium, or both. These changes would then define the extent to which satellite cell replication would occur and the extent to which new fibers would form. In the presence of a mild stimulus, one might expect simple recruitment of satellite cells to old fibers. As the stimulus increases (extreme exercise bordering on damage), one might expect extensive recruitment of satellite cells, and new fiber formation and addition of new nuclei to hypertrophying old fibers. These ideas have the virtue of being testable by experiment in vivo.

Fibroblast Growth Factor in Hypertrophied Muscle

The discovery of heparin-binding growth factors [29, 37, 38] and the demonstration that muscle basal lamina is rich in heparin glycosaminoglycans [39] have advanced our understanding of the mechanisms of satellite cell activation and provided the basis for a theory of satellite cell activation [34]. These studies have led to a further discovery of bFGF in chicken, mouse and rat muscle [19, 40, 41]. Other studies on the regulation of satellite cell and myogenic cell activation by bFGF [17, 20, 21, 37, 41–43] have provided a rational basis for activation mechanisms and have restimulated interest in satellite cell hyperplasia in muscle growth and hypertrophy. It is likely that other growth factors are also involved in regulating satellite cell activation in hypertrophy and regeneration. TFG-β has profound effects on myogenic cell differentiation [23, 27] and IGF-I is expressed in satellite cells during early phases of regeneration in rats [22].

The role of bFGF in activating satellite cell growth is made more plausible by the demonstration that this growth factor is localized in the same extracellular domain as the satellite cell itself.

The present study has proven that both aFGF and bFGF are concentrated in muscle fiber extracellular matrix, or more specifically in the endomysium. Moreover, hypertrophied muscle appears to contain higher levels of FGF activity as compared with controls. The localization of FGF in satellite cells makes sense as long as FGF binding to heparin persists. Under such conditions, the mitogenic activity of FGF on satellite cells may be suppressed by heparin. Localization of bFGF in the extracellular matrix has also been reported for corneal basement membrane in vivo [44] and for endothelial cell extracellular matrix in vitro [45]. Likewise, aFGF mRNA is reportedly present in rat satellite cells. The source of both types of FGFs is not currently known. Production by certain cells related with blood circulation is likely to occur. These cells include, for example, endothelial cells, mast cells, satellite cells, fibroblasts and macrophages. Although bFGF-positive macrophages were confirmed in accordance with other studies, no aFGF was found in this cell type even in hypertrophied muscle. The present immunohistochemical study has shown that blood vessel walls stain positive for bFGF, suggesting the production in endothelial cells. In addition, our preliminary data suggest that bFGF is produced by myogenic cells in culture and/or macrophage in the hypertrophied muscle. It is now clear that FGFs are stored in muscle fiber basal lamina. The modulation of FGF binding to heparin, a component of the basal lamina, may determine the availability of bound FGF to satellite cells. Many possibilities can be listed for the impairment of normal bFGF-heparin interaction, and for satellite cell-heparin interaction within the basal lamina. Such a list may be ranging from the mechanical distortion of muscle fibers which accompanies physical activity [46] to an inflammatory response which accompanies focal or more extensive damage to muscle fibers [36]. Although the present situation is somehow complicated, what is open for at least experimental testing is the exact mechanism by which muscle activity in exercise is translated into changes in basal-lamina-associated FGF or satellite cell responsiveness. We have shown elsewhere, for example, that heparin inhibits the activity of bFGF in cultures of chicken myoblasts or satellite cells from adult mouse skeletal muscle. Furthermore, in this study, the elimination of FGF immunostaining with heparinase and 2 *M* NaCl, both being capable to disrupt FGF binding to heparin, argues strongly for FGF binding to heparin-rich muscle fiber basal lamina.

References

1 Haggmark T, Jansson E, Svane B: Cross-sectional area of the thigh muscle in man measured by computed tomography. Scand J Clin Lab Invest 1978;38:355–360.
2 Gonyea WJ, Sale DG, Gonyea FB, Mikesky A: Exercise induced increases in muscle fiber number. Eur J Appl Physiol 1986;55:137–141.
3 Pette D, Smith ME, Staudte HW, Vrbova G: Effects of long-term electrical stimulation on some contractile and metabolic characteristics of fast rabbit muscles. Pflügers Arch 1973;338:257–272.
4 Kowalski K, Gordon EE, Martinez A, Adamek J: Changes in enzyme activites of various muscle fiber types in rat induced by different exercise. J Histochem Cytochem 1969;17:601–607.
5 Ashmore CR, Summers PJ: Stretch-induced growth of chicken wing muscles: Myofibrillar proliferation. Am J Physiol 1981;241:C93–C97.
6 Binkhorst RA: The effect of training on some isometric contraction characteristics of a fast muscle. Pflügers Arch 1969;309:193–202.
7 Goldberg AL, Etlinger D, Goldspink DF, Jablecki C: Mechanism of work-induced hypertrophy of skeletal muscle. Med Sci Sports 1975;7:248–261.
8 Denny-Brown D: Experimental studies pertaining to hypertrophy, regeneration and degeneration. Neuromusc Dis 1961;38:147–196.
9 Edstrom L, Grimby L: Effect of exercise on the motor unit. Muscle Nerve 1986;9: 104–126.
10 Darr KC, Schultz E: Exercise-induced satellite cell activation in growing and mature skeletal muscle. J Appl Physiol 1987;63:1816–1821.
11 Jablecki CK, Hauser JE, Kaufman S: Autoradiographic localization of new RNA synthesis in hypertrophying skeletal muscle. J Cell Biol 973;57:743–759.
12 Yamada S, Buffinger N, Dimario J, Strohman RC: Fibroblast growth factor is stored in fiber extracellular matrix and plays a role in regulating muscle hypertrophy. Med Sci Sports Exerc 1989;21:S173–180.
13 Kennedy JM, Eisenberg BR, Reid SK, Sweeney LJ, Zak R: Nascent muscle fiber appearance in overloaded chicken slow-tonic muscle. Am J Anat 1988;181:203–215.
14 Gollnick PD, Timson BF, Moore RL, Riedy M: Muscle enlargement and number of fibers in skeletal muscles of rats. J Appl Physiol 1981;50:936–943.
15 Bader D, Masaki IT, Fischman D: Immunochemical analysis of myosin heavy chain during avian myogenesis in vivo and in vitro. J Cell Biol 1982;95:763–770.
16 Goldberg AF: Role of insulin in work-induced growth of skeletal muscle. Endocrinology 1968;83:1071–1073.
17 Allen R, Dodson M, Luiten L: Regulation of skeletal muscle satellite cell proliferation by bovine pituitary fibroblast growth factor. Exp Cell Res 1984;152:154–160.
18 Bischoff R: A satellite cell mitogen from crushed muscle. Dev Biol 1986;115:140–147.
19 Kardami E, Spector D, Strohman RC: Myogenic growth factor present in skeletal muscle is purified by heparin affinity chromatography. Proc Natl Acad Sci USA 1985;82:8044–8047.
20 Olwin BB, Hauschka S: Identification of the fibroblast growth factor receptor of Swiss 3T3 cells and mouse skeletal muscle myoblasts. Biochemistry 1986;25:3487–3492.

21 Florini JR: Hormonal control of muscle growth. Muscle Nerve 1987;10:577–598.
22 Jennische E, Skottner A, Hannson HA: Satellite cells express the trophic factor IGF-I in regenerating skeletal muscle. Acta Physiol Scand 1987;129:9–15.
23 Florini JR, Roberts AB, Ewton DZ, Falen SL, Flanders KC, Sporn MB: Transforming growth factor-beta: A very potent inhibitor of myoblast differentiation identical to the differentiation inhibitor secreted by BRL cells. J Cell Chem 1986;261:16509–16518.
24 Massague J, Cheifetz S, Endo T, Nadal-Ginard B: Type B transforming growth factor is an inhibitor of myogenic differentiation. Proc Natl Acad Sci USA 1986;83: 8206–8210.
25 Weydert A, Daubas P, Caravatti M: Sequential accumulation of mRNAs encoding different myosin heavy chain isoforms during skeletal muscle development in vivo detected with a recombinant plasmid identified as coding for an adult fast myosin heavy chain from mouse skeletal muscle. J Biol Chem 1983;258:13867–13874.
26 Weydert A, Daubas P, Lazarides I, et al: Genes for skeletal muscle myosin heavy chains are clustered and are not located on the same mouse chromosome as a cardiac myosin heavy chain gene. Proc Natl Acad Sci USA 1985;82:7183–7187.
27 March CJ, Mosley B, Larsen A, et al: Cloning, sequence and expression of two distinct human interleukin-1 complementary DNAs. Nature 1985;315:641–647.
28 Bohlen P, Esch F, Baird A, Gospodarowicz D: Acidic fibroblast growth factor (FGF) from bovine brain: Amino-terminal sequence and comparison with basic FGF. Eur Mol Biol Org J 1985;4:1951–1956.
29 Gospodarowicz D, Cheng J, Lui G, Baird A, Bohlen P: Isolation of brain fibroblast growth factor by heparin Sepharose affinity chromatography: Identity with pituitary fibroblast growth factor. Proc Natl Acad Sci USA 1984;81:6963–6967.
30 Schiaffino S, Bormioli SP, Aloisi M: Cell proliferation in rat skeletal muscle during early stages of compensatory hypertrophy. Virchows Arch [B] 1972;11:268–273.
31 Whalen RG, Sell SM, Butler GS, Schwartz K, Bouveret P, Pinse T: Three myosin heavy chain isozymes appear sequentially in rat muscle development. Nature 1981; 292:805–809.
32 Saad AD, Obinata T, Fishman DA: Immunochemical analysis of protein isoforms in thick myofilaments of regenerating skeletal muscle. Dev Biol 1987;119:336–349.
33 Matsuda R, Spector DH, Strohman RC: Regenerating adult chicken skeletal muscle and satellite cell culture express embryonic patterns of myosin and tropomyosin isoforms. Dev Biol 1983;100:478–488.
34 Strohman RC, Kardami V: Muscle regeneration revisited: Growth factor regulation of myogenic cell replication; in Serrero G, Hayashi J (eds): Cellular Endocrinology. New York, Liss, 1986, pp 287–296.
35 Shultz E, Jaryszak D, Valliere CR: Response of satellite cells to focal skeletal muscle injury. Muscle Nerve 1985;8:217–222.
36 Naparstek Y, Cohen IR, Fuks Z, Vlodasky I: Activated T-lymphocytes produce a matrix-degrading heparin sulfate endoglycosidase. Nature 1984;310:241–244.
37 Lobb R, Fett J: Purification of two distinct growth factors from bovine neural tissue by heparin affinity chromatography. Biochemistry 1984;23:6295–6299.
38 Maniatis T, Mehlman T, Friesel R, Schreiber A: Heparin binds endothelial cell growth factor, the principal endothelial cell mitogen in bovine brain. Science 1984; 225:932–935.

39 Bayne EK, Anderson MJ, Famborough DM: Extracellular matrix organization in developing muscle contraction with acetylcholine receptor aggregates. J Cell Biol 1984;99:1486–1501.
40 DiMario J, Buffinger N, Yamada S, Strohman RC: Fibroblast growth factor in the extracellular matrix of dystrophic (mdx) mouse muscle. Science 1989;244;688–689.
41 Kardami E, Spector D, Strohman RC: Selected muscle and nerve extracts contain an activity which stimulates myoblast proliferation and which is distinct from transferrin. Dev Biol 1985;112:353–358.
42 Lathrop BK, Olson EN, Glaster L: Control by FGF of differentiation in the BC3H1 muscle cell line. J Cell Biol 1985;100:1540–1547.
43 Spizz G, Roman D, Strauss A, Olson EN: Serum and fibroblast growth factor inhibit myogenic differentiation through a mechanism dependent on protein synthesis and independent of cell proliferation. J Biol Chem 1986;261:9483–9488.
44 Folkman J, Klagsbrun M, Sasse J, Wadzinski M, Ingber ID, Vlodavsky IV: A heparin-binding angiogenic protein – basic fibroblast growth factor – is stored within basement membrane. Am J Pathol 1988;130:393–400.
45 Vlodavsky I, Folkman J, Sullivan R, et al: Endothelial cell-derived basic fibroblast growth factor: Synthesis and deposition into subendothelial extracellular matrix. Proc Natl Acad Sci USA 1987;84:2292–2296.
46 Irintchev A, Wernig A: Muscle damage and repair in voluntarily running mice: Strain and muscle differences. Cell Tissue Res 1987;249:509–521.

Shigeru Yamada, PhD, University of Tokyo, 3-8-1 Komaba, Meguro-ku,
Tokyo 153 (Japan)

Sato Y, Poortmans J, Hashimoto I, Oshida Y (eds): Integration of Medical and Sports Sciences. Med Sport Sci. Basel, Karger, 1992, vol 37, pp 84–90

Expression of Glucose Transporter Isoforms in Atrophied and Enlarged Muscle

Yoshitomo Oka, Hisamitsu Ishihara, Tomoichiro Asano

Third Department of Internal Medicine, Faculty of Medicine, University of Tokyo, Hongo, Tokyo, Japan

Skeletal muscle comprises a large part of the body mass and is a primary tissue responsible for insulin-induced glucose disposal in vivo. Since glucose transport is the first and probably a rate-limiting step of muscle glucose utilization [1], alteration of the muscle glucose transporter which catalizes glucose transport across the muscle plasma membrane may be responsible for the altered insulin responsiveness of this tissue. Here, we have studied the expression of glucose transporter isoforms in skeletal muscle in two conditions. One is the muscle atrophied by hindlimb suspension and another is the enlarged muscle of aged rats.

Materials and Methods

Animals

Male rats of the Wistar strain (Nisseizai, Tokyo, Japan) were maintained with ad libitum feeding on standard chow for several days before experiments. When body weights reached 180–200 g the animals were randomly assigned to control or suspended groups. Tail-casted hindlimb suspension was performed according to the methods described by Jaspers and Tischler [2] with slight modifications. Briefly, rats were fitted with a Hexcelite orthopedic-tape tail harness. A fish swivel incorporated into the harness allowed the rats to move freely. Experimental rats were tail-suspended to elevate their hindlimbs not to touch the floor of the cage, while the forelimbs could be used for maneuvering, feeding and grooming for 6 days (TCHS group). Control rats were also tail-casted but were allowed normal load bearing of all four limbs by extension of the casting tape (TCWB group). Seven-week- and 20-month-old rats were used as the young and aged rats, respectively.

Preparation of Membranes

Total membrane particulate fractions from rat tissues were prepared as previously described with some modifications [3]. The entire gastrocnemius muscle was excised and homogenized by Polytron (Kinematica, Littau, Switzerland) in 10 mmol/l Tris-HCl, 1 mmol/l EDTA and 250 mmol/l sucrose, pH 7.4, containing 1 mmol/l PMSF. The homogenates were centrifuged at 900 *g* for 10 min at 4 °C to sediment the fraction containing mainly the nuclei and mitochondria. The resulting supernatant was centrifuged at 146,000 *g* for 75 min at 4 °C to yield a pellet designated as the total membrane fraction in this study.

Western Blot Analysis

Production and characterization of antibodies raised against synthetic peptides corresponding to the COOH-terminal domain of the rat GLUT1 glucose transporter (residues 478–492) and of rat GLUT4 (residues 495–509) were described previously [3]. Membrane fractions prepared as described above were suspended in 1% SDS and 50 mmol/l dithiothreitol and subjected to SDS polyacrylamide (10%) gel electrophoresis. Electrophoretic transfer to nitrocellulose paper and detection of the immunocomplex with [^{125}I]-labeled protein A were carried out as described previously using antisera at a final dilution of 1:40. The dried blots were autoradiographed using Fuji RX X-ray film and intensifying screen at −70 °C.

Northern Blot Analysis

Total cellular RNA was prepared from rat tissues by the lithium/urea procedure as described previously [4]. RNA was denatured with formaldehyde, separated by 1.2% agarose gel electrophoresis [5] and transferred onto a nitrocellulose filter. The cDNAs for glucose transporter isoforms were prepared as previously described [4] and the MaeI fragment of rabbit GLUT1 cDNA (nucleotides 111–2230), and DraI-XbaI fragment of rat GLUT4 cDNA (nucleotides 143–1947) were labeled with ^{32}P by random priming and used for the measurement of rat GLUT1 and GLUT4 mRNA, respectively. Hybridization of the filters with the probes was carried out in 50% formamide, 5 × SSC (1 × SSC = 150 mmol/l NaCl, 15 mmol/l sodium citrate), 5 × Denhardt's solution (1 × Denhardt's solution = 0.02% polyvinylpyrrolidone, 0.02% Ficoll, 0.02% bovine serum albumin), 50 mmol/l sodium phosphate buffer, pH 7.0, and 1% SDS at 42 °C. The filters were washed with 0.2% SDS/0.5 × SSC at 42 °C for 30 min and subjected to autoradiography.

Results

There was no difference between TCWB and TCHS rats in final body weight (208 ± 10 vs. 207 ± 10 g) and plasma glucose level (6.38 ± 0.61 vs. 5.61 ± 0.44 mmol/l). However, the weight of the soleus muscle was markedly decreased by 31% in TCHS rats. Although total membrane protein tends to be reduced with hindlimb suspension in soleus muscle when expressed per whole muscle, the differences are not statistically significant ($p = 0.052$). However, total RNA amount in the atrophied soleus muscle was markedly decreased with hindlimb suspension.

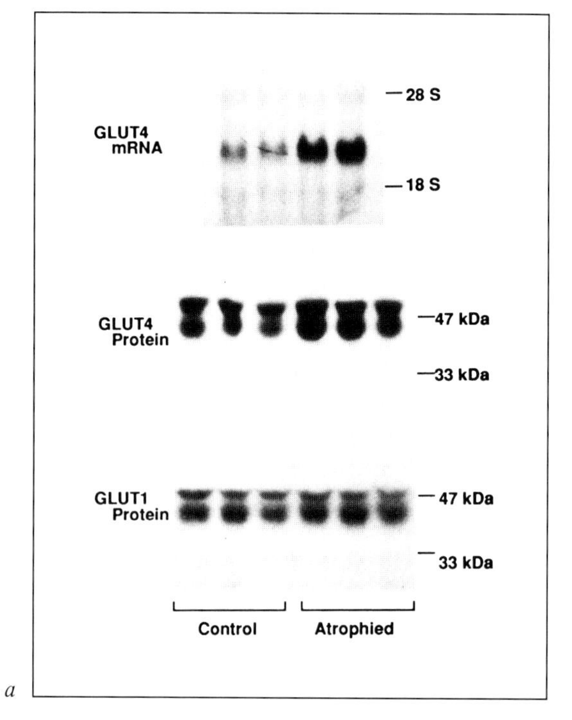

a

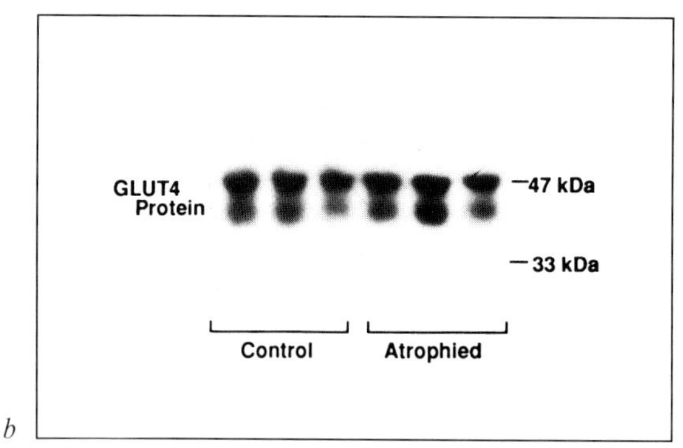

b

Fig. 1. Expression of glucose transporter in control weight-bearing (control) and hindlimb-suspended atrophied (atrophied) soleus muscle. *a* 15 μg of total RNA in each lane were loaded and hybridized with [^{32}P]-labeled rat GLUT4 cDNA (upper panel). 80 μg of total membrane protein was subjected to 8% SDS-PAGE, and immunoblotting was performed using an antiserum raised against the GLUT4 carboxy-terminal domain (middle panel) or against the GLUT1 carboxy-terminal domain (lower panel). *b* 20% of total membrane protein obtained from the same rats shown in *a* was loaded in each lane and immunoblotting was performed using anti-rat GLUT4 carboxy-terminal antisera.

Northern blotting detects mRNA of 2.8 kb in skeletal muscle, the size of which was identical to that of GLUT4 mRNA (fig. 1a, upper panel). When an equivalent amount of total RNA was loaded per lane the amount of GLUT4 mRNA is markedly increased to 265% in atrophied soleus muscle (fig. 1a, upper panel). Western blotting detects GLUT4 protein (fig. 1a, middle panel) and GLUT1 protein (fig. 1a, lower panel), which appeared in doublet forms of 43–50 kDa. A doublet of glucose transporter is sometimes observed as previously reported [6]. The amount of GLUT4 and GLUT1 in the unit weight of the membrane protein was increased in muscle atrophied by hindlimb suspension. However, this way of demonstration may be misleading in the case of muscle atrophy, because membrane protein per one muscle was decreased with atrophy. When 20% of the total membrane protein of soleus muscle was loaded per lane, no significant difference was observed between the atrophied and control muscle (fig. 1b). Thus, the amount of GLUT4 protein in whole soleus muscle is not changed in atrophied muscle compared with control muscle. These data are summarized in figure 2.

The amount of GLUT4 protein per unit weight of membrane of the gastrocnemius muscle was significantly decreased by 30% in the aged rats compared with the young rats (data not shown). The membranes recovered from the gastrocnemius were increased 2-fold in the aged rats compared with the young rats. Thus, the amount of GLUT4 per one gastrocnemius muscle was increased by only 50% in the aged rats compared with the young rats, although the weight of the gastrocnemius was increased approximately 3-fold in the aged rats. GLUT4 mRNA level was also determined by using rat GLUT4 cDNA as a probe. Surprisingly, the amount of 2.8 kb mRNA for GLUT4 per microgram of cellular total RNA was rather increased (1.85-fold) in the aged rats (data not shown). The amount of cellular total RNA per one gastrocnemius muscle was increased 2.5-fold in the aged rats compared with the young rats. Thus, the amount of GLUT4 mRNA per one gastrocnemius of the aged rats was considerably increased (4.6-fold) compared with that of the young rats.

Discussion

Effects other than reduced activity of the muscle might affect the results in the case of the hindlimb suspension model. Some previous studies reported reduction in food intake during tail-casted hindlimb suspen-

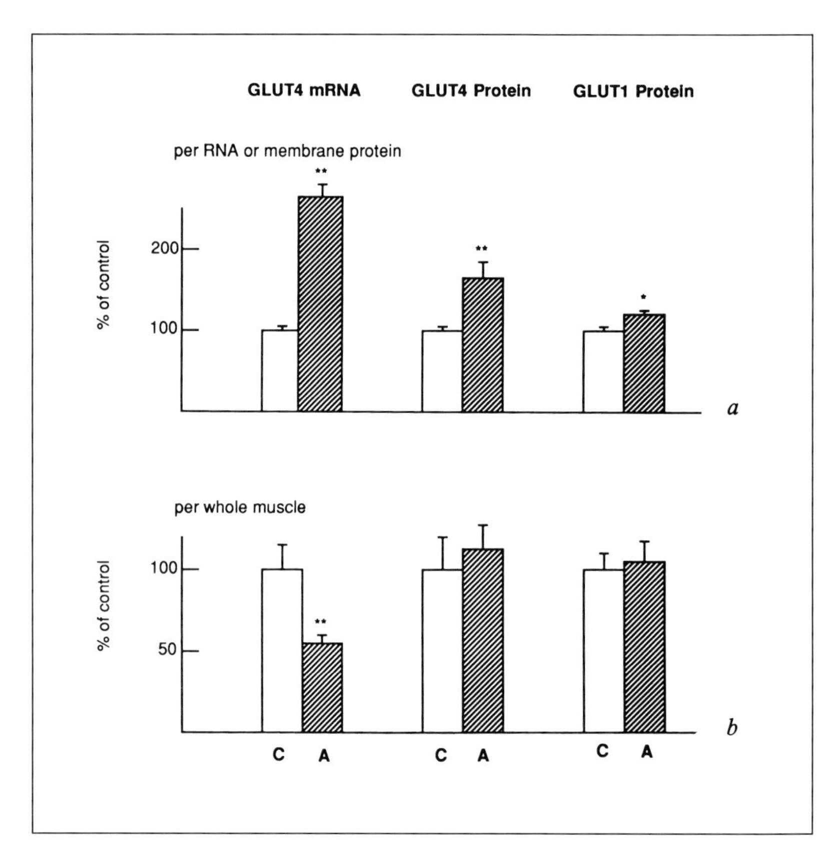

Fig. 2. Amount of GLUT4 mRNA and GLUT4 and GLUT1 proteins from control (C) and hindlimb-suspended atrophied (A) soleus muscle. *a* Amount of GLUT4 mRNA per equal amount of total RNA and amount of GLUT4 and GLUT1 protein per equal amount of total membrane protein. This was determined by quantitative densitometry of Northern and Western blotting from 3 separate experiments. The results are expressed as a percentage of the control and are represented as the mean ± SE. *b* Amount of GLUT4 mRNA and of GLUT4 and GLUT1 protein in whole soleus muscle. Results are the means ± SE of 3 separate experiments. * $p < 0.05$; ** $p < 0.01$: difference from control.

sion [7, 8], and the expression of skeletal muscle glucose transporter is reported to be elevated in the fasting state [9, 10]. To minimize this problem we have chosen TCWB rats as control according to the recommendation of Jaspers and Tischler [2], who reported no difference in food intake within 6 days between TCWB and TCHS rats. In the present study the body weight was not significantly changed and total RNA amount in the extensor digitorum longus muscle was not reduced; this muscle is known to be markedly decreased in the fasting state [11]. Thus, there was almost no difference in feeding status between TCWB and TCHS rats in this study.

An interesting observation is that GLUT4 mRNA level was decreased, while GLUT4 protein level was not decreased in the atrophied soleus muscle. Thus, GLUT4 protein per GLUT4 mRNA was increased in the atrophied soleus muscle, suggesting that the translational efficiency and/or stability of GLUT4 protein is increased in suspended soleus muscle. This is in marked contrast to the results obtained in the enlarged muscle of aged rats. GLUT4 protein per GLUT4 mRNA was markedly decreased by 70% in the enlarged gastrocnemius muscle of aged rats compared with young rats. This suggests that the translational efficiency and/or stability of the GLUT4 protein in the muscle is decreased in the enlarged muscle of aged rats. The results obtained in the enlarged muscle of aged rats might be due to differences in physical activity or age, rather than in tissue size. This issue should be investigated in the future.

References

1 Ziel FH, Venkatesan N, Davidson MB: Glucose transport is rate limiting for skeletal muscle glucose metabolism in normal and STZ-induced diabetic rats. Diabetes 1988;37:885–890.
2 Jaspers S, Tischler ME: Atrophy and growth failure of rat hindlimb muscles in tail-cast suspension. J Appl Physiol 1984;57:1472–1479.
3 Oka Y, Asano T, Shibasaki Y, Kasuga M, Kanazawa Y, Takaku F: Studies with antipeptide antibody suggest the presence of at least two types of glucose transporter in rat brain and adipocyte. J Biol Chem 1988;263:13432–13439.
4 Asano T, Shibasaki Y, Kasuga M, Kanazawa Y, Takaku F, Akanuma Y, Oka Y: Cloning of a rabbit brain glucose transporter cDNA and alteration of glucose transporter mRNA during tissue development. Biochem Biophys Res Commun 1988; 154:1204–1211.
5 Lehrach H, Diamond D, Wozney JM, Boedtker H: RNA molecular weight determination by gel electrophoresis under denaturing conditions: A critical examination. Biochemistry 1977;16:4743–4748.
6 Kahn BB, Charron MJ, Lodish HF, et al: Differential regulation of two glucose transporters in adipose cells from diabetic and insulin-treated rats. J Clin Invest 1989;84:404–411.
7 Babij P, Booth FW: α-Actin and cytochrome C mRNA in atrophied adult rat skeletal muscle. Am J Physiol 1988;254:C651–C656.

8 Steffen JM, Musacchia XJ: Effect of hypokinesia and hypodynamia on protein, RNA, and DNA in rat hindlimb muscles. Am J Physiol 1984;247:R728–R732.
9 Charron MJ, Kahn BB: Divergent molecular mechanisms for insulin-resistant glucose transport in muscle and adipose cells in vivo. J Biol Chem 1990;265:7994–8000.
10 Bourey RE, Koranyi L, James DE, et al: Effect of altered glucose homeostasis on glucose transporter expression in skeletal muscle of the rat. J Clin Invest 1990;86:542–547.
11 Goodman MN, Rudereman NB: Starvation in the rat. I. Effect of age and obesity on organ weights, RNA, DNA and protein. Am J Physiol 1980;239:E269–E276.

Yoshitomo Oka, MD, Third Department of Internal Medicine, Faculty of Medicine, University of Tokyo, 7-3-1 Hongo, Bunkyo-ku, Tokyo 113 (Japan)

Sato Y, Poortmans J, Hashimoto I, Oshida Y (eds): Integration of Medical and Sports Sciences. Med Sport Sci. Basel, Karger, 1992, vol 37, pp 91–108

Health Benefits of Exercise in the Elderly

John O. Holloszy, Robert J. Spina, Wendy M. Kohrt[1]

Section of Applied Physiology, Department of Internal Medicine, Washington University School of Medicine, St. Louis, Mo., USA

As a result of the great success of preventive medicine and public health measures directed against infectious diseases and nutritional deficiencies, the majority of people born in the wealthy, technologically advanced nations are attaining old age. As a consequence, the proportion of the population that is over the age of 65 years is increasing rapidly and the degenerative diseases of middle and old age, the decline in functional capacity and the increased incidence of disability in the elderly have become the major public health problem in these countries.

Evidence from a wide range of studies has accumulated indicating that exercise has beneficial effects in the prevention of some of the major degenerative diseases, and in slowing or partially reversing the decline in physical functional capacity with advancing age. The latter effect is mediated by adaptations that run counter to changes that generally occur with advancing age. There is, therefore, much interest in the possibility that regularly performed exercise could make a major contribution to maintenance of health and functional capacity in the elderly. Before reviewing the adaptations to exercise and their possible health benefits, it may be useful to examine the role of exercise in the context of primary and secondary aging.

Both primary and secondary aging contribute to the morbidity and disability in old age. Primary aging results in a progressive deterioration in structure and function due to intrinsic, i.e.'normal' aging processes such as cross-linking of proteins, loss of postmitotic cells and somatic mutations.

[1] The authors' research is supported by grants AGO5562, AG00425, DK18986 and AG00078.

Secondary aging is the deterioration in structure and function due to cumulative damage from disease processes, injuries, and environmental and life-style factors (such as smoking and overeating). It is extremely difficult, if not impossible, to obtain information regarding the effect of an intervention such as exercise on primary aging in humans, and the available data have come from studies on a short-lived species, the rat.

Exercise and Longevity in Rats

Some of the adaptations to regularly performed exercise run counter to various changes in structure and function that occur with advancing age in sedentary rats. These include partial protection against the increase in body fat content, the increase in fat cell size and the decrease in fat cell insulin sensitivity with aging [1–3]. It is of interest that exercise is more effective in this regard than food restriction sufficient to keep the body weights of sedentary rats the same as those of the exercisers [1, 2]. Exercise-training also reverses the prolongation of cardiac contraction and relaxation [4–7], and the decline in muscle respiratory capacity [8] in old rats. In light of these and other adaptations to exercise that appear to counteract some of the deterioration in structure and function with aging, it seems reasonable that exercise might have a beneficial effect on longevity of laboratory rats. A number of studies have been conducted to evaluate this possibility. The results of these studies show that rats that exercise regularly live longer than freely eating sedentary controls [9–12]. However, although exercise improves survival, i.e. average age at death, it does not appear to increase maximal life span [9, 10]. This is in contrast to food restriction, which is the only intervention that has clearly been shown to prolong life span in rodents [13].

Male rats do not increase their food intake to compensate for the increase in energy expenditure caused by exercise. As a consequence, male rats exercised by means of voluntary wheel running attain a body weight that is only ~66% as great as that of freely eating sedentary rats. When sedentary rats are food-restricted so as to keep their body weight in the same range as the runners', they show a significant extension of maximal life span [9, 10], even though the food-restricted rats are fatter than the runners of similar body weight. The effect of exercise in the form of voluntary wheel running and the effect of food restriction on longevity of male Long Evans rats are illustrated in figure 1.

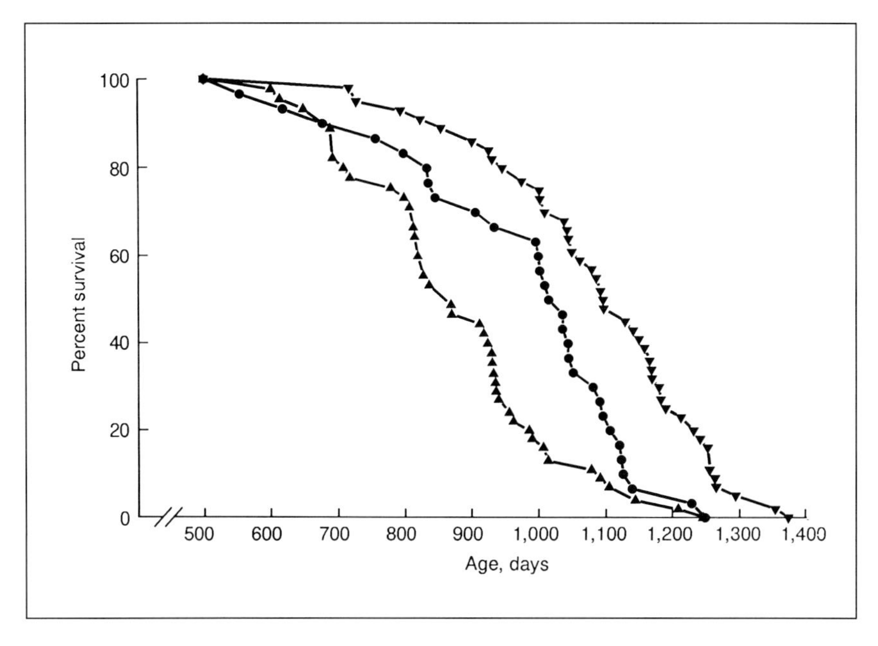

Fig. 1. Effects of voluntary wheel running and of food restriction of sedentary rats on longevity of male Long Evans rats [from ref. 9]. The runners (●) had access to voluntary running wheels beginning at the age of 3 months. The food-restricted sedentary rats (▼) had their food intake reduced so as to maintain their body weights in the same range as those of the runners (~ 30% below ad libitum). The survival curve of the runners shows an increase in average longevity without an extension of maximal life span, indicative of a beneficial effect on secondary aging. The food-restricted sedentary rats' survival curve shows an extension of maximal life span, indicative of a slowing of primary aging. ▲ = Sedentary rats.

Laboratory rats are usually confined to cages that restrict their physical activity and are provided food ad libitum. This is clearly an abnormal situation that generally leads to obesity. It appears from the results of the studies reviewed above that allowing rats to exercise, by giving them access to voluntary running wheels, prevents deleterious effects of a sedentary life combined with overeating that contribute to secondary aging. This makes it possible for a larger proportion of the animals to live to an old age, without slowing primary aging. Food restriction clearly does something more, resulting in a slowing of primary aging as evidenced by an extension of maximal life span.

Studies on Humans

At the same time that the advances in preventive medicine and public health were occurring, technological advances were making it possible for the majority of people living in the wealthy nations to go through life with minimal physical activity which, at least in the Western nations, is frequently associated with consumption of food in excess of energy demands. As in the case of the ad libitum fed, sedentary laboratory rat, this is an abnormal situation to which most of us are not adapted in genetic terms. Evolution occurred in the context of vigorous physical exercise which was necessary for survival. As is now evident, exercise is also necessary for maintenance of normal physiological function.

Chronic exercise deficiency has a number of deleterious effects that are corrected when a sedentary person undertakes a program of regular exercise. The harmful effects of a sedentary life fall into two related, and in some cases overlapping, categories. One of these involves the development of various forms of disuse atrophy, while the other is manifested as increased risk of developing certain disease processes, including atherosclerosis – particularly coronary artery disease (CAD), non-insulin-dependent diabetes (NIDDM) and hypertension.

Exercise and the Prevention and/or Reversal of 'Disuse' Atrophies in Older People

Advancing age is associated with progressive declines in the maximal capacity for aerobic exercise ($\dot{V}O_2max$), endurance, in skeletal muscle mass and strength, flexibility, agility, balance, coordination and movement speed. Thus, a young person outperforms an older individual who is otherwise comparable (i.e. except for age) in tests of these functions. However, similar differences in performance are seen between young sedentary and young well-trained people as exist between 25- and 60-year-old sedentary individuals [14].

Because of the similarities between the disuse atrophies seen with a sedentary life-style and the decline in physical functional capacities with aging, there is increasing interest in the possibilities that (1) much of the decline in functional capacity with advancing age in physically inactive people is due to progressive disuse atrophy rather than to primary aging, and (2) $\dot{V}O_2max$, strength, flexibility and balance can be significantly improved in sedentary older people by means of exercise-training. A number of studies have been conducted to evaluate these possibilities.

Aerobic Power. Evaluation of the effect of regularly performed endurance exercise on the rate of decline in $\dot{V}O_2max$ is difficult. While cross-sectional studies can provide some insights [cf. ref. 15], a definitive answer can only be provided by longitudinal studies. The difficulty of performing a longitudinal study is increased by the need to study individuals who maintain the same relative training stimulus over many years. This is necessary, because a decrease in training will exaggerate, while an increase in training will mask the decline in $\dot{V}O_2max$ with aging. Our approach to this problem has been to study master middle- and long-distance runners who are serious about doing well in races, and who train as hard as they can in order to do well in competition. This study is still in progress, but preliminary results suggest that the rate of decline in $\dot{V}O_2max$ due to primary aging is about 5% per decade up to ~70 years of age, while the rate of decline due to aging plus the disuse atrophy resulting from physical inactivity is 11% per decade [15, 16] (fig. 2).

Although a number of early studies suggested that people over the age of 60 years had little capacity to adapt to training with an increase in $\dot{V}O_2max$, more recent studies have shown that individuals in the 60- to 80-year age range can respond to endurance training with an increase in $\dot{V}O_2max$ similar in relative magnitude to that seen in young people [17–21]. It appears that the reason that $\dot{V}O_2max$ did not increase in the earlier studies on older people was the use of an inadequate training stimulus. As in young people, the increase in $\dot{V}O_2max$ in older people is mediated by cardiovascular adaptations that result in an increased cardiac output and by peripheral adaptations that result in increased extraction of O_2 from the blood by the working muscles.

Skeletal Muscle. There is a decline in skeletal muscle mass and an associated decrease in strength with advancing age. There is also a decline in muscle respiratory capacity in sedentary individuals with advancing age. Recent studies have shown that elderly men and women can adapt to strength training with large increases in muscle mass and strength [22, 23], and even mild exercise can increase strength in deconditioned older people [24]. Master endurance athletes avoid the decline in muscle respiratory capacity with age [25], and older men and women adapt to endurance

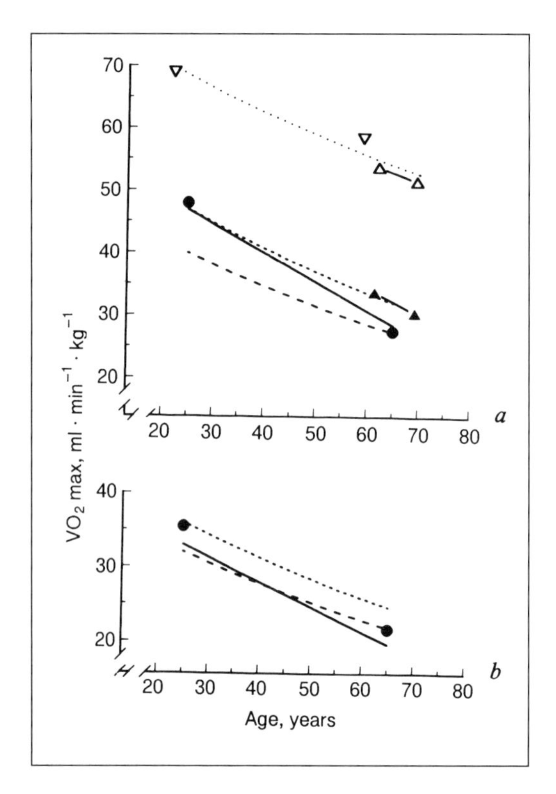

Fig. 2. Decline of $\dot{V}O_2$max with age from cross-sectional [15, 21, 89] and longitudinal [16] studies in groups of sedentary men and women and endurance-trained men. The rate of decline of $\dot{V}O_2$max is 9–13% per decade in sedentary people and 5–6% per decade in endurance-trained athletes who maintain training volume. *a* Men. *b* Women. ---- = Normally active, lean people [15, 21]; – – – = inactive, overweight people [15, 21]; —— = sedentary 25–65 years olds [89]; ● = sedentary young and older people [21]; ▽ = trained young and master athletes [15]; ▲—▲ = lean, sedentary men [16]; △—△ = master athletes [16]; ······ = estimated decline of $\dot{V}O_2$max of endurance-trained athletes.

exercise-training with increases in muscle respiratory enzyme levels similar to those seen in young individuals [20].

Flexibility and Balance. People in the Western nations who are physically inactive and spend much time sitting on chairs or couches undergo a remarkable loss of flexibility, with the decreases in range of motion frequently being sufficiently great to interfere with activities of daily living.

This loss of range of motion can be, at least in part, reversed by means of programs of stretching exercises [24].

Balancing is a major aspect of all standing activities, including walking and rising from a chair. Balance as reflected in tests such as one-legged standing balance time deteriorates, and postural sway increases with aging. The magnitude of these changes correlates with an increased risk of falling. Thus, impaired balance in older people is a major problem, one of the major manifestations of which is hip fractures. Preliminary data suggest that as little as 3 months of a low-intensity, flexibility and strength exercise program that inlcudes some challenge to balancing ability can significantly improve balance in elderly subjects with impaired balance [24].

Physical Benefits and Types of Exercise

The adaptations to exercise of the cardiovascular system, skeletal muscles, ligaments and tendons, and central nervous system can play major roles in enabling older people to remain independent and self-sufficient, and in protecting against falls that result from weakness and poor balance. Furthermore, as discussed in the next section, these adaptations to exercise-training make it possible for previously sedentary older people to exercise sufficiently long at a high enough intensity to help protect against or reverse the harmful effects to health of a chronic exercise deficiency. In recommending or prescribing exercise, it is important to keep in mind that the effects of the endurance, strength, flexibility and coordination-skill types of exercise are quite specific. It is, therefore, necessary to combine these types of exercise in a regularly performed program in order to maintain cardiovascular function, endurance, strength, flexibility, agility and balance.

Exercise and the Prevention or Reversal of Certain Disease Processes

There is enormous genotypic and phenotypic diversity among humans, and the importance of exercise for disease prevention, therefore, varies greatly among individuals. A genotype that is harmless, or may even have a survival advantage, in hunting-gathering or survival farming populations can be responsible for greatly increased risk of morbidity and mortality in sedentary people.

A phenotype that is not seen in hunter-gatherers and subsistence farmers is the 'abdominal obesity' syndrome which, in its full-blown form, includes central obesity, insulin resistance, hyperglycemia, hyperinsulinemia, hyperlipidemia and, sometimes, elevated blood pressure. The severity of the obesity, the degree of insulin resistance, the magnitude of the hyperinsulinemia, the characteristics of the hyperlipidemia, and the presence and degree of hypertension vary considerably among individuals living under similar conditions, probably in large part due to differences in their genetic constitution.

The abdominal obesity syndrome, which is arguably the most serious public health problem in the USA and some of the European countries, is associated with a greatly increased risk of developing atherosclerosis (manifested as CAD, cerebrovascular disease and claudication) and NIDDM with its many complications. This syndrome is preventable and reversible, by regular exercise and/or by moderately severe food restriction. The latter approach, which appears to be successful in Japan, seems unacceptable to a large proportion of people living in other countries, i.e. the wealthy Western nations, as well as the wealthier individuals in the less developed countries, for whom exercise is therefore a necessary health measure.

Individuals with other phenotypes that increase the risk of CAD and/or NIDDM for whom exercise provides health benefits include those with certain lipoprotein abnormalities, with insulin resistance and hyperglycemia in the absence of abdominal obesity and mild essential hypertension. There are also individuals who have no disease risk factors that are reversed by exercise, and who require exercise only to prevent disuse atrophy.

Role of Exercise in Prevention and Treatment of Coronary Artery Disease. In nations such as the US in which people eat diets high in saturated animal fats, cholesterol and calories, and have a high incidence of CAD, sedentary individuals have approximately twice as great a risk of developing clinical CAD as people who exercise regularly [26–29]. Because of the large proportion of middle-aged and elderly people who are physically inactive, lack of exercise is the CAD risk factor that has greatest negative effect on public health in the US [30]. Although the majority of studies have involved middle-aged men, available evidence suggests as strong a protective effect in elderly men up to at least the age of 75 years [27]. Regularly performed exercise has beneficial effects on a number of coronary risk factors. It has been shown that favorable modification of risk factors by other interventions, i.e. diet and medications, can prevent progression and even bring about regression of advanced coronary atherosclerosis [31, 32]. It, therefore, seems reasonable that exercise mediates its protective effect, at least in part, by its beneficial effects on coronary risk factors.

There is a negative correlation between CAD and plasma high-density lipoprotein (HDL) cholesterol [33]. This protective effect of plasma HDL appears to be mediated by the HDL_2 fraction [34]. People who regularly perform vigorous endurance exercise have a higher plasma HDL concentration than sedentary persons [35], and short periods of exercise training result in modest increases in plasma HDL [35]. A longer period of training resulted in a large increase in HDL in 9 coronary patients who were retested after 1 year and again after an average of 7 years of training [36]. At the end of the initial 12 months of training, these patients were jogging 29 ± 5 km/week. They continued to exercise regularly for an additional 6 ± 0.4 years and were running about 42 km/week at an intensity of 85 ± 4% of $\dot{V}O_2max$ during the last year. In the untrained state, HDL cholesterol was 38 ± 3 mg/dl, after 12 months it was 45 mg/dl and after 7 years it was 53 ± 5 mg/dl. Over the 7-year period, the ratio of low-density lipoprotein to HDL cholesterol decreased from 3.8 to 2.4. The increase in HDL cholesterol between the end of the 1st year and the 6th year of training occurred in the absence of additional weight loss [36].

It is likely that the exercise-induced increase in HDL_2 is mediated by an increase in the half-life of the HDL_2 fraction [37]. The lowering of the rate of HDL_2 degradation may result from an increase in lipoprotein lipase (LPL) activity and enhanced triglyceride (TG) clearance induced by exercise [37]. Exercise has a potent plasma TG-lowering effect that is evident within hours after a bout of exercise [38, 39]. Exercise lowers both fasting [38, 39] and postprandial [40] plasma TG. This effect is cumulative, so that daily bouts of exercise result in a progressive decrease in fasting plasma TG concentration [39]. The TG lowering is not mediated by negative calorie balance [39], but by exercise- and exercise-training-induced increases in skeletal muscle, adipose tissue and postheparin plasma LPL activities [41, 42]. Elevation of plasma TG results in enrichment of HDL_2 with TG, which makes the HDL_2 more susceptible to conversion to HDL_3 and degradation by hepatic lipase [43, 44]. Raising LPL activity by means of exercise lowers plasma TG levels and the TG content of HDL_2, and, thus, probably plays a major role in bringing about the increase in HDL_2. The best evidence for an important role of LPL in determining plasma HDL_2 concentration comes from studies in which inhibition of LPL resulted in accumulation of TG-rich lipoproteins and a decrease in HDL_2 [45, 46]. A second factor that may contribute to the lowering of plasma TG and the increase in HDL_2 is the decrease in insulin secretion brought about by exercise [47], resulting in a decrease in the stimulus for hepatic TG synthesis [48, 49].

Hyperinsulinemia is common in older people, particularly those with abdominal obesity. In addition to playing important roles in the development of other risk factors, including hyperlipidemia, a low HDL_2 level and hypertension, hyperinsulinemia may play a direct role in the development of atherosclerosis [50, 51] manifested as CAD [52]. As short a period as 7 days of vigorous exercise-training brings about a marked lowering of the plasma insulin response to glucose feeding [53], and regular exercise prevents the increase in plasma insulin levels with advancing age [54, 55].

Even after atherosclerosis has progressed sufficiently to result in clinical CAD, exercise-training can improve myocardial oxygenation in some patients. In early studies, exercise was found to improve exercise capacity, endurance and the minimal exercise intensity required to induce ischemia in patients with CAD [56–58]. These effects were a consequence of adaptations in the skeletal muscles and autonomic nervous system that result in smaller increases in heart rate and systolic blood pressure and, therefore, a reduced myocardial oxygen requirement at a given submaximal exercise intensity. However, in these studies, the training did not improve myocardial ischemia or contractile function at the same myocardial O_2 requirement [56–58]. The exercise-training in these early studies was brief and mild. Studies in experimental animals have shown that intense training improves myocardial blood supply with a reduction in ischemia and improvement in left-ventricular function [59–62], suggesting that the lack of improvement in the early studies on patients with CAD was due to an insufficient exercise stimulus. The results of studies performed during the past 10 years, of the effects of prolonged, high-intensity endurance exercise-training, support this view.

These studies have shown that, in addition to peripheral adaptations, long-term exercise training of progressively increasing duration, frequency and intensity can induce further adaptations indicative of improvements in myocardial ischemia and left-ventricular function. These findings include: (1) less ST segment depression at the same rate pressure product [63, 64], and even during maximal exercise despite attainment of a higher heart rate and systolic blood pressure after training [36]; (2) a higher stroke volume at a comparable heart rate and peripheral vascular resistance [65]; (3) a decrease in stress-induced myocardial ischemia as assessed by thallium-201 scintigraphy [66], and (4) an improvement in myocardial contractile function assessed by equilibrium radionuclide ventriculography [64]. The improvement in myocardial contractile function, measured during maximal supine cycle exercise, was reflected in an increase in ejection fraction from rest to maximal exercise after training, whereas before training, and in control patients, ejection fraction did not change or decreased with exercise. Exercise-induced regional wall motion abnormalities also improved in the training group [64]. An encouraging finding is that these improvements can occur in patients who are in their 60s and 70s. It remains to be determined whether these manifestations of improved myocardial oxygenation induced by exercise-training are due to regression of atherosclerosis, an increase in diameter of the coronary arteries and/or an improvement in collateral circulation.

Role of Exercise in the Prevention and Treatment of Insulin Resistance, Glucose Intolerance and Non-Insulin-Dependent Diabetes mellitus. Glucose tolerance frequently deteriorates with advancing age [67, 68]. This deterioration in glucose tolerance is largely due to development of resistance to insulin [68]. Exercise has the opposite effect and increases insulin sensitivity [69–72]. People who exercise vigorously on a regular basis appear to be protected against development of insulin resistance and a reduction in glucose tolerance as they age [54, 55, 73]. In some individuals, probably those with a genetic predisposition to develop NIDDM, exercise prevents manifestation of the tendency for insulin resistance, which is unmasked when they stop exercising for a few days [73]. In others, regular exercise appears to prevent the development of insulin resistance and the decline in glucose tolerance; this is evidenced by the finding that when they stop exercising for long enough to permit the short-term effects of exercise on insulin sensitivity and secretion to wear off, their glucose and insulin responses to a glucose load are similar to those of young, lean sedentary men [73].

Regularly performed exercise has both long-term and short-term beneficial effects in the prevention and treatment of insulin resistance, glucose intolerance and NIDDM. The long-term effect involves the prevention or reversal of abdominal obesity and is mediated by increased caloric expenditure. One of the short-term effects of exercise is a suppression of insulin secretion, which is manifested as a markedly blunted insulin response not only to a glucose load [47, 74–78] but also to arginine and a fat meal [47]. The blunted insulin response compensates for another short-term adaptation to exercise, an increase in the sensitivity of muscle to the action of insulin [79, 80]. Since skeletal muscle is the major site of insulin-mediated glucose disposal, the increased muscle insulin sensitivity is manifested as increased glucose disposal during a euglycemic clamp [71, 72, 81] or an

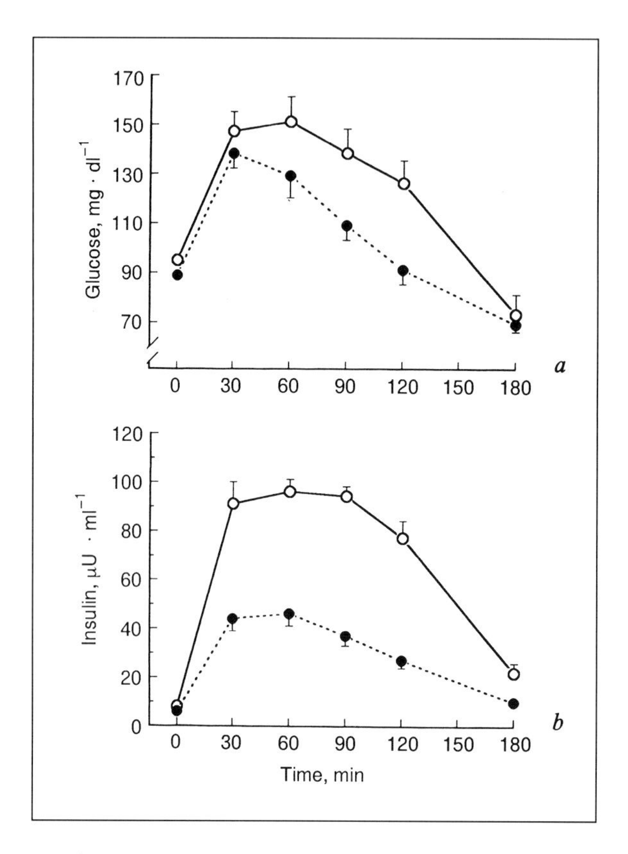

Fig. 3. Plasma glucose *(a)* and insulin *(b)* concentrations (mean ± SE) during a 75-gram oral glucose tolerance test in 7 men and 5 women, aged 63 ± 1 years, before (○) and after (●) 9 months of endurance exercise-training (4 days/week, 45 min/day, at ~ 80% of maximal heart rate). Body weight decreased from 69.1 ± 10 to 67.4 ± 10 kg, and $\dot{V}O_2$max increased from 1.84 ± 0.4 to 2.19 ± 0.7 liters/min in response to the exercise-training.

intravenous glucose tolerance test [82] and excellent tolerance to oral glucose despite the blunted insulin response [55, 74–76, 83] (fig. 3).

Exercise in the Prevention and Treatment of Osteoporosis. The decision to include osteoporosis under the heading of 'disease' rather than 'disuse' was an arbitrary one, as disuse atrophy is certainly a major factor. However, it is not the only one, as anabolic hormone (estrogen, dehydroepiandrosterone, growth hormone) deficiencies, inadequate calcium and vitamin D intake, and various other factors also play roles.

Exercise is necessary for bone growth and maintenance of bone mass. Bone growth ceases and bone atrophy develops in limbs in which the muscles are paralyzed. Bed rest [84] and space flight [85] result in rapid loss of bone. The findings on the effects of space flight have been taken as evidence that weight-bearing, i.e. antigravity, exercise is necessary for maintenance of bone mass, but this has not yet been firmly established. Evidence is beginning to accumulate that exercise-training can induce significant increases in both appendicular [86] and axial [87] bone mineral content in postmenopausal women. An adequate intake of calcium and vitamin D is, of course, necessary for the anabolic effect of exercise on bone [88]. Research on the effects of exercise on osteoporosis in elderly women is still at an early stage, and relatively little is known regarding what represents an optimal exercise program for reversing osteoporosis or about the interactions between exercise and hormone replacement.

Conclusion

Human beings are not genetically adapted for a sedentary life. Regular exercise is necessary for maintenance of normal functioning of the cardiovascular system, the skeletal muscles, the skeleton and the autonomic nervous system. Chronic physical inactivity plays an important role in the deterioration of functional capacity, the increase in body fat content and the loss of lean body mass – particularly the skeletal muscle and bone atrophy – that occur with advancing age in sedentary people. These 'secondary aging' effects of physical inactivity are preventable and, also, to a considerable extent reversible even in people in their 60s and 70s. Lack of exercise is also a major risk factor for development of CAD and NIDDM, which are responsible for much of the morbidity and mortality among older people in the wealthy nations.

References

1 Craig BW, Garthwaite SM, Holloszy JO: Adipocyte insulin resistance: Effects of aging, obesity, exercise and food restriction. J Appl Physiol 1987;62:95–100.

2 Garthwaite SM, Cheng H, Bryan JE, Craig BW, Holloszy JO: Ageing, exercise and food restriction: Effects on body composition. Mech Ageing Dev 1986;36:187–196.

3 Lawrence JC, Colvin J, Cartee GD, Holloszy JO: Effects of aging and exercise on insulin action in rat adipocytes are correlated with changes in fat cell volume. J Gerontol 1989;44:B88–B92.

4 Li Y, Lincoln T, Mendelowitz D, Grossman W, Wei JW: Age-related differences in effect of exercise training on cardiac muscle function in rats. Am J Physiol 1986;251: H12–H18.

5 Spurgeon HA, Steinbach MF, Lakatta EG: Chronic exercise prevents characteristic age-related changes in cardiac contraction. Am J Physiol 1983;244:H513–H518.

6 Starnes JW, Beyer RE, Edington DW: Myocardial adaptations to endurance exercise in aged rats. Am J Physiol 1983;245:H560–H566.

7 Tate CA, Taffet GE, Hudson EK, Blaylock SL, McBride RP, Michael LH: Enhanced calcium uptake of cardiac sarcoplasmic reticulum in exercise-trained old rats. Am J Physiol 1990;258:H431–H435.

8 Cartee GD, Farrar RP: Muscle respiratory capacity and $\dot{V}O_2max$ in identically trained young and old rats. J Appl Physiol 1987;63:257–261.

9 Holloszy JO, Schechtman KB: Interactions between exercise and food restriction: Effects on longevity of male rats. J Appl Physiol 1991;70:1529–1535.

10 Holloszy JO, Smith EK, Vining M, Adams S: Effect of voluntary exercise on longevity of rats. J Appl Physiol 1985;59:826–831.

11 Goodrick CL: Effects of long-term voluntary wheel exercise on male and female Wistar rats 1. Longevity, body weight and metabolic rate. Gerontology 1980;26: 22–23.

12 Goodrick CL, Ingram DK, Reynolds MA, Freeman JR, Cider NL: Differential effects of intermittent feeding and voluntary exercise on body weight and lifespan in adult rats. J Gerontol 1983;38:36–45.

13 Weindruch R, Walford RL: The Retardation of Aging and Disease by Dietary Restriction. Springfield, Thomas, 1988, pp 3–436.

14 Hagberg JM, Seals DR, Yerg JE, Gavin J, Gingerich R, Premachandra B, Holloszy JO: Metabolic responses to exercise in young and older athletes and sedentary men. J Appl Physiol 1988;65:900–908.

15 Heath GW, Hagberg JM, Ehsani AA, Holloszy JO: A physiological comparison of young and older endurance athletes. J Appl Physiol 1981;51:634–640.

16 Rogers MA, Hagberg JM, Martin WH, Ehsani AA, Holloszy JO: Decline in $\dot{V}O_2max$ with aging in master athletes and sedentary men. J Appl Physiol 1990;68:2195–2199.

17 Seals DR, Hagberg JM, Hurley BF, Ehsani AA, Holloszy JO: Endurance training in older men and women. 1. Cardiovascular responses to exercise. J Appl Physiol 1984; 57:1024–1029.

18 Hagberg JM, Graves JE, Limacher M, Woods DR, Leggett SH, Cononie C, Gruber JJ, Pollock ML: Cardiovascular responses of 70- to 79-yr-old men and women to exercise training. J Appl Physiol 1989;66:2589–2594.

19 Makrides L, Heigenhauser GJF, Jones NL: High-intensity endurance training in 20- to 30- and 60- to 70-yr-old healthy men. J Appl Physiol 1990;69:1792–1798.

20 Meredith CN, Frontera WR, Fisher EC, Hughes VA, Herland JC, Edwards J, Evans WJ: Peripheral effects of endurance training in young and old subjects. J Appl Physiol 1989;66:2844–2849.

21 Kohrt WM, Malley MT, Coggan AR, Spina RJ, Ogawa T, Ehsani AA, Bourey RE,

Martin WH, Holloszy JO: Effects of gender, age, and fitness level on the response of $\dot{V}O_2max$ to training in 60 to 71 yr-olds. J Appl Physiol 1991;71:2004–2011.
22 Fiatarone MA, Marks EC, Ryan ND, Meredith CN, Lipsitz LA, Evans WJ: High intensity strength training in nonagenarians. JAMA 1990;263:3029–3034.
23 Frontera WR, Meredith CN, O'Reilly KP, Knuttgen HG, Evans WJ: Strength conditioning in older men: Skeletal muscle hypertrophy and improved function. J Appl Physiol 1988;64:1038–1044.
24 Brown M, Holloszy JO: Effects of a low intensity exercise program on selected physical performance characteristics of 60- to 71-year olds. Aging 1991;3:129–139.
25 Coggan AR, Spina RJ, Rogers MA, King DS, Brown M, Nemeth PM, Holloszy JO: Histochemical and enzymatic characteristics of skeletal muscle in master athletes. J Appl Physiol 1990;68:1896–1901.
26 Leon AS, Connett J, Jacobs DR Jr, Rauramaa R: Leisure-time physical activity levels and risk of coronary heart disease and death. JAMA 1987;258:2388–2395.
27 Morris JN, Everitt MG, Pollard R, Chave SPW, Semmence AM: Vigorous exercise in leisure-time: Protection against coronary heart disease. Lancet 1980;ii:1207–1210.
28 Paffenbarger RS, Hyde RT, Wing AL, Hsieh CC: Physical activity, all cause mortality and longevity of college alumni. N Engl J Med 1986;314:605–613.
29 Paffenbarger RS, Hyde RT, Wing AL, Steinmetz CH: A natural history of athleticism and cardiovascular health. JAMA 1984;252:491–495.
30 Centers for Disease Control: Protective effect of physical activity on coronary heart diesase. MMWR 1987;36:426–430.
31 Blankenhorn DH, Nessin SA, Johnson RL: Beneficial effects of combined colestipol-niacin therapy on coronary atherosclerosis and coronary venous bypass grafts. JAMA 1987;257:3233–3240.
32 Ornish D, Brown SE, Scherwitz LW, Billings JH, Armstrong WT, Ports TA, McLanahan SM, Kirkeeide RL, Brand RJ, Gould KL: Can lifestyle changes reverse coronary heart disease? Lancet 1990;336:129–133.
33 Miller GJ: High density lipoproteins and atherosclerosis. Annu Rev Med 1980;31:97–108.
34 Miesenböck G, Patsch JR: Relationship of triglyceride and high-density lipoprotein metabolism. Atheroscler Rev 1990;21:119–127.
35 Haskell WL: The influence of exercise on the concentrations of triglyceride and cholesterol in human plasma. Exerc Sport Sci Rev 1984;12:205–244.
36 Rogers MA, Yamamoto C, Hagberg JM, Holloszy JO, Ehsani AA: The effect of 7 years of intense exercise training on patients with coronary artery disease. J Am Coll Cardiol 1987;10:321–326.
37 Herbert PN, Bernier DN, Cullinane EM, Edelstein L, Kantor MA, Thompson PD: High-density lipoprotein metabolism in runners and sedentary men. JAMA 1984;252:1034–1037.
38 Holloszy JO, Skinner JS, Toro G, Cureton TK: Effects of a six month program of endurance exercise on the serum lipids of middle-aged men. Am J Cardiol 1964;14:752–760.
39 Gyntelberg F, Brennan R, Holloszy JO, Schonfeld G, Rennie MJ, Weidman SW: Plasma triglyceride lowering by exercise despite increased food intake in patients with type IV hyperlipoproteinemia. J Clin Nutr 1977;30:716–720.

40 Weintraub MS, Rosen Y, Otto R, Eisenberg S, Breslow JL: Physical exercise conditioning in the absence of weight loss reduces fasting and postprandial triglyceride-rich lipoprotein levels. Circulation 1989;79:1007–1014.
41 Lithell H, Cedermark M, Fröberg J, Tesch P, Karlsson J: Increase of lipoprotein-lipase activity in skeletal muscle during heavy exercise: Relation to epinephrine excretion. Metabolism 1981;30:1130–1134.
42 Nikkilä EA, Taskinen M, Rehunen S, Härkönen M: Lipoprotein lipase activity in adipose tissue and skeletal muscle of runners: Relation to serum lipoproteins. Metabolism 1978;27:1661–1671.
43 Patsch JR, Prasad S, Gotto AM Jr, Patsch W: High density $lipoprotein_2$: Relationship of the plasma levels of this lipoprotein species to its composition, to the magnitude of postprandial lipemia, and to the activities of lipoprotein lipase and hepatic lipase. J Clin Invest 1987;80:341–347.
44 Tall AR: Plasma high density lipoproteins: Metabolism and relationship to atherogenesis. J Clin Invest 1990;86:379–384.
45 Behr SR, Patsch JR, Forte T, Bensadoun A: Plasma lipoprotein changes resulting from immunologically blocked lipolysis. J Lipid Res 1981;22:443–451.
46 Goldberg IJ, Blaner WS, Vanni TM, Moukides M, Ramakrishnan R: Role of lipoprotein lipase in the regulation of high density lipoprotein apolipoprotein metabolism: Studies in normal and lipoprotein lipase-inhibited monkeys. J Clin Invest 1990;86:463–473.
47 King DS, Staten MA, Kohrt WM, Dalsky GP, Elahi D, Holloszy JO: Insulin secretory capacity in endurance-trained and untrained young men. Am J Physiol 1990; 259:E155–E181.
48 Reaven GM, Lerner RL, Stern MP, Farquhar JW: Role of insulin in endogenous hyperglyceridemia. J Clin Invest 1967;46:1756–1767.
49 Tobey TA, Greenfield M, Kraemer F, Reaven GM: Relationship between insulin resistance, insulin secretion, very low density lipoprotein kinetics and plasma triglyceride levels in normotriglyceridemic man. Metabolism 1981;30:165–171.
50 DeFronzo RA, Ferrannini E: A multifaceted syndrome responsible for NIDDM, obesity, hypertension, dyslipidemia and atherosclerotic cardiovascular disease. Diabetes Care 1991;14:173–194.
51 Stout RW: The relationship of abnormal circulating insulin levels to atherosclerosis. Atherosclerosis 1977;27:1–13.
52 Fontbonne AM, Eschwege EM: Insulin and cardiovascular disease: Paris prospective study. Diabetes Care 1991;14:461–469.
53 Rogers MA, Yamamoto C, King DS, Hagberg JM, Ehsani AA, Holloszy JO: Improvement in glucose tolerance after one week of exercise in patients with mild NIDDM. Diabetes Care 1988;11:613–618.
54 Holloszy JO, Schultz J, Kusnierkiewicz J, Hagberg JM, Ehsani AA: Effects of exercise on glucose tolerance and insulin resistance. Acta Med Scand [Suppl] 1986;711: 55–65.
55 Seals DR, Hagberg JM, Allen WK, Hurley BF, Dalsky GP, Ehsani AA, Holloszy JO: Glucose tolerance in young and older athletes and sedentary men. J Appl Physiol 1984;56:1521–1525.
56 Detry J-M, Bruce RA: Effects of physical training on exertional ST segment depression in coronary heart disease. Circulation 1971;44:390–396.

57 Hellerstein HK, Burlando A, Hirsch EZ, Plotkin FH, Feil GH, Winkler O, Marik S, Margolis N: Active physical reconditioning of coronary patients. Circulation 1965; 32:11–110.
58 Redwood DR, Rosing DR, Epstein SE: Circulatory and symptomatic effects of physical training in patients with coronary-artery disease and angina pectoris. N Engl J Med 1972;286:959–965.
59 Heaton WH, Marr KC, Capurro NL, Goldstein RE, Epstein SE: Beneficial effect of physical training on blood flow to myocardium perfused by chronic collaterals in the exercising dog. Circulation 1978;57:575–581.
60 McElroy CL, Gissen SA, Fishbein MC: Exercise-induced reduction in myocardial infarct size after coronary occlusion in the rat. Circulation 1978;57:958–962.
61 Wyatt HL, Mitchell J: Influences of phyisical conditioning and deconditioning on coronary vasculature in dogs. J Appl Physiol 1978;45:619–625.
62 Scheuer J, Tipton CM: Cardiovascular adaptations to physical training. Annu Rev Physiol 1977;39:221–251.
63 Ehsani AA, Hagberg JM, Heath GW, Sobel BE, Holloszy JO: Effects of twelve months of intense exercise training on ischemic ST-segment depression in patients with coronary artery disease. Circulation 1981;64:1116–1124.
64 Ehsani AA, Biello D, Schultz J, Sobel B, Holloszy JO: Improvement of left ventricular contractile function by exercise-training in patients with coronary artery disease. Circulation 1986;74:350–358.
65 Hagberg JM, Ehsani AA, Holloszy JO: Effect of 12 months of intense exercise training on maximum stroke volume in patients with coronary artery disease. Circulation 1983;67:1194–1199.
66 Schuler G, Schlierf G, Wirth A, Mautner H-P, Scheurlen H, Thumm M, Roth H, Schwarz F, Kohlmeier M, Mehmel HC, Kubler W: Low-fat diet and regular, supervised physical exercise in patients with symptomatic coronary artery disease: Reduction of stress-induced myocardial ischemia. Circulation 1988;77:172–181.
67 Davidson MB: The effect of aging on carbohydrate metabolism: A review of the English literature and a practical approach to the diagnosis of diabetes mellitus in the elderly. Metabolism 1979;28:686–705.
68 DeFronzo RA: Glucose intolerance and aging: Evidence for tissue insensitivity to insulin. Diabetes 1979;28:1095–1101.
69 Burstein R, Polychronakos C, Toews CJ, MacDougall JD, Huyda HJ, Posner BI: Acute reversal of the enhanced insulin action in trained athletes: Association with insulin receptor changes. Diabetes 1985;34:756–760.
70 Hollenbeck CB, Haskell W, Rosenthal M, Reaven GM: Effect of habitual physical activity on regulation of insulin-stimulated glucose disposal in older males. J Am Geriatr Soc 1984;33:273–277.
71 King DS, Dalsky GP, Clutter WE, Young DA, Staten MA, Cryer PE, Holloszy JO: Effects of exercise and lack of exercise on insulin sensitivity and responsiveness. J Appl Physiol 1988;64:1942–1946.
72 King DS, Dalsky GP, Staten MA, Clutter WE, Van Houten DR, Holloszy JO: Insulin action and secretion in endurance-trained and untrained humans. J Appl Physiol 1987;53:2247–2252.
73 Rogers MA, King DS, Hagberg JM, Ehsani AA, Holloszy JO: Effect of 10 days of inactivity on glucose tolerance in master athletes. J Appl Physiol 1990;68:1833–1837.

74 Bjorntorp P, Fahlen M, Grimby G, et al: Carbohydrate and lipid metabolism in middle-aged physically well trained men. Metabolism 1972;21:1027–1044.
75 Lohman D, Liebold F, Heilman W, Singer H, Pohl A: Diminished insulin response in higly trained athletes. Metab Clin Exp 1978;27:521–524.
76 LeBlanc J, Nadeau A, Richard D, Tremblay A: Studies on the sparing effect of exercise on insulin requirements in human subjects. Metabolism 1981;30:1119–1124.
77 Galbo H, Hedeskov CJ, Capito K, Vinten J: The effect of physical training on insulin secretion of rat pancreatic islets. Acta Physiol Scand 1981;11:75–79.
78 King DS, Dalsky GP, Clutter WE, Young DA, Staten MA, Cryer PE, Holloszy JO: Effects of exercise and lack of exercise on insulin secretion. Am J Physiol 1988;254: E537–E542.
79 Richter EA, Garetto LP, Goodman MN, Ruderman NB: Muscle glucose metabolism following exercise in the rat: Increased sensitivity to insulin. J Clin Invest 1982;59: 785–793.
80 Cartee GD, Young DA, Sleeper MD, Zierath J, Wallberg-Henriksson H, Holloszy JO: Prolonged increase in insulin-stimulated glucose transport in muscle after exercise. Am J Physiol 1989;256:E494–E499.
81 Mikines KJ, Sonne B, Farrell PA, Tronier B, Galbo H: Effect of physical exercise on sensitivity and responsiveness to insulin in humans. Am J Physiol 1988;254:E248–E259.
82 Kahn SE, Larson VG, Beard JC, Cain KC, Fellingham GW, Schwartz RS, Veith RC, Stratton JR, Cerqueira MD, Abrass IB: Effect of exercise on insulin action, glucose tolerance, and insulin secretion in aging. Am J Physiol 1990;259:E937–E943.
83 Heath GW, Gavin JR III, Hinderliter JM, Hagberg JM, Bloomfield SA, Holloszy JO: Effects of exercise and lack of exercise on glucose tolerance and insulin sensitivity. J Appl Physiol 1983;55:512–517.
84 LeBlanc AD, Schneider VS, Evans HJ, Engelbretson DA, Krebs JM: Bone mineral loss and recovery after 17 weeks of bed rest. J Bone Miner Res 1990;5:843–850.
85 Morey ER, Baylink DJ: Inhibition of bone formation during space flight. Science 1978;201:1138–1141.
86 Margulies JY, Simkin A, Leichter I: Effect of intense physical exercise on the bone-mineral content in the lower limbs of young adults. J Bone Joint Surg 1986;68A: 1090–1093.
87 Dalsky GP, Stocke KR, Ehsani AA, Slatopolsky E, Lee WC, Birge SJ: Weight-bearing exercise training and lumbar bone mineral content in postmenopausal women. Ann Intern Med 1988;108:824–828.
88 Kanders B, Dempster D, Lindsay R: Interaction of calcium nutrition and physical activity on bone mass in young women. J Bone Miner Res 1988;3:145–149.
89 Hossack KF, Bruce RA: Maximal cardiac function in sedentary normal men and women: Comparison of age-related changes. J Appl Physiol 1982;54:799–804.

John O. Holloszy, MD, Washington University School of Medicine,
2nd Floor, West Building, Campus Box 8113, 4566 Scott Avenue,
St. Louis, MO 63110 (USA)

Sato Y, Poortmans J, Hashimoto I, Oshida Y (eds): Integration of Medical and Sports Sciences. Med Sport Sci. Basel, Karger, 1992, vol 37, pp 109–125

Ageing, Skeletal Muscle Contractile Properties and Enzyme Activities with Exercise

A.W. Taylor[a, b], *E.G. Noble*[a, b], *D.A. Cunningham*[a, b], *D.H. Paterson*[a], *P. Rechnitzer*[a, c]

[a] Centre for Activity and Ageing, Faculty of Kinesiology, and
[b] Department of Physiology, Faculty of Medicine,
The University of Western Ontario;
[c] St. Joseph's Health Centre, London, Ont., Canada

It is an accepted fact that physical work capacity declines with advancing age [1]. Recently, a great deal of scientific literature has been published which suggests that exercise and physical training may result in increased ability of skeletal muscle to carry out work even into the 9th decade [2]. The majority of this research involves cross-sectional sampling and often it is difficult to parcel out the effects of ageing from disease-related processes and physical inactivity [1]. Unfortunately, few longitudinal studies are available [3]. The fields of muscle physiology and biochemistry are also inhibited by the sampling techniques available. Nonetheless, a great deal of progress in studying the ageing process and the adaptation to regular exercise has been made.

Hayflick [4] has noted the 8 theories of biogerontology, the study of how and why living organisms age. The following review will deal primarily with skeletal muscle and it is left to individual readers to 'pigeon-hole' the cause of ageing in this organ and to surmise whether or not regular exercise affects any of these theories or merely retards the rate of application.

Contractile Properties and Motor Unit Composition

The ultimate expression of skeletal muscle function is the manner in which the muscle contracts. Numerous authors have previously demonstrated an age-related change in contractile properties (time to peak ten-

Table 1. Relationship of ageing to contractile properties in rat skeletal muscle

Age months	Muscle	TPT ms	½RT ms	V_{max} m/s	Reference
6	TA	17	14		13
20–24	TA	21	14		13
6	S	24	46		13
20–24	S	34	46		13
9	EDL			5.4	14
20–30	EDL			3.5	14
9	SOL			0.99	14
29–30	SOL			0.97	14
3–4	P	16.4	9.8		15
30–34	P	17.7	12.0		16

TA = Tibialis anterior; S = sartorius; EDL = extensor digitorum longus; SOL = soleus; P = plantaris.

sion = TPT, 1/2 relaxation time = 1/2 RT, maximal shortening velocity = V_{max} and maximal torque) in humans [5–12] and subhuman mammals [13–15] for several muscles (tables 1, 2) [16]. In addition to a general slowing of the twitch contraction times (table 2), which may be associated with age-related changes in sarcoplasmic reticulum function [17], muscle force output also declines (table 3). While the reason(s) for some of these changes may be related to factors discussed in the following sections, changes in motor unit composition may play a role.

The number of functioning human motor units has been shown to decline with age and is thought to reflect primary loss of spinal motor neurons. Doherty et al. [18] have recently investigated the influence of this age-associated loss on the contractile function of the muscles of the arm. Using the spike-triggered averaging technique of Brown et al. [19], the motor unit estimate in older adults (68 years) was decreased by 45% in comparison to younger controls (29 years). The force of the electrically evoked isometric twitch contraction and maximal voluntary contraction of the elbow flexor was reduced some 33% in older subjects. These results suggest a strong relationship between loss of motor units and reduced voluntary strength in older adults. In addition to motor unit loss, there is

Table 2. Relationship of ageing to contractile properties in human skeletal muscle

Age years	Sex	TPT ms	½RT ms	Muscle	Reference
24.7	M	105	91	TS	6
20–36	M, F	101	94	TS	7
63–76	M	127	110	TS	5
66.3	M	128	92	TS	6
66.9	M	137	123	G	11
69.1	M	151	98	TS	10
80–100	M	113	–	LG	8
80–100	F	145	–	LG	8
80–100	M	116	–	MG	8
80–100	F	128	–	MG	8
80–100	M	166	–	SOL	8
80–100	F	222	–	SOL	8

TS = Triceps surae; G = gastrocnemius; LG = lateral gastrocnemius; MG = medial gastrocnemius; SOL = soleus.

Table 3. Comparison of force output in young and elderly subjects

Age years	Sex	Pt N	Po^{10} N	Po^{20} N	Po^{50} N	Muscle	Reference
22.4	M	–	936	1,456		TS	10
24.7	M	87	493	832	1,027	TS	6
66.3	M	64	405	649	784	TS	6
69.1	M	92	553	748		TS	10
63–76	M	81	388	618	760	TS	5

Pt = Twitch tension; Po = tetanic tension; TS = triceps surae.

probably a rearrangement in the motor unit pool. For example, the significant enlargement of the motor unit action potential in aged humans [18] suggests that nerve sprouting and reinnervation of 'orphaned' muscle fibres may provide an adaptive mechanism to maintain muscle mass despite substantial motor unit loss.

Table 4. Motor unit proportions (% of total) in the plantaris muscles of ageing rats

Age	FF	FInt	FR	S	Transitional
7 weeks	41	22	10	16	11
6 months	36	41	9	12	2
14 months	14	40	13	18	15
20 months	21	27	11	30	11

FF = Fast fatigable; FInt = fast intermediate-fatigable; FR = fast fatigue-resistant; S = slow. [Adapted in part from ref. 20.]

Animal models have demonstrated similar findings. Pettigrew and Gardiner [15] observed a significant decline in motor unit number in the rat plantaris, concomitant with an increase in the proportion and size of the slow motor units. It appears that, in the rat, these changes start at middle age and progress as the animal grows older [20, 21]. One interesting observation [20] was that, as the rats aged, motor units which displayed unusual or 'transitional' properties were observed (table 4). The transitional characteristics of these units could presumably arise when multiple isoforms of the proteins involved in muscle contraction are found within a single motor unit/muscle fibre. Such observations have been noted with ageing in human muscles [22].

It would be an advantage to the muscle, and in fact the total entity, to find ways to prevent or retard the change in contractile properties noted above. Klitgaard et al. [23], trained old rats to endurance swim and to weight lift. The strength training groups did not demonstrate the normal age-related decreases in twitch and tetanic tension. Furthermore, this training decreased TPT and ½RT in both the soleus and plantaris muscles. On the other hand, the swimming affected only the endurance of the skeletal muscles.

In our laboratories we have considered the effects of regular endurance exercise on the contractile properties of human triceps surae. The data [5] suggest that, while the muscle of elderly subjects is slower and weaker than the muscle of younger subjects and older training subjects, aerobic physical activity has little effect on the depressed muscle force observed with ageing. We continue our studies using a weight training programme for subjects in their 7th decade.

Table 5. Skeletal muscle fibre distribution with age in different animals

Species	Age	Muscle	Type I %	Type IIA %	Type IIB %	Reference
Pigeon	6 w	ALD	99.55	0.45		26
	6 y	ALD	99.53	0.47		26
	6 w	ABVC	12.60	87.40		26
	6 y	ABVC	10.95	89.05		26
Barrier-reared rat	3 m	EDL	3.4	12.8	83.8	27
	30 m	EDL	3.0	11.1	85.9	27
	3 m	SOL	84.9	9.2	5.9	27
	30 m	SOL	94.1	2.8	3.1	27
Rat	6 m	EDL	3	22	75	13
	20–24 m	EDL	3	17	80	13
	6 m	S	88	10	0	13
	20–24 m	S	99	0	0	13
Human	20–29 y	VL	57	25	12	28
	73–83 y	VL	52	31	17	29
	73–83 y	BB	51	28	20	29

W = Weeks; ALD = anterior latissimus dorsi; y = years; ABVC = anterior biventer cervicis musculi; m = months; EDL = extensor digitorum longus; SOL = soleus; S = sartorius; VL = vastus lateralis; BB = biceps brachii.

Skeletal Muscle Fibre Number, Distribution and Size

Motor units are composed of muscle fibres with distinct morphological and biochemical characteristics. There is little doubt that, in conjunction with a loss of motor units, total muscle fibre numbers decrease with age. Direct counting techniques have been utilized in human muscle [24] and other mammalian species [25] to verify this finding. No report has been published to demonstrate or even suggest that fibre number is maintained with exercise and ageing.

Distribution of fibre types, however, has proven to be more controversial. In several animal models (table 5) [26–29], ageing may be associated with changes in muscle fibre distribution, with an apparent increase in the proportion of slow-twitch fibres [15, 21, 30] Larsson [31] was one of the first to suggest that ageing, itself, results in progressive increases in the percent-

Table 8. Skeletal muscle fibre areas with age in the exercising rat

Age months	Muscle	Condition	Type I μm^2	Type IIA μm^2	Type IIB μm^2	Reference
9	S	sedentary	4.618	5.291		34
29	S	sedentary	3.960	3.264		34
29	S	strength-trained	4.857	5.038	3.725 IM	34
29	S	swim-trained	4.180	3.536		34
9	Pl	sedentary	2.427	2.932	5.206	34
29	Pl	sedentary	1.950	2.080	3.471	34
29	Pl	strength-trained	2.245	2.155	4.206	34
29	Pl	swim-trained	1.991	1.965	3.762	34
4	S	sedentary	1.0 ($\mu m^2 \cdot 10^3$)	1.0		35
24	S	sedentary	3.0	NR		35
24	S	run-trained	2.8	NR		35
4	RF	sedentary	0.8	2.0	3.8	35
24	RF	sedentary	2.0	2.6	4.6	35
24	RF	run-trained	2.0	3.2	5.2	35

S = Sartorius; Pl = plantaris; RF = rectus femoris; NR = not recorded.

observed no age-associated change in skeletal muscle fibre composition or myosin ATPase activity. Similarly, Brooks and Faulkner [43] observed no age-related changes in V_{max} and in the force-velocity relationship in mouse soleus and extensor digitorum longus muscles. In contrast, decreases in calcium activated myosin ATPase activity have been noted for minor pectoral muscles in humans with age [42] (table 9). Additionally, as mentioned, an increase in the relative proportion of both myosin heavy and light chains associated with type I fibres has been observed. The age-associated increases in type I myosin may be influenced by exercise training, particularly strength training, however [16] (table 6).

Muscle Enzyme Activities

A great deal of research has been carried out on enzyme activities in muscle from aged individuals (tables 10, 11) [44–53]. Of particular interest has been the effects of exercise and training on skeletal muscle enzymes in

Table 9. Age-related myosin ATPase activity of rat and human skeletal muscle

Age	Condition	Muscle	Ca-ATPase mmol Pi/min/mg protein	Reference
Rat, months				
8	sedentary	S	172	41
16	sedentary	S	209	41
24	sedentary	S	175	41
8	sedentary	EDL	811	41
16	sedentary	EDL	867	41
24	sedentary	EDL	784	41
			Ca/Mg-ATPase OD600/h/g fresh weight	
Human, years				
30–39	sedentary	PM	38	42
40–49	sedentary	PM	28	42
50–59	sedentary	PM	24	42
60–69	sedentary	PM	21	42
70–79	sedentary	PM	19	42

S = Sartorius; EDL = extensor digitorum longus; PM = pectoralis minor.

Table 10. Effects of ageing and physical activity on selected mammalian skeletal muscle enzyme activities

Enzyme	Species	Age effect	Exercise effect	Reference
Hexokinase	rat	↓	↑	34, 44
Phosphofructokinase	rat	→↓	→	40
Lactate dehydrogenase	rat	↓	→↓	34, 44
CS	mouse	→	→	45
CS	rat	↓	↑	34, 44
Cytochrome oxidase	rat	↓	↑	46
Creatine phosphokinase	mouse	↓	↑	47
HAD	rat	↓	↑	44
SDH	rat	↓	↑	40
Acid proteases	mouse	↑	→↑	45
Alkaline phosphatase	mouse	↓	↑	48

CS = Citrate synthase; HAD = β-hydroxyacyl CoA dehydrogenase; SDH = succinic dehydrogenase.

Table 11. Effects of ageing and physical activity on selected human skeletal muscle enzyme activities

Enzyme	Age effect	Exercise effect	Reference
Hexokinase	↓	→	49–51
Phosphofructokinase	↑→↓	→	5, 39, 52
Lactate dehydrogenase	↓	↑	39, 49–51
Succinic dehydrogenase	↓	→↑	5, 50–53
CS		→	39
Creatine phosphokinase		→↑	50, 51

CS = Citrate synthase.

the energy-producing pathways [54] (tables 10, 11). Generally speaking, the activity levels of such primary enzymes as hexokinase, lactate dehydrogenase, β-hydroxyacyl CoA dehydrogenase (HAD) cytochrome oxidase and creatine phosphokinase have been shown to decrease with age [34, 44–47]. PFK appears to plateau or decrease in activity depending upon the age and physical ability of the subjects (fig. 1) [55–59].

Regular physical activity, primarily of an endurance nature, affects most of the metabolic enzymes measured to date (tables 10, 11). Nonetheless, with ageing the activity of the enzymes is reduced (for example, see fig. 1 to observe the effect of ageing and exercise on succinic dehydrogenase). The longitudinal study of Aniansson et al. [29] confirms the age-related reductions in body cell mass, size of fibres and muscle strength, and our work [5] demonstrates the concomitant decrease in enzyme activity. Exercise training can attenuate this decrease in enzyme activity so that the rate of decline is reduced [39, 50, 51, 58]. However, training is unable to halt the age-associated loss in enzyme activity [5].

Other Metabolic Parameters to Consider

This review has been restricted to the four general areas mentioned previously. However, there are numerous other parameters that play a considerable role in the basic metabolism of skeletal muscle and which are particularly affected by exercise and training (tables 12, 13) [60–73]. These

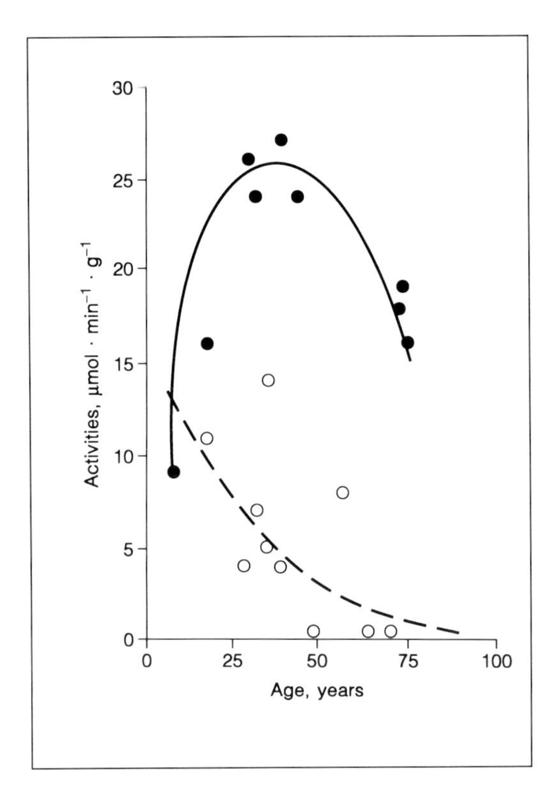

Fig. 1. Trend of succinic dehydrogenase (○) and phosphofructokinase (●) activities with age. [Data points are taken from ref. 5, 12, 39, 50–53 and 55–59.]

include substrate concentrations [60, 61], transport across the muscle membranes, capillarization, myoglobin content [62, 63], metabolite concentrations [64, 65], mitochondrial content [66], Na/K pump concentration [67], collagen metabolism [68], connective tissue thickness [69], number of satellite cells in muscle [70] and infiltration of the muscle by nonmuscle tissue [71, 72].

This latter topic has been studied by means of computed tomography in our laboratory [71, 72] and is worthy of comment. We have observed increased infiltration by nonmuscle tissue. The muscles of old subjects demonstrate a great deal of infiltration when compared to muscles of young subjects. However, and perhaps more importantly, we observed less

Table 12. Substrate concentration changes in mammalian skeletal muscle with age and/or exercise training

Substrate	Species	Age-related changes	Exercise effects with age	Reference
Fatty acids	human	→		60
Cholesterol	human	→		60
Glycogen	rat	→	→	61
Energy-rich phosphagens	human	→	→	62
Lactate	rat	→	estim. →	64
Pcr/Pi	human	↓	→	65

Table 13. Effects of ageing and/or physical training on selected mammalian skeletal muscle parameters

Parameter	Species	Age-related changes	Exercise effects with age	Reference
Na/K pump con^{n}	human	↓	↑	67
Collagen metabolism	rat	→↓	↑	68
Perimysium thickness	rat	↑		69
Endomysium thickness	rat	↑		69
Satellite cells	rat	↓		70
Fatty infiltration	human	↑	↓	71, 72
Fibre size	human	↓	↑	38
State 3 respiration rates	rat	↓		66
Myoglobin	rat	→	↑	63
Cytochrome c	rat	↓		73

infiltration in old muscle with regular exercise. In fact, other than size, it is difficult to differentiate between old exercised muscle and young sedentary or exercised muscle by computed-tomographic scan. The force-generating capacity of the contractile material in old age was maintained with training, but with decreased tetanic force relative to muscle weight, and may be a result of a gradual increase in the infiltration of fat and connective tissue in relation to the number of muscle fibres [24].

Concluding Remarks

Few studies have investigated the relationship between the effects of ageing on morphological, biochemical and contractile properties of skeletal muscle [5]. Reduced force generation per cross-sectional area [6] may result from a decrease in fibre number [24], area [29] or fatty infiltration [71, 72]. Slowed contractile speed may be caused by a change in fibre distribution [31], reduced ATPase activity [42] or changes in myosin isozyme pattern [16]. Decreased metabolic capacity might also play a role [17, 28, 34, 44, 46, 49, 64–66] in the reduced contractile properties observed with ageing [6, 10].

It is relatively well accepted that cell replication is related to and slowed by the ageing process [73]. As well, a general decline in gene expression, translation and transcription has been observed to occur with increasing age in a wide variety of organisms and tissues [74]. In particular, protein turnover declines with increasing age and this could lead to an age-related decrease in the response of inducible enzymes to stimuli, and explain the molecular basis for the decline in the ability of ageing organisms to respond to a variety of environmental factors [74]. Thus, it would appear that the process of ageing is not reversible but regular exercise stimulates the appropriate mechanisms to retard the ageing effect in skeletal muscles.

References

1 Green HJ: Characteristics of aging human skeletal muscles; in Sutton JR, Brock RH (eds): Sports Medicine for the Mature Athlete. Indianapolis, Benchmark Press, 1986, pp 17–26.
2 Fiatrone MA, Marks EC, Ryan ND, Meredith CN, Lipsitz LA, Evans WJ: High-intensity strength training in nonagenarians. JAMA 1990;263:3029–3034.
3 Saltin B: The aging endurance athlete; in Sutton JR, Brock RM (eds): Sports Medicine for the Mature Athlete. Indianapolis, Benchmark Press, 1986, pp 59–80.
4 Hayflick L: Theories of biological aging. Exp Gerontol 1985;20:145–159.
5 Keh-Evans L, Rice CL, Noble EG, Paterson DH, Cunningham DA, Taylor AW: Comparison of histochemical, biochemical and contractile properties of triceps surae of trained aged subjects. Can J Ageing, submitted.
6 Klein C, Cunningham DA, Paterson DH, Taylor AW: Fatigue and recovery contractile properties of young and elderly men. Eur J Appl Physiol 1988;57:684–690.
7 Rice CL, Cunningham DA, Taylor AW, Paterson DH: Comparison of the histochemical and contractile properties of human triceps surae. Eur J Appl Physiol 1988;58:165–170.

43 Brooks SV,Faulkner JA: Contractile properties of skeletal muscles from young, adult and aged mice. J Physiol 1988;404:71–82.
44 Sanchez J, Bastien C, Monod H: Enzymatic adaptations to treadmill training in skeletal muscle of young and old rats. Eur J Appl Physiol 1983;52:69–74.
45 Salminen A, Vihko V: Effects of age and prolonged running on proteolytic capacity in mouse cardiac skeletal muscles. Acta Physiol Scand 1981;112:89–95.
46 Cartee GD, Farrar RP: Muscle respiratory capacity and VO_2 max in identically trained young and old rats. J Appl Physiol 1987;63:257–261.
47 Steinhagen-Thiessen E, Reznick AZ: Effect of short- and long-term endurance training on creatine phosphokinase activity in skeletal and cardiac muscles of CW-1 and C57-BL mice. Gerontology 1987;33:14–18.
48 Reznick AZ, Steinhagen-Thiessen E, Silbermann M: Alkaline phosphatase activity in striated muscle: The effect of aging and long-term training in female mice. Arch Gerontol Geriatr 1989;9:59–65.
49 Roch Norlund AE, Borrebaek B: The decrease with age in the activities of enzymes of human skeletal muscle: Some observations on palmityl-carnitine formation, hexokinase activity, and lactate dehydrogenase activity. Biochem Med 1978;20: 378–381.
50 Suominen H, Heikkinen E: Enzyme activities in muscle and connective tissue of M. vastus lateralis in habitually and sedentary 33–70 year old men. Eur J Appl Physiol 1975;34:249–254.
51 Suominen H, Heikkinen E, Liesen H, Michel D, Hollman W: Effects of 8 weeks' endurance training on skeletal muscle metabolism in 56–70 year old sedentary men. Eur J Appl Physiol 1977;37:173–180.
52 Gollnick PD, Armstrong RB, Saltin B, Saubert CW IV, Sembrowich WL, Shepherd RE: Effect of training on enzyme activity and fiber composition of human skeletal muscle. J Appl Physiol 1973;34:107–111.
53 Houston ME, Benzen H, Larsen H: Interrelationships between muscle adaptation and performance by detraining and retraining. Acta Physiol Scand 1979;105:163–170.
54 Taylor AW, Lavoie S, Lemieux G, Dufresne C, Skinner JS, Vallée J: Effects of endurance training on the fibre area and enzyme activities of skeletal muscle of French-Canadians; in Landry F, Orban W (eds): 3rd Int Symp on Biochemistry of Exercise. Miami, Symposium Specialists, 1978, pp 267–278.
55 Aniansson A, Grimby G, Rundgren A, Svanborg A: Physical training in old men. Age Aging 1980;9:186–187.
56 Essen B, Jansson E, Henriksson J, Taylor AW, Saltin B: Metabolic characteristics of fibre types in human skeletal muscle. Acta Physiol Scand 1975;95:153–165.
57 Essen B, Henriksson J: Metabolic characteristics of human type 2 skeletal muscle fibers. Muscle Nerve 1980;3:263.
58 Eriksson BD, Gollnick PD, Saltin B: Muscle metabolism and enzyme activities after training in boys 11–13 years old. Acta Physiol Scand 1973;87:485–497.
59 Orlander J, Kiessling KH, Larsson L, Karlsson J, Aniansson A: Skeletal muscle metabolism and ultrastructure in relation to age in sedentary men. Acta Physiol Scand 1978;104:249–261.
60 Thomas TR, Londeree BR, Gerhardt KO, Gehrke CW: Fatty acid pattern and cholesterol in skeletal muscle of men aged 22 to 73. Mech Ageing Dev 1978;8:429–434.

61 Cartee GD, Farrar RP: Exercise training induces glycogen sparing during exercise in old rats. J Appl Physiol 1988;64:259–265.
62 Moller P, Brandt R: The effects of physical training on elderly subjects with special reference to energy-rich phosphagens and myoglobin in leg skeletal muscle. Clin Physiol 1982;2:307–314.
63 Beyer RE, Fattore JE: The influence of age and endurance exercise on the myoglobin concentration of skeletal muscle of the rat. J Gerontol 1984;39:525–530.
64 Dudley GA, Fleck SJ: Metabolite changes in aged muscle during stimulation. J Gerontol 1984;39:183–186.
65 McCully KK, Forciea MA, Hack LM, Donlon E, Wheatley RW, Oatis CA, Goldberg T, Chance B: Muscle metabolism in older subjects using magnetic resonance spectroscopy. Can J Physiol Pharmacol 1981;69:576–580.
66 Murfitt RR, Rao Sanadi D: Evidence for increased degeneration of mitochondria in old rats: A brief note. Mech Ageing Dev 1978;8:197–201.
67 Klitgaard H, Clausen T: Increased total concentration of Na-K pumps in vastus lateralis muscle of old trained human subjects. J Appl Physiol 1989;67:2491–2499.
68 Kovanen V, Suominen H: Age- and training-related changes in the collagen metabolism of rat skeletal muscle. Eur J Appl Physiol 1989;58:765–771.
69 Alnaqeeb MA, Al Zaid NS, Goldspink G: Conncetive tissue changes and physical properties of developing and ageing skeletal muscle. J Anat 1984;139:677–689.
70 Gibson MC, Schultz E: Age-related differences in absolute numbers of skeletal muscle satellite cells. Muscles Nerve 1983;6:574–580.
71 Rice CL, Cunningham DA, Paterson DH, Lefcoe MS: A comparison of anthropometry with computed tomography in limbs of young and aged men. J Gerontol Med Sci 1990;45:M175–179.
72 Rice CL, Cunningham DA, Paterson DH, Lefcoe MS: Arm and leg composition determined by computed tomography in young and elderly men. Clin Physiol 1989; 9:207–220.
73 Schneider EL: Cell replication and aging: In vitro and in vivo studies. Fed Proc 1979;38:1857–1861.
74 Richardson A, Cheung HT: The relationship between age-related changes in gene expression, protein turnover, and the responsiveness of an organism to stimuli. Life Sci 1987;31:605–613.

Dr. Albert W. Taylor, Faculty of Kinesiology, University of Western Ontario, London, ON N6A 3K7 (Canada)

Sato Y, Poortmans J, Hashimoto I, Oshida Y (eds): Integration of Medical and Sports Sciences. Med Sport Sci. Basel, Karger, 1992, vol 37, pp 126–136

Plasma Lipoprotein and Apolipoprotein Profiles in Aged Japanese Athletes

M. Higuchi[a], *T. Tamai*[b], *S. Kobayashi*[a], *T. Nakai*[b]

[a]Division of Health Promotion, National Institute of Health and Nutrition, Tokyo; [b]Third Department of Internal Medicine, Fukui Medical School, Fukui, Japan

Japanese men and women have the highest longevities in the world. In 1988 the life expectancies in Japanese men and women were 75.9 and 81.8 years, respectively. These longevities could be partly due to the remarkably lower mortality rate ($41.5/10^5$ persons including men and women) from coronary heart disease (CHD) when compared to those of the other developed countries.

Even though the percentage of the people aged over 65 years is already 12% in Japan, the population of aged Japanese people has been rapidly increasing toward the coming century. According to the 1988 health statistics in Japan, the mortality rate caused by CHD was 138 per 10^5 persons in men aged 65–69 years, and 62 in women aged 65–69 years. These values are 10 and 20 times higher than those in men and women aged 45–49 years, respectively.

Epidemiological studies have suggested that at older ages the most potent lipid risk factor for the development of CHD is high-density lipoprotein (HDL) cholesterol, and that a weaker association with the incidence of CHD is observed for low-density lipoprotein (LDL) cholesterol [1].

HDL is responsible for carrying cholesterol away from the arterial wall back to the liver. Cholesterol is metabolized and excreted in the liver. On the other hand, LDL is formed as a consequence of the catabolism of intermediate-density lipoproteins.

A benefit of an active life-style appears to be protection against CHD. Farrell et al. [2] summarized HDL cholesterol values for various groups

Table 4. Plasma lipid and lipoprotein concentrations for the male and female subjects

	n	Cholesterol			TG
		total	LDL	HDL	
Male subjects					
Middle-aged					
Runners	30	203±27	114±24[a]	74±12[b]	75±22[b]
Untrained	30	201±27	127±24	54±13	103±41
Elderly					
Runners	30	219±33[c]	127±29[c]	77±16[b]	73±17[b]
Swimmers	5	184±20	97±26	70±15	89±49
Untrained	15	197±38	119±33	57±16	110±56
Female subjects					
Middle-aged					
Premenopausal					
Runners	17	182±28	86±25	85±12	56±14[a]
Untrained	26	173±26	81±23	77±14	72±29
Postmenopausal					
Runners	16	191±39	96±32	81±16	68±28
Untrained	15	194±28	104±23	76±15	72±32
Elderly					
Runners	15	195±23	105±23	76±16	71±24[a]
Untrained	28	212±33	117±31	74±14	101±32

Values are means ± SD and are expressed as milligrams per deciliter.
[a] $p < 0.05$, [b] $p < 0.01$ vs. untrained subjects; [c] $p < 0.05$, [d] $p < 0.01$ vs. swimmers.

more, there was no significant difference in the ratio of saturated/unsaturated fat consumed among the groups.

Plasma lipid and lipoprotein concentrations of all of the groups are summarized in table 4. Total cholesterol concentration was higher (8–19%) for the elderly male runners compared to the other 4 groups of men. LDL cholesterol level in the elderly runners was similar to that in the untrained age-matched individuals. In the middle-aged and elderly groups of men, HDL cholesterol concentrations of the runners were markedly higher than those of the sedentary subjects. The runners had lower concentrations of TG than the untrained men in both age groups. No difference was observed in HDL cholesterol concentration between the elderly and middle-aged runners. There were no differences in body composition, smoking and dietary habits between the middle-aged and older runners.

Thus, it appears that elderly men who regularly engage in endurance running have an elevation of HDL cholesterol similar to that found in middle-aged, trained individuals. The fact that the elderly runners had significantly higher HDL cholesterol and lower TG concentrations suggests that endurance running may have a favorable effect on the prevention of CHD in elderly men.

Apolipoproteins play an important role in the regulation of lipoprotein metabolism. ApoAI is the major protein constituent of HDL. ApoAI plays an important role in the activation of lecitin:cholesterol acyltransferase in the reverse cholesterol transport system. On the other hand, apoB is the major protein constituent of LDL, and apoB is recognized by specific high-affinity receptors that mediate clearance of LDL particles from plasma [18].

Elderly runners had higher levels of apoAI (156 ± 16 mg/dl; mean ± SD) and AII (35 ± 4 mg/dl) than age-matched untrained men (apoAI: 144 ± 13 mg/dl; apoAII: 32 ±2 mg/dl), whereas no difference was observed in apoB between the trained (77 ± 15 mg/dl) and untrained (80 ± 17 mg/dl) groups [19].

Based upon these findings, we concluded that in elderly men regularly performed endurance running can bring about favorable changes of the lipid component (in terms of cholesterol) and the protein component (in terms of apoAI and apoAII) in HDL.

Middle-Aged and Elderly Female Runners

We compared plasma lipid and lipoprotein concentrations between female runners and untrained subjects in both middle-aged and elderly groups [20, 21]. All premenopausal women menstruated monthly and did not use any oral contraceptives. All postmenopausal middle-aged and elderly women had gone through a natural menopause. They did not take any form of hormonal replacements.

There were significant differences in the percent fat between the trained and untrained women in both pre- and postmenopausal middle-aged groups (table 1).

Both pre- and postmenopausal middle-aged runners had substantially higher $\dot{V}O_2max$ values than the untrained middle-aged women (table 1). The postmenopausal middle-aged women had markedly lower plasma estradiol concentrations than the premenopausal women (premenopausal untrained:

123 ± 96 pg/ml; premenopausal runners: 98 ± 64 pg/dl; postmenopausal untrained: 11 ± 6 pg/ml; postmenopausal runners: 9 ± 2 pg/dl).

Total caloric intakes of the runners were higher than those of untrained women in both the middle-aged and elderly groups, whereas there were generally no differences in the proportions of different energy nutrients (table 3).

Plasma lipid and lipoprotein profiles in middle-aged women are shown in table 4. No significant differences were observed in plasma total cholesterol. The runners, both before and after menopause, had slightly higher levels of HDL cholesterol than the untrained women. LDL cholesterol in the postmenopausal trained and untrained women tended to be higher than in the premenopausal groups. The difference in the plasma lipid and lipoprotein profiles between males and females could be partly associated with the different influences of sex hormones [22, 23].

There were significant differences in body weight and percent fat between the trained and untrained elderly women (table 1). The elderly female runners had lower training distances and lower race times than the middle-aged female runners, as shown in table 2. The elderly runners had a 38% higher $\dot{V}O_2max$ compared to age-matched untrained women, but their $\dot{V}O_2max$ was 12% lower than that of the postmenopausal middle-aged runners (table 1). Although the elderly female runners were less active than the middle-aged runners, their levels of daily physical activity are substantially high for elderly populations in industrialized countries [24].

Plasma lipid and lipoprotein profiles in elderly women are shown in table 4. Plasma TG levels in the elderly female runners were lower than in the age-matched untrained, whereas no significant difference was observed in total cholesterol between the two groups. No significant differences were observed in HDL cholesterol between the trained and untrained elderly women. It has been suggested that Japanese have higher levels of HDL cholesterol than Americans because they consume relatively high amounts of fish, shellfish, soybeans and cereals [25]. In this context, it seems reasonable to infer that running of 35 km/week does not cause much increase in HDL cholesterol in elderly nonobese women who are habitually eating a typical Japanese diet and already have a higher HDL cholesterol level than elderly Japanese men.

LDL cholesterol in the elderly female runners tended to be lower, but not statistically significantly, than in the untrained elderly women. The levels of LDL cholesterol in the trained and untrained elderly women were higher by 22 and 44% than in the comparable groups of premenopausal middle-aged women. This finding suggests that regularly performed endurance running could partly prevent an age-related increase in LDL cholesterol level in women.

The plasma apolipoprotein profile in the elderly female runners was compared to that of age-matched untrained women. ApoAI (n = 7; 162 ± 29 mg/dl) and apoAII (n = 7; 39 ± 19 mg/dl) in the female runners tended to be higher than in the untrained (n = 28; apoAI: 147 ± 20 mg/dl; apoAII: 35 ± 9 mg/dl), but this difference was not statistically significant. No difference was observed in apoB between the groups (runners: 74 ± 30 mg/dl; untrained: 77 ± 19 mg/dl).

Elderly Male Swimmers

The elderly male swimmers had been swimming regularly for 15 years; their swimming distances were 12 km weekly in the latest year. They were leaner than untrained men (table 1). $\dot{V}O_2max$ of the elderly swimmers was higher than that of untrained elderly men, and lower than that of runners in the same age group. The elderly swimmers tended to have lower total and LDL cholesterol and TG than untrained men, whereas their HDL cholesterol levels were slightly higher than those of the age-matched untrained men.

These results suggest that regularly performed swimming can induce favorable effects on cardiovascular and lipid metabolic functions in elderly men.

Elderly Runner and Swimmer Aged over 80 Years

We obtained data on excellent Japanese master athletes aged over 80 years. One is a master runner whose $\dot{V}O_2max$ was 48 ml/kg/min. The other is a master swimmer who set two world records in his age group in 1991, and had a $\dot{V}O_2max$ of 32 ml/kg/min. They had slightly higher levels of total cholesterol (runner: 236 mg/dl; swimmer: 212 mg/dl) than untrained men aged over 60 years, as shown in table 4, but they had remarkably higher HDL cholesterol (runner: 92 mg/dl; swimmer: 83 mg/dl) than untrained and trained men aged over 60 years, as demonstrated in table 4. TG levels of the athletes aged over 80 years were 97 mg/dl in the runner and 83 mg/dl in the swimmer.

These results suggest that, in addition to their natural physical endowments, their excellent health status was in part due to their vigorous exercise training for many years.

In conclusion, these studies indicate that regularly performed aerobic exercise favorably modifies the plasma lipoprotein profile, and that the changes may thus reduce the risk of developing CHD in elderly individuals, particularly in men.

References

1 Gordon T, Castelli WP, Hjortland MC, Kannel WB, Dawber TB: High density lipoprotein as a protective factor against coronary heart disease: The Framingham study. Am J Med 1977;62:707–714.

2 Farrell PA, Maksud MG, Pollock ML, Anholm J, Hare J, Leon AS: A comparison of plasma cholesterol, triglycerides, and high density lipoprotein-cholesterol in speed skaters, weightlifters and non-athletes. Eur J Appl Physiol 1982;48:77–82.

3 Hurley BF, Hagberg JM, Seals DR, Ehsani AA, Goldberg AP, Holloszy JO: Glucose tolerance and lipid-lipoprotein levels in middle-aged powerlifters. Clin Physiol 1987;7:11–19.

4 Seals DR, Allen WK, Hurley BF, Dalsky GP, Ehsani AA, Hagberg JM: Elevated high-density lipoprotein cholesterol levels in older endurance athletes. Am J Cardiol 1984;54:390–393.

5 Rainville S, Vaccaro P: The effects of menopause and training on serum lipids. Int J Sports Med 1984;5:137–141.

6 Foster VI, Hume GJE, Byrnes WC, Dickinson AI, Chatfield SJ: Endurance training for elderly women: Moderate vs low intensity. J Gerontol 1989;44:M184–188.

7 Lokey EA, Tran ZV: Effects of exercise training on serum lipid and lipoprotein concentrations in women: A meta-analysis. Int J Sports Med 1989;10:424–429.

8 Nagamine S, Suzuki S: Anthropometry and body composition of Japanese young men and women. Hum Biol 1964;36:8–15.

9 Saltin B, Astrand PO: Oxygen uptake in athletes. J Appl Physiol 1967;53:353–358.

10 Bruce RA, Horsten TR: Exercise stress testing in evaluation of patients with ischemic heart disease. Prog Cardiovasc Dis 1969;11:371–390.

11 Lipid Research Clinics Program: Manual of Laboratory Operation, Lipid and Lipoprotein Analysis. DHEW Publication (NIH). 1974, pp 75–628.

12 Friedewald WT, Levy RI, Fredrickson DS: Estimation of the concentration of low-density lipoprotein cholesterol in plasma, without use of the preparative ultracentrifuge. Clin Chem 1972;18:499–509.

13 Goto Y, Akanuma Y, Harano Y, Hata Y, Itakura H, Kajiyama G, Kawade M, Koga S, Kuzuya F, Maruhama Y, Matsuzawa Y, Murai A, Murase T, Naito C, Nakai T, Noma A, Saitoh Y, Sasaki J, Takeuchi N, Tamachi H, Uzawa H, Yamamoto A, Yamazaki S, Yasugi T, Yukawa S: Determination by the SRID method of normal values of serum apolipoproteins (A-I, A-II, B, C-II, C-III, and E) in normolipidemic healthy Japanese subjects. J Clin Biochem Nutr 1986;1:73–88.

14 Hartung GH, Foreyt JP, Mitchell RE, Vlasek I, Gotto AM Jr: Relation of diet to high-density-lipoprotein cholesterol in middle-aged marathon runners, joggers and inactive men. N Engl J Med 1980;302:357–361.

15 Kiens B, Lithell H, Vessby B: Further increase in high density lipoprotein in trained males after enhanced training. Eur J Appl Physiol 1984;52:426–430.
16 Higuchi M, Fuchi T, Iwaoka K, Yamakawa K, Kobayashi S, Tamai T, Takai H, Nakai T: Plasma lipid and lipoprotein profile in elderly male long-distance runners. Clin Physiol 1988;8:137–145.
17 Heath GW, Hagberg JM, Ehsani AA, Holloszy JO: A physiological comparison of young and older endurance athletes. J Appl Physiol 1981;51:634–640.
18 Breslow JL: Apolipoprotein genetic variation and human disease. Physiol Rev 1988; 68:85–132.
19 Tamai T, Nakai T, Takai H, Fujiwara R, Miyabo S, Higuchi M, Kobayashi S: The effects of physical exercise on plasma lipoprotein and apolipoprotein metabolism in elderly men. J Gerontol 1988;43:M75–79.
20 Higuchi M, Iwaoka K, Ishii K, Matsuo S, Kobayashi S, Tamai T, Takai H, Nakai T: Plasma lipid and lipoprotein profiles in pre- and post-menopausal middle-aged runners. Clin Physiol 1990;10:69–76.
21 Higuchi M, Oishi K, Ishii K, Iwaoka K, Matsuo S, Kobayashi S,Tamai T, Takai H, Nakai T: Plasma lipid and lipoprotein profile in elderly female runners. Clin Physiol 1991;11:545–552.
22 Krauss RM, Lindgren FT, Wingerd J, Bradley DD: Effects of estrogens and progestins on high density lipoproteins. Lipid 1979;14:113–118.
23 Nordoy A, Aakvaag A, Thelle D: Sex hormones and high density lipoproteins in healthy males. Atherosclerosis 1979;34:431–436.
24 Shephard RJ: Activity patterns of the elderly; in Shephard RJ (ed): Physical Activity and Ageing. London, Croom Helm, 1978, pp 146–154.
25 Ueshima H, Iida M, Shimamoto T, Konishi M, Tanigaki M, Takayama Y, Ozawa H, Kojima S, Komachi Y: High-density lipoprotein-cholesterol levels in Japan. JAMA 1982;247:1985–1987.

Dr. Mitsuru Higuchi, Division of Health Promotion, National Institute of Health and Nutrition, 1-23-1 Toyama, Shinjuku-City, Tokyo 162 (Japan)

Sato Y, Poortmans J, Hashimoto I, Oshida Y (eds). Integration of Medical and Sports Sciences. Med Sport Sci. Basel, Karger, 1992, vol 37, pp 137–141

Effect of Habitual Physical Activity on Glucose Tolerance and Peripheral Insulin Action in the Elderly

Kunio Yamanouchi[a], *Kiwami Chikada*[a], *Katsumi Kato*[a], *Yoshiharu Oshida*[b], *Yuzo Sato*[b]

[a] First Department of Internal Medicine, Aichi Medical University, Aichi;
[b] Research Center of Health, Physical Fitness and Sports, Nagoya University, Nagoya, Japan

It has been shown over and over that the prevalence of glucose intolerance and non-insulin-dependent diabetes increases with age [1, 2]. Two mechanisms considered responsible for this are the reduced insulin secretion whereby beta cells exhibit diminished sensitivity to glucose and the increased insulin resistance. In vivo studies using the euglycemic insulin clamp procedure have further narrowed the cause of insulin resistance to two categories: insulin sensitivity and responsiveness [3]. However, the results of studies in the elderly have thus far been inconclusive on these points [4–8]. Discrepancies in results from studies in the elderly may be due to the existence of some secondary factors associated with aging [2]. One of the most important factors mentioned may be the decreased physical activity in the elderly.

Therefore, we undertook the following study to explore this possibility in more detail to find out the role of physical activity in modulating the aging process as it relates to glucose intolerance.

Subjects and Methods

There were 26 elderly subjects aged 65–92 years and young students aged 19–22 years. The elderly were divided into three groups according to the amount of physical activity they participated in on a daily basis. Young subjects were similarly divided into two groups. Eight aged, bedridden individuals (aged bedridden group) were selected from patients at a hospital with specializes in caring for the elderly. They had all been bedridden for at least 3 months. As a control, we also monitored 7 other aged individuals (aged control group) and 11 young students (young control group) who lived normal everyday lives without participating in any particular type of physical activity. The 11 aged athletes (aged athlete group) jogged more than 10 km/day at least 4 days/week and also participated in such activities as endurance races. The 11 young athletes (young athlete group) were long-distance runners from a local university track team. All subjects were nonobese and had no evidence of diabetes as defined by the criteria of the National Diabetes Association at 75 g oral glucose tolerance test (OGTT) [9]. They had no family history of diabetes and took no medication that would affect glucose tolerance. They were all informed of the nature, purpose and possible risks of the study before giving voluntary consent to participate.

Glucose tolerance was examined with 75 g OGTT whereby plasma glucose and insulin were measured at 30-min intervals up to 120 min. SIRI/SBS was calculated by taking the total area under the insulin curve SIRI and dividing it by the total area under the glucose curve SBS which was obtained during OGTT. This was thought to represent an index for insulin secretion ability. Peripheral insulin action was estimated by the euglycemic insulin clamp procedure at two different insulin infusion rates: 40 and 400 mU/m^2/min. Glucose metabolic clearance rate (MCR) was calculated by dividing the glucose infusion rate by the steady-state plasma glucose concentration [10]. This index thus provided a quantitative estimate for insulin action. All data are expressed as means ± SE. Statistical analysis was performed by analysis of variance for one- and two-way observations where appropriate. Significant mean differences were determined using the Newman-Keuls multiple-comparison test.

Results

Plasma Glucose and Insulin Concentrations during 75 g OGTT in Aged Subjects

There was no significant difference between the aged control and aged athlete group in terms of plasma glucose or insulin levels. After 2 h, however, plasma glucose in aged bedridden subjects was significantly higher than in aged controls and aged athletes ($p < 0.05$). Aged bedridden subjects also tended to have a delayed response of insulin with levels significantly low at 30 min, compared to that in aged controls and in aged athletes ($p < 0.05$), but rose continuously throughout the full 2 h of monitoring.

SIRI/SBS during 75 g OGTT in Aged Subjects

SIRI/SBS was significantly lower at 30 min in the aged bedridden group (0.13 ± 0.04) than in aged controls (0.24 ± 0.02) and aged athletes (0.27 ± 0.03; $p < 0.05$). At other times, it continued to be significantly lower in aged bedridden subjects when compared with aged athletes ($p < 0.05$).

Plasma Glucose and Insulin Concentrations at Basal and during Euglycemic Clamp Procedure

There were no statistical differences in basal insulin concentrations between any of the groups. Steady-state plasma insulin ranged from 70 to 100 μU/ml at 40 $mU/m^2/min$. It reached more than 1,000 μU/ml at 400 $mU/m^2/min$ in all groups, but there were no significant differences between the groups. The basal glucose level was almost the same for all groups and was well maintained throughout the entire procedure.

Glucose Metabolic Clearance Rate at Two Insulin Infusion Rates in Young and Aged Subjects

At 400 $mU/m^2/min$, MCR remained at about the same level among young controls, young athletes and aged athletes with no statistical differences. In aged controls and aged bedridden subjects, however, there was no statistical difference in MCR between the groups, but together they were significantly lower than the other three groups ($p < 0.001$).

Although there was no significant difference in MCR at 400 $mU/m^2/min$ between young controls and young athletes and between aged controls and aged bedridden subjects, MCR at 40 $mU/m^2/min$ in young athletes was higher than in young controls ($p < 0.001$) and that in aged bedridden subjects was lower than in aged controls ($p < 0.01$).

Discussion

The fact that a significant rise in glucose in OGTT was seen only in the aged bedridden suggests that one factor influencing decreased glucose tolerance may be the absence of all physical activity in this group. And it also suggests that physical inactivity could have an influence on the glucose intolerance usually seen with aging. Decreased glucose tolerance as a result of prolonged bed rest was first reported by Blotner [11] in 1945. Furthermore, the fact that insulin secretion as shown by OGTT followed a delayed response pattern in the aged bedridden, combined with the significantly lower 30 min and shown by SIRI/SBS for this group in comparison with the aged controls, indicates that lack of physical activity in the elderly also exerts an effect on insulin secretion. Although maximal MCR in the aged bedridden did not differ statistically from that in the aged controls, MCR at the lower insulin infusion rate was statistically lower in the aged bedridden than in the aged controls ($p < 0.01$). This might be taken as showing that the dose-response curve is shifted toward the right, that is to say, that as the elderly become less active physically they develop greater decrease of sensitivity. Meanwhile, judging from the fact that there was no significant difference between aged controls and aged athletes in terms of blood glucose and insulin in OGTT, it seems unlikely that physical training exerts a direct effect on the capability to process sugars from oral intake. Nevertheless, a clear difference in glucose MCR was seen between the two groups when the euglycemic clamp procedure was performed at an insulin infusion rate of 400 mU/m^2/min. The aged controls showed significantly lower levels of MCR at 400 mU/m^2/min than all young groups, but that for the aged athletes was almost the same as the latter. At this time, insulin concentration exceeded 1,000 μU/ml, which is thought to equal the maximal effect insulinemia described by Olefsky et al. [12]. In other words, this suggests that marked improvements in insulin responsiveness may be effected through physical training in aged individuals, and moreover that there may be many variations in the effect of insulin responsiveness depending on the amount and type of physical activity. This may be the factor that explains why there are so many discrepancies in results from other reports [7, 8] dealing with maximal responsiveness of the euglycemic clamp procedure in the elderly.

It should be noted, however, that the effects of training in young individuals are quite different from aged people. There was no difference in maximal MCR between the young controls and the young athletes. But at the lower insulin infusion rate, MCR in the young athletes was higher than in the young controls, indicating that the dose-response curve shifted to the left in the young athletes. In other words, the effects of training on the young could improve insulin sensitivity. Similar results have been reported by King et al. [13], where there was no significant difference in maximally stimulated glucose disposal, but glucose disposal rates were higher in the trained than in the untrained group at an insulin infusion rate of 40 mU/m^2/min. In animal experiments, James et al. [14] also showed that 7 weeks of moderate exercise training in rats resulted in a significant shift to the left of the insulin dose-response curve.

These findings indicate that the aging process might cause a reduction in insulin responsiveness. A lack of exercise over the long term in aged individuals not only induces impairments to glucose tolerance with deteriorations in the insulin secretion response, but also possibly causes poorer insulin sensitivity. All is not lost, however, because there is still a possibility that physical training in the elderly individuals may serve to improve insulin responsiveness or may protect this from further deterioration.

References

1 Spence JC: Some observations on sugar tolerance with special reference to variations found at different ages. Q J Med 1920/21;14:314–326.
2 Davidson MB: The effect of aging on carbohydrate metabolism: A review of the English literature and a practical approach to the diagnosis of diabetes mellitus in the elderly. Metab Clin Exp 11979;8:688–705.
3 Kahn CR: Insulin resistance, insulin sensitivity and insulin unresponsiveness: A necessary distinction. Metabolism 1978;27:1893–1902.
4 DeFronzo RA: Glucose intolerance and aging. Evidence for tissue insensitivity to insulin. Diabetes 1979;28:1095–1101.
5 Fink RI, Griffin J, Olefsky JM: Effect of age on carbohydrate metabolism in man (abstract). Diabetes 1982;31:64a.
6 Andres R, Tobin JD: Endocrine system; in Finch CA, Hay-Flick L (eds): Handbook of the Biology of Aging. New York, Van Nostrand Reinhold, 1977, pp 375–378.
7 Rowe JW, Minaker KL, Pallotta JA: Characteristics of the insulin resistance of aging. J Clin Invest 1983;71:1581–1587.
8 Fink RI, Kolterman OG, Griffin J, Olefsky JM: Mechanism of insulin resistance in aging. J Clin Invest 1983;71:1523–1535.
9 National Diabetes Data Group: Classification and diagnosis of diabetes mellitus and other categories of glucose intolerance. Diabetes 1979;28:1039–1057.
10 Doberne L, Greenfield MS, Rosenthal M, Widstrom A, Reaven GM: Effect of variation in basal plasma glucose utilization (M) and metabolic clearance (MCR) rates during insulin clamp studies in patients with non-insulin-dependent diabetes mellitus. Diabetes 1982;31:396–400.
11 Blotner H: Effect of prolonged physical inactivity on tolerance of sugar. Arch Intern Med 1945;75:39–44.
12 Olefsky JM, Kolterman OG, Scarlett JA: Insulin action and resistance in obesity and noninsulin-dependent type II diabetes mellitus. Am J Physiol 1982;243:E15–E30.
13 King DS, Dalsky GP, Staten MA, Clutter WE, Houten DR, Holloszy JO: Insulin action and secretion in endurance-trained and untrained humans. J Appl Physiol 1987;63:2247–2252.
14 James DE, Kraegen EW, Chisholm DJ: Effect of exercise training on whole-body insulin sensitivity and responsiveness. J Appl Physiol 1984;56:1217–1222.

Dr. Kunio Yamanouchi, First Department of Internal Medicine,
Aichi Medical University, 21 Karimata, Nagakute-cho, Aichi, 480-11 (Japan)

Sato Y, Poortmans J, Hashimoto I, Oshida Y (eds): Integration of Medical and Sports Sciences. Med Sport Sci. Basel, Karger, 1992, vol 37, pp 142–149

Molecular Mechanisms of Muscle Disuse Atrophy (and Strategies of Prevention)

Frank W. Booth[a], *Jon K. Linderman*[b], *Christopher R. Kirby*[a,1]

[a] Department of Physiology and Cell Biology, University of Texas Medical School, Houston, Tex.; [b] Life Sciences Division, NASA-Ames Research Center, Moffett Field, Calif., USA

Atrophy of skeletal muscle diminishes the quality of life of a human or animal by reducing their freedom of movement. Unfortunately, muscle atrophy occurs all too commonly. Atrophy of skeletal muscle often is initiated, or enhanced, by decreased physical activity associated with senescence [1], postoperative repair from orthopedic surgeries [2], spinal cord injury [3], cancer cachexia [4] or chronic bed rest resulting from disease [5]. Although a clinical condition, itself, may produce some atrophy of skeletal muscle, the common factor among the above afflictions is that they reduce muscle usage. To develop a scientifically rational method for preventing muscle atrophy it will be necessary to understand the mechanism(s) by which reduced load bearing triggers the loss of muscle mass.

Experimental Models of Muscle Atrophy

Two experimental animal models of skeletal muscle atrophy (limb immobilization and hindlimb non-weight bearing) have been studied extensively. Each model produces characteristic responses and has inherent advantages and limitations. Limitations of limb immobilization in-

[1] We thank Ms. Grace Belcher for assistance in typing, Dr. Richard Grindeland for stimulating discussions, and NASA grants NAG 2-239 and NAGW-70 for research funding.

clude the absence of isotonic muscle contractions and the difficulty of removing plaster casts for the purpose of exercising during limb fixation. An advantage of limb immobilization in animals is that it faithfully mimics the immobilization of human limbs. In addition, since fast-twitch and slow-twitch muscles atrophy similarly [6] in immobilized, but not in non-weight-bearing animals, immobilization is the preferred model to study reduced-use atrophy of fast-twitch muscle. An important characteristic of limb immobilization is that the length of the muscle, relative to resting length, determines its atrophic response. For example, a muscle fixed at or less than resting length atrophies [7], while a muscle fixed at greater than resting length atrophies less (i.e. fast-twitch muscle) or hypertrophies (i.e. slow-twitch muscle) [6]. In addition, when muscle is fixed at greater than resting length muscle fiber diameter decreases and fiber length increases [7] due to the addition of sarcomeres at the ends of the muscle fiber [8].

A second popular animal model of skeletal muscle atrophy is hindlimb non-weight bearing. Two procedures have been introduced which prevent the hindlimbs from touching the ground and bearing the load of the body weight. In the non-weight-bearing model devised by Musacchia et al. [9], the whole rat or mouse is suspended in a harness with the front legs in contact with a platform. The model developed by Morey [10] elevates the hindlimbs of the rat via a cast placed around the tail. A disadvantage of the non-weight-bearing model is the limited atrophy of fast-twitch muscle (15%) [11] as compared to the marked atrophy (50%) occurring in fast-twitch muscles fixed at less than resting length. Advantages of the non-weight-bearing model are the ability of soleus muscles to perform unloaded isotonic contractions [9] and the ease of removal and return of the rat to the non-weight-bearing position following an exercise bout. In addition, responses of skeletal muscle from non-weight-bearing hindlimbs on Earth are qualitatively similar to muscle responses to microgravity [9, 12]. This suggests that the absence of weight bearing is the common cause of slow-twitch muscle atrophy. Thus, the non-weight-bearing model on Earth is, to date, an excellent model of spaceflight.

Mechanisms of Skeletal Muscle Atrophy

Decreased synthesis of protein is an early (i.e. in the first few hours of reduced use) response of skeletal muscle during limb immobilization and non-weight bearing. Significant decreases (16–66%) in skeletal muscle protein synthesis have been reported in the first 5–6 h of decreased use. For example, fixation of fast-twitch muscle at less than resting length for 6 h reduced the synthesis rates of mixed proteins (37%) [13], actin protein (66%) [14] and cytochrome *c* protein (27%) [15]. The first 5 h of non-weight bearing in slow-twitch muscle reduced synthesis rates of mixed and myofibrillar protein by 16 and 22%, respectively [16]. Thus, a reduction in muscle protein synthesis is an early response to a decrease in muscle usage. The absence of changes in specific mRNAs in the initial 5–6 h of decreased muscle use suggests that these initial reductions in muscle protein synthesis are not triggered by decreases in pretranslational control of gene expression. For example, no changes in the concentrations of α-actin mRNA, cytochrome *c* mRNA or β-myosin heavy-chain mRNA are observed in the first 5–6 h of either limb immobilization or non-weight bearing [14–16]. Such changes occur only after several days of either limb immobilization or non-weight bearing [14–17]. The identity of the cellular factor(s) which alter protein synthesis during reduced muscle use remains to be determined.

Estimated rates of protein degradation in skeletal muscle do not increase during the first day of either limb immobilization [18] or non-weight bearing [16]. Thus, although protein losses from muscle after a week of atrophy are, in part, due to an increase in protein degradation rate, decreased protein synthesis, not increased protein degradation, initiates the loss of muscle protein.

Countermeasures to Skeletal Muscle Atrophy

The old adage 'use it or lose it' describes the most effective countermeasure to muscle atrophy induced by decreased use. This concept is confirmed by the ability of continuous platform support of rat hindlimbs during harness suspension to prevent the loss of soleus muscle mass [19]. In addition, the onset of reductions in protein synthesis is rapid after muscle activity decreases [13–16] and the absolute quantity of protein lost is greatest in the first week of either limb immobilization or non-weight bearing [7, 20]. However, since it is not feasible to provide continuous contractile activity during reduced muscle use resulting from limb immobilization or spaceflight, an adequate exercise countermeasure to muscle atrophy is needed. At present, it seems likely that the product of the intermittency of exercise within a day and the quantity of weight bearing during the exercise

Table 1. Summary of various countermeasures to soleus muscle atrophy during non-weight bearing

Treatment	Percentage of atrophy prevented	Reference
Ground support		
2 h/day × 4 weeks	28	20
2 h/day × 8 weeks	33	20
4 × 10 min/day × 1 week	70[a]	21
2 h/day × 1 week	52	22
4 × 15 min/day × 1 week	61	23
Treadmill running		
1.5 h/day last 4 weeks of 8 weeks (speed = 20 m/min)	29	20
1.5 h/day × 4 weeks (speed = 20 m/min)	50	26
1.5 h/day × 4 weeks (speed = 20 m/min)	57	27
4 × 10 min/day × 1 week (speed = 5 m/min)	46	24
4 × 10 min/day × 1 week (speed = 5 m/min)	71[a]	21
4 × 10 min/day × 1 week (speed = 20 m/min)	95[a]	21
Ten daily contractions upon impact from a fall of 58 cm × 4 weeks	0	25
Ladder climbing		
4 × 10 reps/day × 1 week; 50% load, 1 m at 85° grade	89[a]	21
4 × 8 reps/day × 1 week; 75% load, 1 m at 85° grade	43	21
4 × 8 reps/day × 1 week; 75% load, 1 m at 85° grade	75[a]	21
2 × 10 reps/day × 1 week; 75% load, 1 m at 85° grade	55[a]	21
Centrifugation		
2 h/day × 1 week, 1.5 *g*	44	22
4 × 15 min/day × 1 week, 1.2 *g*	39	23

The percentage of atrophy prevented was calculated by the equation: percentage atrophy prevented = [(muscle weight of countermeasure group – muscle weight of non-weight-bearing only group) ÷ (muscle weight of control group – muscle weight of non-weight-bearing only group)], where muscle weight is the wet weight of the soleus muscle.
[a] In these examples above, data in table 4 of ref. 21 presented as percentage weight lost were used in the following equation: Percentage atrophy prevented = (% decrease in non-weight-bearing soleus weight/body weight ratio – % decrease in countermeasure soleus weight/body weight ratio) ÷ (% decrease in non-weight-bearing soleus weight/body weight ratio).

is the most important determinant of an effective countermeasure (table 1).

Intermittent, short bouts of exercise throughout the day may be more efficient on a time basis than is one single, long bout of exercise during the day as a countermeasure to atrophy during non-weight bearing (table 1). We speculate that a possible reason for the efficacy of intermittent bouts of exercise is the maintenance of protein synthesis at control values during periods of reduced muscle use. For example, if protein synthesis in non-weight-bearing slow-twitch muscle declines significantly 3 h after a 2-hour exercise bout, then protein synthesis rates will be less than control for the remaining 19 h in that day. Alternately, if a 15-min bout of exercise occurs every 6th hour, then protein synthesis rates would be less than control for 12 h each day. This speculation would explain why equivalent quantities of contractile activity during a 1-hour total duration of exercise (intermittent) would be more effective than a 2-hour duration (continuous exercise). However, to date the effect of exercise on protein synthesis rates during reduced use muscle atrophy is not known. Table 1 further demonstrates a proportionality between the quantity of weight bearing during intermittent exercise and the percentage of atrophy prevented. For example, in experiments employing 4 intermittent bouts of weight bearing for 10–15 min per bout, the amount of atrophy prevented was greater in 20-m/min as compared to 5-m/min running (table 1). Although neither EMG activity nor force output were quantified during these exercise bouts, we assume that either more muscle fibers were recruited or more total work was done by a similar number of fibers in the 20-m/min as compared to the 5-m/min run. It must be noted, however, that these speculations do not take into account possible reductions of protein synthesis and/or increases of protein degradation in the first hour following an exercise bout [28].

The previous discussion suggests that the following equation might be useful in planning future experiments designed to evaluate the efficacy of potential countermeasures to muscle atrophy:

Maintenance of muscle protein content = frequency of weight-bearing bouts each day × total load on muscle (i.e. number of repetitions at a given load).

In normal adult rats a strong positive correlation exists between soleus muscle weight and body weight. This is due to the fact that the functional demand placed on this antigravity muscle is proportional to rat body weight. In contrast, since non-weight bearing removes the load of the body weight from the soleus muscle, soleus muscle weight is independent of body weight during the non-weight-bearing procedure in adult rats. Because of this independence from body weight, normalization of soleus muscle weight to body weight may not be the optimal method to determine the percentage of atrophy prevented by a countermeasure. As body weight decreases in adult rats during non-weight bearing, expression of soleus muscle weight to body weight ratio will result in an overestimation of the true prevention of atrophy by a given countermeasure. This contention is supported by the discrepancy in the percentage of atrophy prevented when soleus weights are expressed in absolute versus relative terms (table 1). For example, the reduction of muscle atrophy atrophy by treadmill running (4 × 10 min/day, 5 m/min) was 71% when calculated from normalized soleus weights (i.e. mg/100 g body weight) [21], as compared to 46% when calculated from absolute muscle wet weights [24]. Additionally, in growing animals, comparisons between absolute muscle weights from postcontrol and non-weight-bearing groups will likely underestimate the percentage of atrophy prevented by a given countermeasure. This is due to the combined effects of the increase in muscle weight between pre- and postcontrol animals as well as growth failure in non-weight-bearing muscle during the course of the experiment. This emphasizes the importance of using adult, nongrowing animals or the necessity of making comparisons to precontrol animals of similar body weight when using growing animals. Furthermore, the use of muscle wet weights cannot account for potential changes in muscle water or collagenous protein content. Increases in muscle wet weight due to alterations in these constituents would not maintain muscle functional integrity. Thus, the optimal expression of atrophy prevention is absolute noncollagenous protein content per soleus muscle in adult rats with relatively stable body weights. Finally, the influence of countermeasures on the functional capacity of muscle must also be addressed.

References

1 Grimby G, Saltin B: The ageing muscle. Clin Physiol 1983;3:209–218.
2 Young A, Hughes I, Round JM, Edwards RHT: The effect of knee injury on the number of muscle fibres in the human quadriceps femoris. Clin Sci 1982;62:227–234.
3 Douglas AJ, Walsh EG, Wright GW, Edmond P: Muscle tone around the human knee in paraplegia. Q J Exp Physiol 1989;74:897–905.
4 Young VR: Energy metabolism and requirements in the cancer patient. Cancer Res 1977;37:2336–2347.

5 LeBlanc A, Gogia P, Schneider V, Krebs J, Schonfeld E, Evans H: Calf muscle area and strength changes after five weeks of horizontal bed rest. Am J Sports Med 1988; 16:624–629.
6 Booth FW: Time course of muscular atrophy during immobilization of hindlimbs in rats. J Appl Physiol 1977;43:656–661.
7 Spector SA, Simard CP, Fournier M, Sternlicht E, Edgerton VR: Architectural alterations of rat hind-limb skeletal muscles immobilized at different lenghts. Exp Neurol 1982;76:94–110.
8 Williams PE, Goldspink G: Longitudinal growth of striated muscle fibres. J Cell Sci 1971;9:751–767.
9 Musacchia XJ, Deavers DR, Meininger GA, Davis TP: A model for hypokinesia: Effect on muscle atrophy in the rat. J Appl Physiol 1980;48:479–486.
10 Morey ER: Spaceflight and bone turnover: Correlation with a new rat model of weightlessness. Bioscience 1979;29:168–172.
11 Thomason DB, Booth FW: Atrophy of the soleus muscle by hindlimb unweighting. J Appl Physiol 1990;68:1–12.
12 Miu B, Martin TP, Roy RR, Oganov V, Ilyina-Kakueva E, Marini JF, Leber JJ, Bodine-Fowler SC, Edgerton VR: Metabolic and morphologic properties of single muscle fibers in the rat after spaceflight, Cosmos 1887. FASEB J 1990;4:64–72.
13 Booth FW, Seider MJ: Early change in skeletal muscle protein synthesis after limb immobilization of rats. J Appl Physiol 1979;47:974–977.
14 Watson PA, Stein JP, Booth FW: Changes in actin synthesis and α-actin-mRNA content in rat muscle during immobilization. Am J Physiol 1984;247:C39–C44.
15 Morrison PR, Montgomery JA, Wong TS, Booth FW: Cytochrome c protein synthesis rates and mRNA contents during atrophy and recovery in skeletal muscle. Biochem J 1987;241:257–263.
16 Thomason DB, Biggs RB, Booth FW: Protein metabolism and β-myosin heavy-chain mRNA in unweighted soleus muscle. Am J Physiol 1989;257:R300–R305.
17 Babij P, Booth FW: α-Actin and cytochrome c mRNAs in atrophied adult rat skeletal muscle. Am J Physiol 1988;254:C651–C656.
18 Goldspink DF: The influence of immobilization and stretch on protein turnover of rat skeletal muscle. J Physiol 1977;264:267–282.
19 Stump CS, Overton JM, Tipton CM: Influence of single hindlimb support during simulated weightlessness in the rat. J Appl Physiol 1990;68:627–634.
20 Thomason DB, Herrick RE, Baldwin KM: Activity influences on soleus muscle myosin during rodent hindlimb suspension. J Appl Physiol 1987;63:138–144.
21 Herbert ME, Roy RR, Edgerton VR: Influence of one-week hindlimb suspension and intermittent high load exercise on rat muscles. Exp Neurol 1988;102:190–198.
22 D'Aunno DS, Thomason DB, Booth FW: Centrifugal intensity and duration as countermeasures to soleus muscle atrophy. J Appl Physiol 1990;69:1387–1389.
23 D'Aunno DS, Robinson RR, Smith GS, Thomason DB, Booth FW: Intermittent acceleration as a countermeasure to soleus muscle atrophy. J Appl Physiol 1992;72: 428–433.
24 Hauschka EO, Roy RR, Edgerton VR: Periodic weight support on rat soleus fibers after hindlimb suspension. J Appl Physiol 1988;65:1231–1237.

25 Hauschka EO, Roy RR, Edgerton VR: Size and metabolic properties of single muscle fibers in rat soleus after hindlimb suspension. J Appl Physiol 1987;62:2338–2347.
26 Shaw SR, Zernicke RF, Vailas AC, DeLuna D, Thomason DB, Baldwin KM: Mechanical, morphological and biochemical adaptations of bone and muscle to hindlimb suspension and exercise. J Biomech 1987;20:225–234.
27 Graham SC, Roy RR, West SP, Thomason DB, Baldwin KM: Exercise effects on the size and metabolic properties of soleus fibers in hindlimb-suspended rats. Aviat Space Environ Med 1989;60:226–234.
28 Booth FW, Watson PA: Control of adaptations in protein levels in response to exercise. Fed Proc 1985;44:2293–2300.

Frank Booth, PhD, Department of Physiology, University of Texas
Medical School, PO Box 20708, Houston, TX 77225 (USA)

Sato Y, Poortmans J, Hashimoto I, Oshida Y (eds): Integration of Medical and Sports Sciences. Med Sport Sci. Basel, Karger, 1992, vol 37, pp 150–161

Effect of Endurance Training on Disuse Muscle Atrophy Induced by Body Suspension in Rats

A Structural and Biochemical Study

Toshitada Yoshioka, Hiroaki Takekura, Katsumasa Yamashita

Department of Physiology, St. Marianna University School of Medicine, Kawasaki, Kanagawa, Japan

It has been reported that histochemical and biochemical changes occur in skeletal muscle as a result of regular endurance and sprint exercise-training [1, 2]. Glycolytic and oxidative enzyme activities increased remarkably following both types of training [3, 4]. However, glycolytic and oxidative enzyme activities in fast muscle and oxidative enzyme activity in slow muscle clearly decreased following experimental body suspension hypodynamia/hypokinesia [5–8]. Similar results were obtained from the three types of single muscle fibers, with the conclusion that the structural and metabolic changes differed according to fiber type [7–10]. The present study was designed to prevent degenerative alterations in structural and functional properties following suspension disuse muscle atrophy in rats trained previously by endurance running exercise.

Materials and Methods

Classification of Single Muscle Fibers

Wistar strain male rats were used. Single muscle fibers were dissected from both slow soleus and fast extensor digitorum longus (EDL) muscle in a mammalian relaxing solution (120 m*M* KCl, 0.5 m*M* EGTA, 4 m*M* $MgCl_2$, 10 m*M* PIPES, 10 m*M* ATP, pH 6.8). Three or four small fragments were cut from a single fiber. Each tiny fragment was split longitudinally along the myofilament direction and placed on a microglass slide with the cell membrane against the glass. They were stained histochemically with succinate

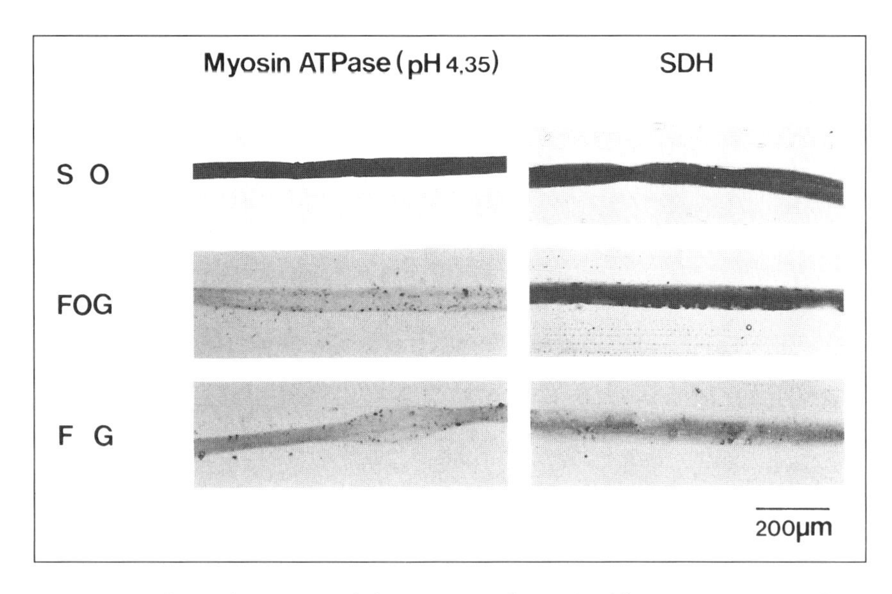

Fig. 1. Light micrographs of single muscle fibers classified by ATPase and SDH staining.

dehydrogenase (SDH) and actomyosin ATPase after preincubation at pH 4.35, in order to classify each single muscle fiber (fig. 1). Based on the staining characteristics, all single muscle fibers were classified as slow-twitch oxidative (SO) fiber, fast-twitch oxidative and glycolytic (FOG) fiber or fast-twitch glycolytic (FG) fiber according to Peter et al. [11]. SO, FOG and FG fibers were basically equal to type I, type IIA and type IIB fibers, respectively. The remaining parts of the single muscle fibers which were classified according to type were used for analysis of enzyme activity and for electron-microscopic measurements. A piece from the fragments was fully homogenized with a microglass homogenizer in Tris-HCl buffer for analysis of creatine kinase (CK), lactate dehydrogenase and phosphofructokinase (PFK) as glycolytic enzymes, and analysis of malate dehydrogenase and SDH as oxidative enzymes.

CK isoenzymes were again separated into cytoplasmic (CK-MM, CK-MB, CK-BB) and mitochondrial (m-CK) isoenzymes using ion-exchange column chromatography [12]. The concentration of total fiber protein was also measured.

Another method for classification of fiber types was used in this study. The soleus and EDL muscles were cut in cross-sections at the muscle belly level. Serial sections were obtained from one side of the rapidly frozen muscle and stained histochemically to classify fiber types. The other side of the muscle specimen, the opposite side of the frozen muscle, was processed for electron-microscopic study. A 1-µm-thick section of a plastic-embedded preparation corresponded to the section stained histochemically by SDH and ATPase (fig. 2). Mitochondrial volumes were obtained stereologically on the micrographs [13].

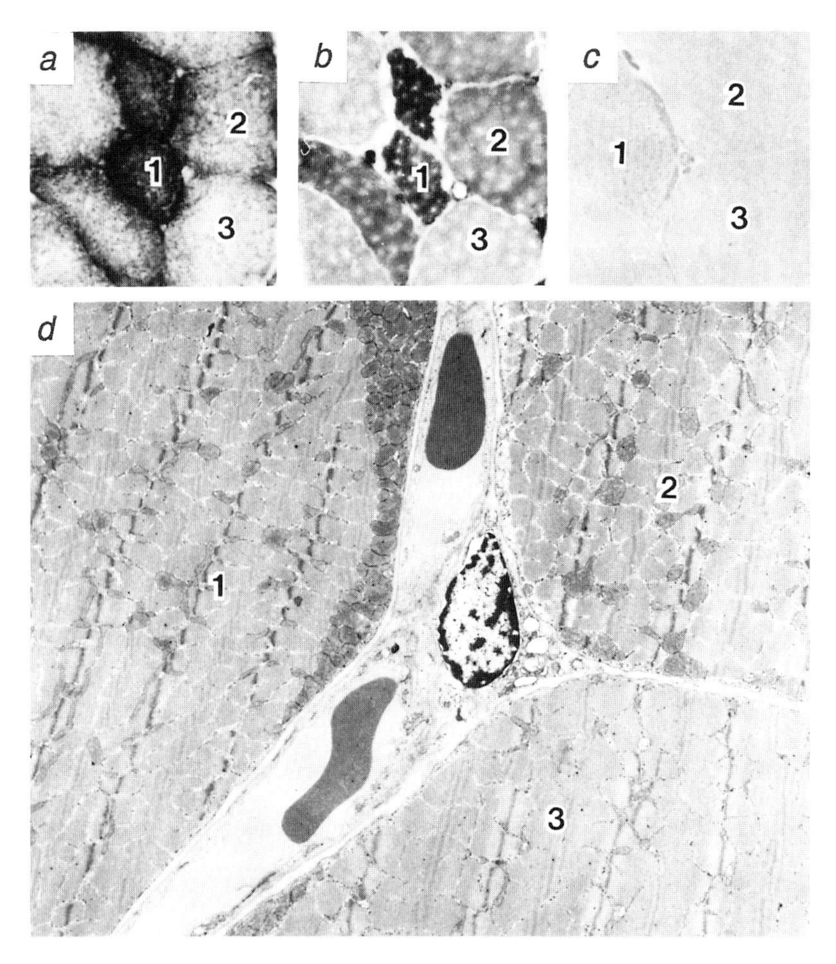

Fig. 2. Light (*a–c;* × 230) and electron (*d;* × 4,300) micrographs from the EDL muscle of an endurance-trained rat. The micrographs were taken from the same portion of the muscle. 1 = SO fiber; 2 = FOG fiber; 3 = FG fiber. *a* SDH staining. *b* Actomyosin ATPase staining (preincubation at pH 4.35). *c* Toluidine blue staining of a 1-µm section of a plastic-embedded preparation.

Animal Care, Training Program and Suspension Procedure

Rats were divided into four groups: sedentary control (C) group; endurance exercise-training (EX) group; hindlimb-suspended (body suspension; BS) group, and body suspension after training (EX + BS) group. The rats of the EX group were subjected to continuous running for 120 min at a speed of 40 m/min [2]. The intensity of the running training used in this study was considered to result in 85% of maximum O_2 uptake [14], effecting an increase of oxidative capacity not only in soleus, but also in EDL muscles. Body suspension was performed by the modified method reported by Musacchia et al. [15]. After endurance training, rats were subjected to suspension for 2 weeks [16]. Electrical stimuli were also applied to suspended muscle via the sciatic nerves by implanted electrodes.

Results and Discussion

Body Weight and Muscle Weight

The mean body weight of the C, BS, EX and EX+BS groups were approximately the same at the end of the experiments. The wet weight of the soleus muscle in the BS group was about 40% lower than that of the C group, but the weight of the EDL was approximately the same in the two groups. A recovery in the soleus muscle weight of the EX+BS group was obtained in comparison to the BS group, suggesting that disuse atrophy due to suspension mainly affected the slow muscle, as has been reported earlier [5, 7, 12].

Structural Features and Mitochondrial Volumes of Exercised and Suspended Muscle Fibers

Figure 3 shows electron micrographs of three types of single muscle fiber after suspension and after application of electrical stimuli during body suspension. Myofibrillar disruptions, Z band streaming and ruptures of mitochondria and sarcoplasmic reticulum were often observed in the treated muscles. We measured the mitochondrial and lipid droplet volumes stereologically on the three types of fibers from the control and electrically stimulated soleus and EDL muscles during suspension (fig. 4, 5). Remarkable decreases in mitochondrial volumes were apparent in all three kinds of fibers following 2 weeks suspension. The volume of lipid droplets, which was significantly elevated by suspension, was markedly decreased after stimulation of 50 Hz, 0.1 ms duration, for 8 h, via the sciatic nerve.

One factor causing increased lipid droplets could be the low rate of the lipid metabolism cycle probably effected by suspension. Only in FG fibers did the mitochondrial volume recover to the control level after application of electrical stimuli. Therefore, the chronic stimulation used in this study was not effective in preventing the degenerative alterations of suspension-induced atrophy.

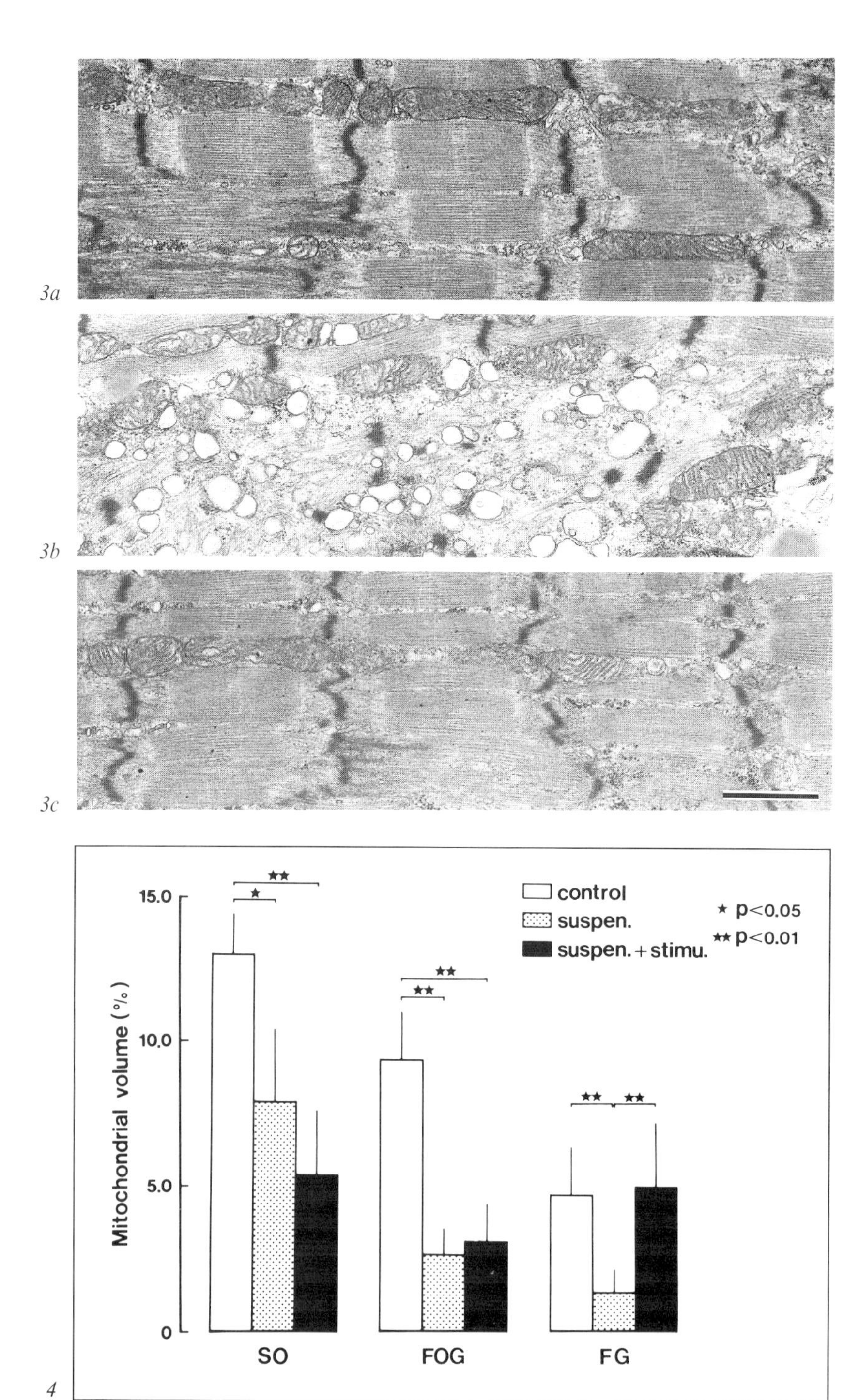
3a
3b
3c
Mitochondrial volume (%)
15.0
10.0
5.0
0
SO
FOG
FG
control
suspen.
suspen. + stimu.
★ p<0.05
★★ p<0.01
4

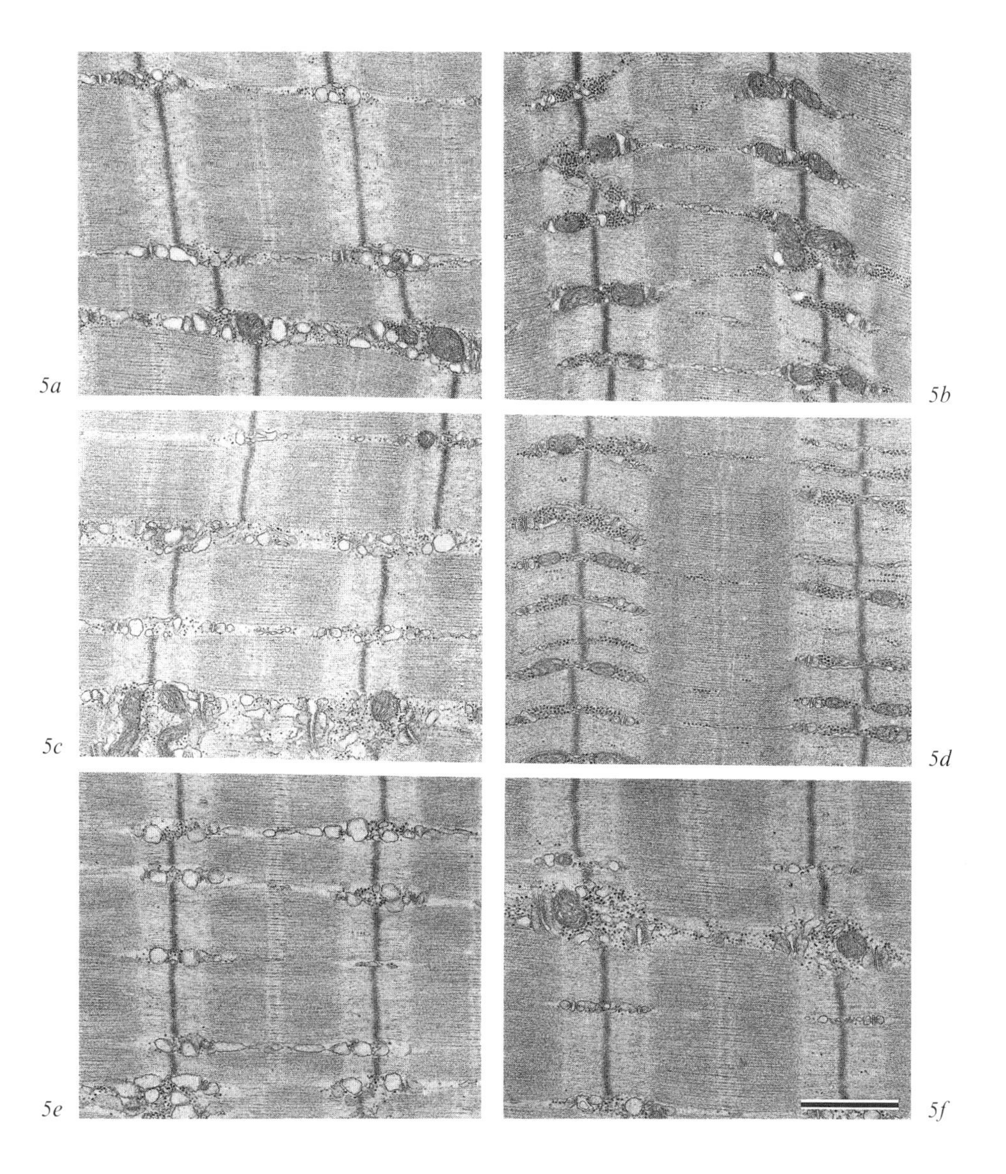

Fig. 3. Electron micrographs of single muscle fibers. *a* Suspended FOG fiber. *b* Suspended SO fiber with electrical stimulation. *c* Suspended FG fiber with electrical stimulation. Bar = 1 μm.

Fig. 4. Mitochondrial volumes of three types of fibers from sedentary control (control) muscle, suspended (suspen.) muscle and suspended muscle with electrical stimulation (suspen. + stimu.).

Fig. 5. Electron micrographs of SO fibers *(a, b)*, FOG fibers *(c, d)* and FG fibers *(e, f)*. *a*, *c* and *e* are from suspended muscles; *b*, *d* and *f* are from exercised and body-suspended muscles.

Table 1. Activities of total CK and CK isoenzymes in single muscle fibers of different types

Fiber type	n	Total CK	m-CK	CK-MM	CK-MB	CK-BB
SO	38	30.90±27.89	4.34±7.22	23.57±22.73	2.24±3.67	0.74±1.67
FOG	49	43.40±42.87	2.41±4.93	37.50±39.17	2.29±4.81	1.20±2.92
FG	38	61.65±42.17**	1.90±2.73	55.65±37.92*,**	2.59±4.89	1.55±6.45

Values are means ± SD and are expressed as international units per milligram protein. Significant difference: * $p < 0.05$ vs. FOG fiber type; ** $p < 0.01$ vs. SO fiber type.

Creatine Kinase Isoenzyme Compositions in Single Muscle Fibers

The activities of total CK and of the 4 isoenzymes in each type of single muscle fiber are given in table 1. Total CK activity was highest in FG, lowest in SO and intermediate in FOG fibers. CK-MM activity resembled the total activity. These results suggested that total CK activity could be referred to as an intracellular energy storage function for muscle contraction. Total CK and CK isoenzyme compositions in single fibers may be influenced by exercise-training [12].

Oxidative and Glycolytic Enzyme Activities of Trained and Suspended Muscles

The effects of different types of runnung training on the glycolytic and oxidative enzyme activities and mitochondrial content in the different kinds of single muscle fibers were studied. Glycolytic enzyme activities were increased in FG and FOG fibers following sprint training. Oxidative enzyme activities were increased in all fiber types following both types of training. Mitochondrial volumes in the single muscle fibers were significantly increased from 12.8 to 19.9% ($p < 0.05$) in SO fibers and from 7.9 to 9.6% ($p < 0.05$) in FOG fibers of soleus muscle following endurance training. Glycolytic and oxidative enzyme activities tended to decrease in whole-muscle homogenates following hindlimb suspension [18]. When we examined a single muscle fiber, we noted that the PFK activity did not change except for FG fibers (fig. 6). The results show that the glycolytic enzyme activities in SO and FOG fibers did not alter at all, even when the

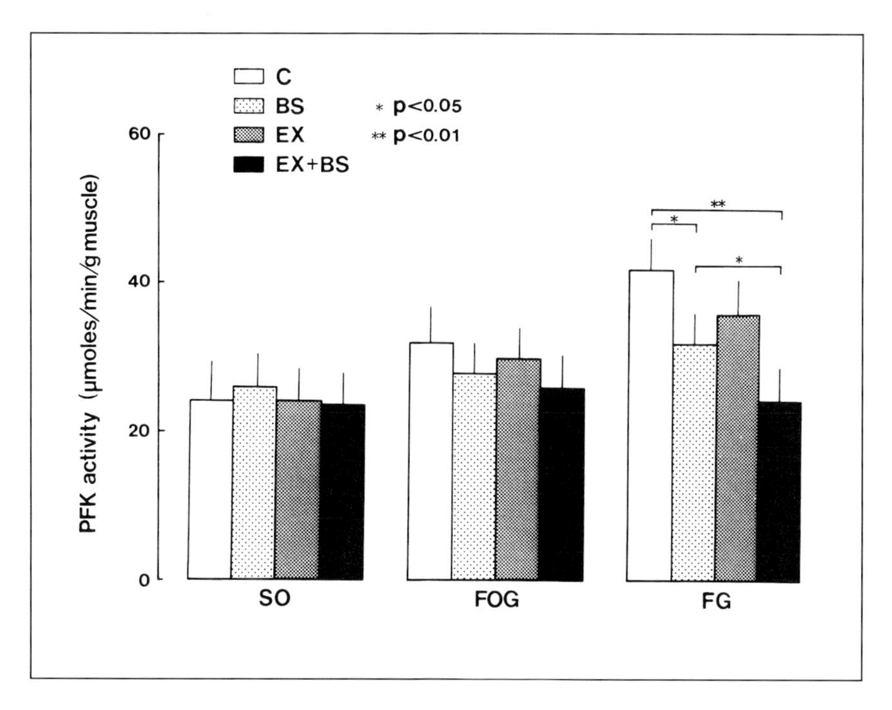

Fig. 6. PFK activities of three types of fibers (SO, FOG, FG) from muscles of the C, BS, EX and EX+BS groups.

soleus muscle underwent suspension. However, a small decrease in activity in FG fibers of the BS or EX+BS group suggested a reduction in glycolytic enzyme capacity in the entire muscle, because EDL muscle contains many FG fibers, while soleus muscle contains relatively high amounts of SO and FOG fibers.

The SDH activity was apparently elevated in all three types of fibers following endurance training, but not in the BS group. Significantly higher activities of oxidative enzyme in the EX+BS group were obtained when comparing them to the C and BS groups. Therefore, the oxidative activity was effectively maintained in the EX+BS group (fig. 7). Figure 8 depicts histograms of the mitochondrial volumes in all three types of muscle fiber under the various conditions. Extremely high volumes were obtained in the SO fiber type of the EX group. When we compared these results with those of the BS group, the EX group was recognized to show an increase in

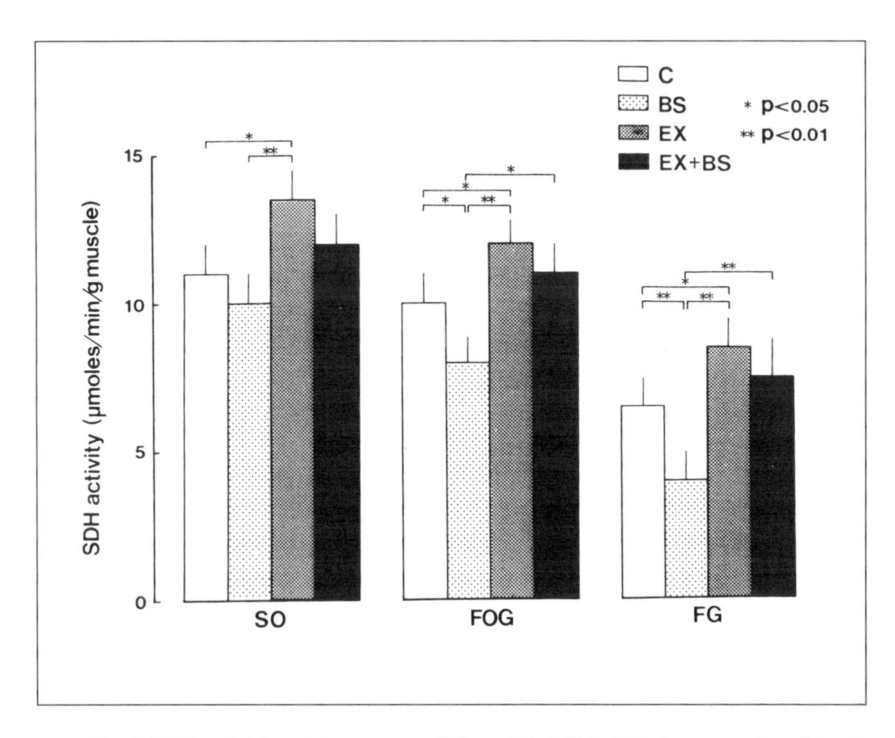

Fig. 7. SDH activities of three types of fibers (SO, FOG, FG) from muscles of the C, BS, EX and EX+BS groups.

mitochondrial volume in all types of muscle fiber. The volume recovered effectively in previously trained muscle (EX+BS group) from the decrease in mitochondrial volume in response to suspension [19, 20]. These results suggest a decrease in the contractile energy metabolism of the muscle, reflecting hypokinesia due to suspension.

Oxidative Activity and Mitochondrial Volume in Single Muscle Fibers

The relationship between mitochondrial volume and SDH activity for each type of fiber of the control, sprint training and endurance training groups were compared. In SO and FOG fibers, but not in FG fibers, of the endurance exercise group, there was a significant relationship between relative mitochondrial volume and SDH activity, suggesting that the increase

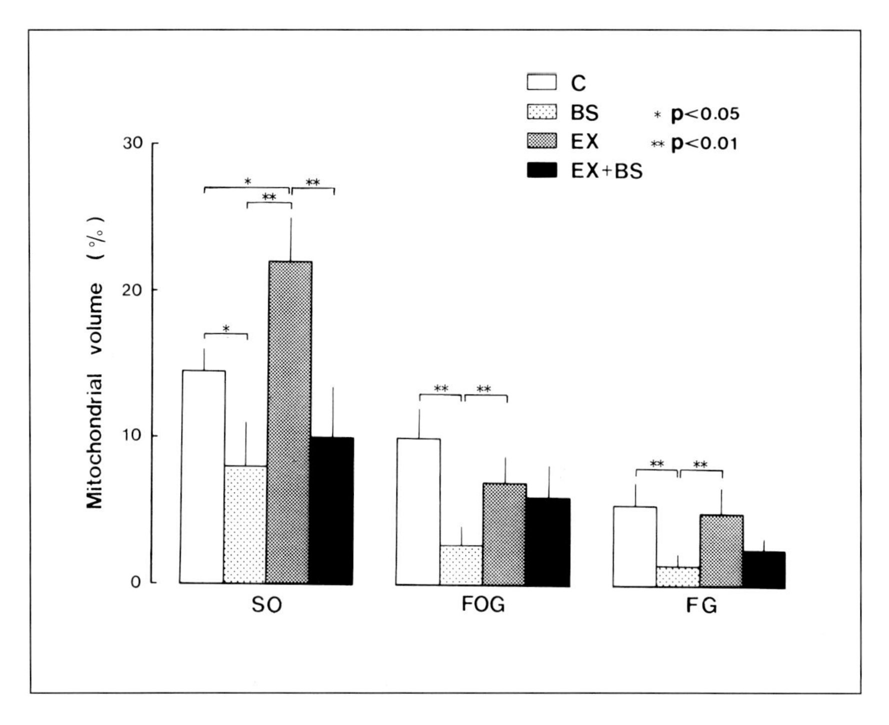

Fig. 8. Mitochondrial volumes of three types of muscle fibers (SO, FOG, FG) from the C, BS, EX and EX + BS groups.

in oxidative capacity in skeletal muscle fibers is not always due to the increase in mitochondrial volume. This tells us that there may be various types of adaptation profiles related to exercise intensity and fiber type. Relative mitochondrial volumes in all types of single muscle fiber of the C group were significantly higher than those of the BS group. However, oxidative enzyme activity did not show the same tendency following hindlimb suspension. The results suggest that the changes in oxidative capacity in response to suspension did not strictly depend on mitochondrial alterations.

In conclusion, the oxidative capacities in three types of muscle fibers dissected from slow and fast muscle were effectively maintained in suspended muscle which was trained previously by a proper protocol of endurance running exercise. Increased mitochondrial volumes were recognized in all types of fibers in treated muscle, resulting in increased oxidative capacity in the entire muscle. However, glycolytic enzyme activity was not generally influenced by suspension and/or endurance exercise.

References

1 Benzi G, Panceri P, Barnard MD, Villa R, Arcelli E, D'Angelo L, Arrigoni E, Berte F: Mitochondrial enzymatic adaptation of skeletal muscle to endurance training. J Appl Physiol 1975;38:565–569.
2 Takekura H, Yoshioka T: Specific mitochondrial responses to running training are induced in each type of rat single muscle fibers. Jpn J Physiol 1989;39:497–509.
3 Holloszy JO, Booth FW: Biochemical adaptations to endurance exercise. Annu Rev Physiol 1976;38:273–291.
4 Saubert CWIV, Armstrong RB, Shepherd RE, Gollnick PD: Anaerobic enzyme adaptations to sprint training in rats. Pflügers Arch 1973;341:305–312.
5 Desplanches D, Mayet MH, Sempore B, Flandoris R: Structural and functional responses to prolonged hindlimb suspension in rat muscle. J Appl Physiol 1987;63: 558–563.
6 Simard C, Lacaille M, Vallieres J: Enzymatic adaptations to suspension hypokinesia in skeletal muscle of young and old rats. Mech Ageing Dev 1985;33:1–9.
7 Takekura H, Yoshioka T: Ultrastructural and metabolic profiles on single muscle fibers of different types after hindlimb suspension in rats. Jpn J Physiol 1989;39: 385–396.
8 Yoshioka T, Takekura H, Ohira Y, Saiki H: The mitochondrial volume and fiber type transition of skeletal muscle after suspension hypokinesia in rat (in Japanese). Jpn J Aerospace Environ Med 1988;25:87–96.
9 Templeton GH, Sweeney HL, Timson BF, Padalino M, Dudenhoeffer GA: Changes in fiber composition of soleus muscle during rat hindlimb suspension. J Appl Physiol 1988;65:1191–1195.
10 Yoshioka T, Takekura H: Effect of body suspension hypokinesia on skeletal muscle trained previously by endurance exercise (in Japanese). Jpn J Aerospace Environ Med 1989;26:47–57.
11 Peter JB, Barnard RJ, Edgerton VR, Gillespie CA, Stempel KE: Metabolic profiles of three fiber types of skeletal muscle in guinea pig and rabbits. Biochemistry 1972; 11:2627–2633.
12 Yamashita K, Yoshioka T: Profiles of creatine kinase isoenzyme compositions in single muscle fibres of different types. J Muscle Res Cell Motil 1991;12:37–44.
13 Weibel ER: Stereological principles for morphometry in electron mircroscopic cytology. Int Rev Cytol 1969;26:235–302.
14 Shepherd RE, Gollnick PD: Oxygen uptake of rats at different work intensities. Pflügers Arch 1976;362:219–222.
15 Musacchia XJ, Deavers DR, Meininger GA, Davis TP: A model for hypokinesia: Effects on muscle atrophy in rat. J Appl Physiol 1980;48:479–486.

16 Elder GCB, McComas AJ: Development of rat muscle during short- and long-term hindlimb suspension. J Appl Physiol 1987;62:1917–1923.
17 Ilyva-Kakueva EI, Portugalou VV, Krivenkova NP: Space flight effects on the skeletal muscles of rats. Aviat Space Environ Med 1976;47:700–703.
18 Nemeth PM, Meyer D, Kark RAP: Effects of denervation and simple disuse on rats of oxidation and on activitics of four mitochondrial enzymes in type I muscle. J Neurochem 1980;35:1351–1360.
19 Graham SC, Roy RR, West SP, Thomason D, Baldwin KM: Exercise effects on the size and metabolic properties of soleus fibers in hindlimb-suspended rats. Aviat Space Environ Med 1989;60:226–234.
20 Herbert ME, Roy RR, Edgerton VR: Influence of one-week hindlimb suspension and intermittent high load exercise on rat muscles. Exp Neurol 1988;102:190–198.

Toshitada Yoshioka, MD, Department of Physiology,
St. Marianna University School of Medicine, Kawasaki, Kanagawa 216 (Japan)

Sato Y, Poortmans J, Hashimoto I, Oshida Y (eds): Integration of Medical and Sports Sciences. Med Sport Sci. Basel, Karger, 1992, vol 37, pp 162–170

Association of Cardiac Ventricular Myosin Isoforms with Hemodynamic Factors[1]

C.D. Ianuzzo, B. Li, N. Hamilton, L. D'Costa, S.E. Ianuzzo, C.A.M. Barrozo, T.A. Salerno, M.H. Laughlin

Departments of Physical Education and Biology, and Center for Health Studies, York University, Toronto, Ont.; Division of Cardiovascular and Thoracic Surgery, St. Michael's Hospital, and University of Toronto, Ont., Canada; Departments of Veterinary Biomedical Sciences and Medical Physiology, and the Dalton Research Center, University of Missouri, Columbia, Mo., USA

The myosin protein is of primary importance in establishing myocardial performance and maintaining normal hemodynamics. The expression of this protein can be altered by changing the hemodynamic demands on the cardiac ventricles. When the hemodynamic load on the myocardium is altered it responds not only by hypertrophy or atrophy, but also by invoking an additional compensating maneuver of switching the ventricular myosin isoform, which incurs energetic benefits. The actual hemodynamic factor(s) that signals the switch in myosin expression is (are) not known. In this paper we explored the possibility that the double product [i.e. rate pressure product (RPP) = heart rate × peak systolic pressure], which has a high correlation with myocardial oxygen consumption [1], may be a correlate of myosin expression. The individual parameters of heart rate and peak systolic pressure were also examined to see if they exhibited a correlation with ventricular myosin type. This paper has used findings from our previous and recent studies, with the goal of providing some insights into whether the above-mentioned hemodynamic factors relate to cardiac myosin expression. The association between ventricular myosin phenotype and heart rate, RPP and systolic pressure has been discussed using the findings

[1] Supported by Canada NSERC grant 0404, an Ontario Heart and Stroke Foundation grant and National Heart, Lung and Blood Institute grant HL36531

from several different animal models. These experimental models include: comparison of differently sized mammalian hearts, experimentally imposed tachycardia, exercise-training, cultured cardiac myocytes, pressure-induced cardiac hypertrophy and heterotopic heart transplants.

Methods

The *comparative* heart approach used mammals consisting of cattle, swine, dogs, rabbits, guinea pigs, rats and mice [2–4]. The resting heart rates ranged from 50 to 475 bpm from the cow to the mouse, respectively. *Myocardial tachycardia* was experimentally produced in Yorkshire pigs for a period of 35–42 days by implanting Medtronic cardiac pacemakers set at 180 pulses per minute [5]. The normal resting heart rate for the pig is between 90 and 120 bpm. Female Yucatan miniature swine were *exercise-trained* using an alternating endurance with an intermittent sprint training protocol for 85 min/day, 5 days per week for 12 weeks [6]. *Cultured rat cardiac myocytes* that were spontaneously beating or arrested and treated with 3,3′,5-triiodo-*L*-thyronine (T_3) were grown in serum-free, 2% ULTROSER-G (Gibco)-supplemented, defined medium for a period of 14 days [7, 8]. The spontaneous beating of the cultured cells was arrested using 50 m*M* KCl and the untreated cultured myocytes contracted spontaneously at 133 ± 5 bpm in the defined medium. Arrested and beating cultures were treated with 10 n*M* T_3. *Cardiac hypertrophy* was studied in Lewis rats by coarctation of the abdominal aorta using neurosurgical hemoclips closed to the diameter of a 25-gauge needle [9]. In the *heart transplant* model, hearts from an inbred Lewis strain of rats were heterotopically transplanted onto the abdominal aorta and inferior vena cava of recipient Lewis rats [10]. This resulted in a coronary perfused heart that was beating at about 230–260 bpm, but only occasionally ejecting blood. The lesser cardiac work accomplished by this heart leads to myocardial atrophy. The native heart had a resting heart rate of 320 bpm. Both hypertrophy and atrophy were allowed to develop for a period of at least 30 days.

The cardiac myosin isoforms were separated electrophoretically using the method described by Hoh et al. [11]. Myosin and myofibrillar ATPases were assayed using the procedures of Solaro et al., Pagani and Solaro, Baldwin et al. and Nakaniski et al., as referenced in Hamilton and Ianuzzo [3], Hornby et al. [10] and Laughlin et al. [6].

Results and Discussion

When comparing the myosin isoforms present in the heart ventricles of mammals with different heart rates, it was observed that the hearts that had beating rates of less than 300 bpm expressed primarily the V3 isoform, whereas those with rates above 300 bpm expressed the V1 form (fig. 1) [4]. From this observation it was postulated that a threshold heart rate of approximately 300 bpm may be the hemodynamic correlate of ventricular

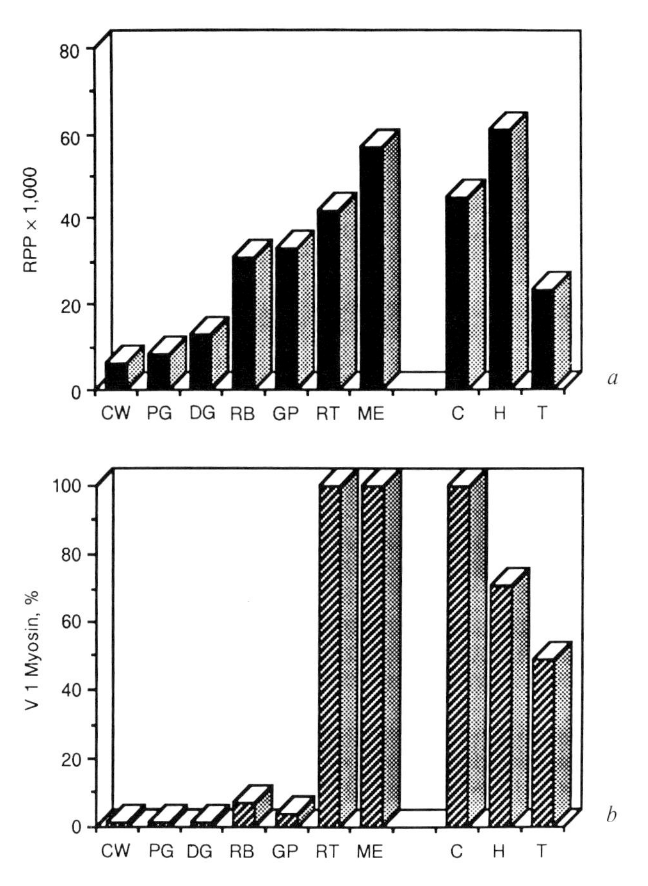

Fig. 1. a RPP of the left ventricles from differently sized mammalian hearts and for control (C), hypertrophied (H) and transplanted (T) hearts. *b* Percentage of V1 myosin isoform in left ventricles of differently sized mammalian hearts. RPP = Rate-pressure product; CW = cow; PG = pig; DG = dog; RB = rabbit; GP = guinea pig; RT = rat; ME = mouse.

myosin expression. Since among these differently sized mammals a number of parameters that could possibly alter cardiac demands and myosin expression remain constant, namely, arterial blood pressure, stroke work index, ventricular wall stress, oxygen transport capacity and plasma thyroid hormone concentrations, this leaves heart rate difference as a main hemodynamic variable that could be associated with myosin type. The RPP among these 7 differently sized mammals ranged from 6,000

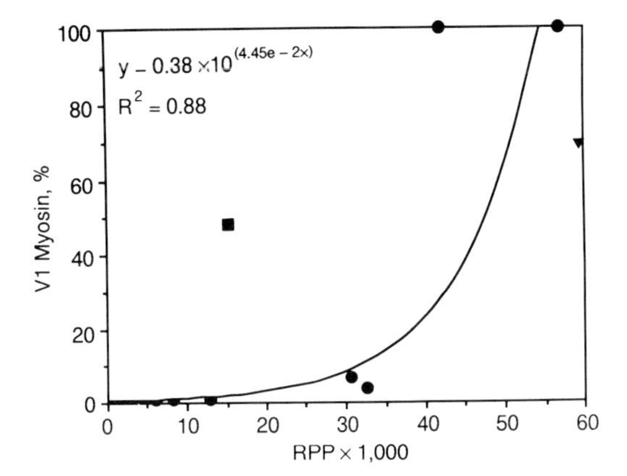

Fig. 2. Exponential relationship between RPP and percentage of V1 myosin isoform in the left ventricle of differently sized mammalian hearts. The RPP and percentage of V1 myosin from the hypertrophied (▼) and transplanted (■) hearts are also shown to illustrate the poor predictability of the equation for these experimental perturbations of RPP and myosin changes.

mm Hg/min for the cow to 60,000 mm Hg/min for the mouse (fig. 1). Although there was a 10-fold difference in the RPP among these mammalian hearts, no significant incremental step was observed between the guinea pig and the rat, to correspond with the shift in myosin expression, as would be expected if RPP was a meaningful correlate of myosin expression. Furthermore, the exponential equation, as shown in figure 2 for percent V1 and RPP, fit the data points from the different mammalian hearts with an $r^2 = 0.88$ ($p < 0.01$), but this equation was not useful as a predictor of the myosin changes that occurred in the atrophied and hypertrophied hearts (see below).

In the chronic tachycardia study, the hearts of 9 pigs were atrial-paced at a rate of 180 bpm for a period of 35–42 days [5]. One purpose was to determine if a chronic increase in heart rate would result in altered myosin expression, with the underlying intent to further test the hypothesis that heart rate is the hemodynamic correlate of myosin expression. We initially attempted to impose a rate of approximately 300 bpm, but were unsuccessful because of partial atrioventricular block at this high rate. The findings from the native myosin gels and Ca^{2+}-activated myosin ATPase at pH 7.2 and 9.8 for the paced compared to the nonpaced hearts strongly indicated that myosin expression was not altered by chronic pacing at 180 bpm

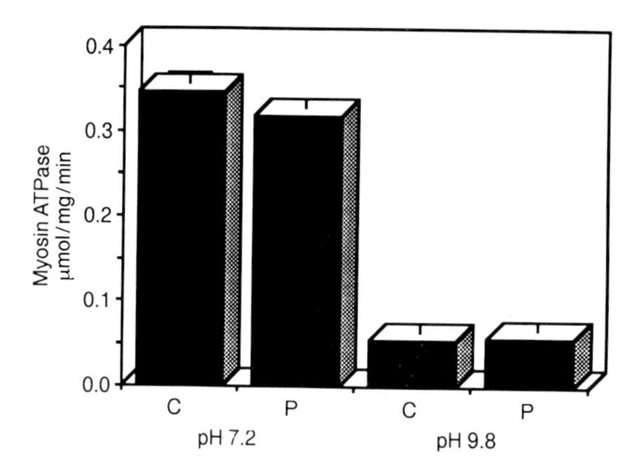

Fig. 3. Myosin ATPase activity of the myocardium from paced (P) and nonpaced (C) pig hearts at pH 7.2 and 9.8.

(fig. 3). The RPP was 19,000 mm Hg/min for the paced compared to 16,000 mm Hg/min for the nonpaced hearts. The RPP for the paced pig hearts was only 50% of that of rat hearts that expressed primarily V1 myosin; therefore, based on the RPP, myosin expression would not be expected to change. The findings from this study are consistent with the findings and hypothesis from the comparative study.

Exercise-training results in a daily pulse of tachycardia and an acute increase in blood pressure and RPP [6]. The heart rates during the 85-min exercise bout reached a maximal level of 250–300 bpm with a peak systolic pressure of 133 ± 7 mm Hg. Therefore, during exercise the RPP attained levels of 33,000–40,000 mm Hg/min, which was 2.6–3.1 times that from the nonexercised pigs. Neither the cardiac myosin isoforms nor the myofibrillar or myosin ATPases in the pig ventricle were affected by this daily pulse of increased RPP resulting mainly from the exercise tachycardia. These findings show that exercise-training of a large mammal heart does not alter myosin expression. This study adds further insight to the assumption that the pulsatile nature of repeated bouts of exercise with numerous transient perturbations, such as increased heart rate, blood pressure, RPP, and altered humoral and hormonal milieu, does not appear to influence the expression of ventricular myosin.

Additional proof that RPP and myosin expression are not associated was obtained by comparing the myosin type and RPP of the left and right ventricle. In normal control hearts from Lewis rats both ventricles con-

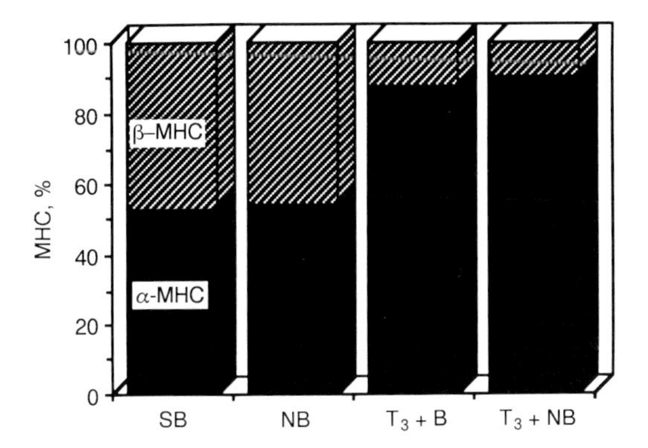

Fig. 4. Percentage of α-and β-MHC for spontaneously beating (SB), nonbeating (NB), T_3-treated beating (B) and T_3-treated nonbeating cultured cardiac myocytes.

tained 100% V1 myosin, yet the RPP for the right ventricle was 8,000 mm Hg/min, whereas for the left ventricle it was 38,000 mm Hg/min. This nearly 5-fold difference in RPP between the two ventricles with the same percentage of myosin type supports the conclusion that ventricular RPP per se is not a hemodynamic correlate of myosin expression.

By altering the spontaneous beating rates of cultured cardiac myocytes, additional information on the influence of chronotropism on myosin expression has been provided. Neonatal rat cardiac myocytes were cultured in serum-free medium for a period of about 2 weeks [1, 7]. Some cultures were allowed to beat at their normal spontaneous rates (133 ± 4 bpm) for 2 weeks, while other cultures had their beating rates arrested by placing 50 m*M* KCl (0 bpm) in the culture medium. The percentage of α-myosin heavy chain (α-MHC), which is the MHC that comprises the V1 isoform, was similar for the arrested (65 ± 5%) and spontaneously beating cultures (64 ± 5%; fig. 4). Both the arrested and beating cultures responded to 10 n*M* T_3 in the medium by expressing 100% α-MHC. These findings indicate that beating rates ranging from 0 to 133 bpm do not alter the myosin expression in culture. The effects of T_3 on myosin expression [12, 13, 14] further confirm that T_3 is a powerful inducer of α-MHC gene transcription and that this T_3 effect does not interact with beating rates to influence myosin regulation. These findings are consistent with the hypothesis that contraction frequencies below 300 bpm do not result in changes in myosin expression. It does show, however, that α-MHC can be expressed in significant amounts in the absence of contractile activity, at least in the cultured environment.

Afterload and RPP were increased in rat hearts by constriction of the abdominal aorta above the renal arteries [9]. The hearts hypertrophied an average of 35% with a range of 10–90% within 30 days. By the end of the postcoarctation period, the hypertrophied hearts had 70% V1, 18% V2 and 12% V3 myosin isotypes in the left ventricle, in contrast to the non-hypertrophied control hearts that possessed 100% V1. The hypertrophied left ventricle had an RPP of 60,000 mm Hg/min, which was more than 50% greater than the nonhypertrophied left ventricle (fig. 1). This further argues against RPP being associated with myosin type, since RPP attained a level similar to that of the mouse, which was 100% V1, but myosin expression moved towards the V3 type, i.e. opposite of what might be expected if high RPP was associated with V1 expression.

Hearts were partially unloaded by using the heterotopic heart transplant model. Hearts that were anastomosed onto the abdominal aorta and vena cava of a recipient rat atrophied by 35% within 30 days after transplantation. The myosin type profile in the transplanted hearts was 50% V1, 23% V2 and 27% V3 compared to 100% V1 in the native hearts from the same rat [9, 10]. The RPP of the left ventricle was 16,000 mm Hg/min (fig. 1). A recent study [13] has shown similar RPP for this transplanted heart model. This is only about 25% of the RPP of the left ventricle from the hypertrophied hearts, yet both the atrophied and hypertrophied hearts had nearly similar myosin profiles. This also verifies that RPP has no association with ventricular myosin type.

Summary and Conclusions

The purpose of this paper has been to explore the possibility that some common hemodynamic factors are associated with ventricular myosin phenotypes and, therefore, may be the initial signal altering in myosin expression. The result of examining our previous work, which has employed various experimental models, with this viewpoint has led to the conclusion that RPP is definitely not a correlate of myosin expression in any of these models. The hypothesis formed from our previous studies [4] stating that 300 bpm is the threshold that leads to a switch in myosin expression still stands. Myocardial afterload, per se, is not the sole factor resulting in ventricular myosin switches since all of these 7 mammalian hearts studied were facing the same arterial pressures yet had different ventricular myosin types. However, when the afterload was chronically increased in the rat heart, myosin expression was altered. This could be interpreted as resulting from an increased amount of wall stress that is translated to the individual cardiac myocytes, which then compensated by myocyte hypertrophy and re-expression of the myosin type. This would be consistent for both the right and left ventricles, which face different absolute amounts of afterload, but have similar relative amounts of stress per cardiac myocyte and also similar myosin isoform profiles. In contrast, when the afterload was reduced, such as in atrophy and in the cultured myocytes, myosin expression was also changed to V3, indicating that increased hemodynamic demand is not the sole factor in inducing β-MHC expression. The hemodynamic parameters that are associated with myosin gene expression remain an enigma and need continued investigative attention.

References

1 Sarnoff SJ, Braunwald E, Welch GH Jr, Case RB, Stainsby WN, Macruz R: Hemodynamic determinants of oxygen consumption of the heart with special reference to the tension-time index. Am J Physiol 1958;192:148–156.
2 Blank S, Chen V, Hamilton N, Salerno TA, Ianuzzo CD: Metabolic characteristics of mammalian myocardia. J Mol Cell Cardiol 1989;21:367–373.
3 Hamilton N, Ianuzzo CD: Contractile and calcium regulating capacities of myocardia of different sized mammals scale with resting heart rate. Mol Cell Biochem 1991; 106:133–141.
4 Ianuzzo CD, Blank S, Hamilton N, O'Brien P, Chen V, Brotherton S, Salerno TA: The relationship of myocardial chronotropism to the biochemical capacities of mammalian hearts; in Taylor AW, Gollnick PD, Green HJ, Ianuzzo CD, Noble EG, Métivier G, Sutton JR (eds): Biochemistry of Exercise VII. Champaign, Human Kinetics, 1990, pp 593–605.
5 Ianuzzo CD, Brotherton S, O'Brien P, Salerno TA, Laughlin MH: Myocardial biochemical and hemodynamic adaptations to chronic tachycardia. J Appl Physiol 1991;70:907–913.
6 Laughlin MH, Hale C, Novela L, Gute D, Hamilton N, Ianuzzo CD: Biochemical characterization of exercise-trained porcine myocardium. J Appl Physiol 1991;71: 229–235.
7 Li B, Ianuzzo CD: Role of contractile activity and thyroid hormone in regulating the biochemical character of cultured neonatal rat heart myocytes. Submitted.
8 Williams H, Ianuzzo CD: The effects of triiodothyronine on cultured neonatal rat cardiac myocytes. J Mol Cell Cardiol 1988;20:689–699.
9 Ianuzzo CD, Li B, Ianuzzo SE, Barrozo CAM, Salerno TA: Cardiac myosin phenotype changes with atrophy and hypertrophy (abstract). Physiologist 1991;34:227.

10 Hornby L, Hamilton N, Marshall D, Salerno TA, Laughlin MH, Ianuzzo CD: Role of cardiac work in regulating biochemical characteristics. Am J Physiol 1990;258: H1482–H1490.
11 Hoh JFY, McGrath PA, Hale PT: Electrophoretic analysis of multiple forms of rat cardiac myosin: Effects of hypophysectomy and thyroxine replacement. J Mol Cell Cardiol 1978;10:1053–1076.
12 Nag A, Cheng M: Expression of myosin isoenzymes in cardiac muscle cells in culture. Biochem J 1984;221:21–26.
13 Korecky B, Masika M: Direct effect of increased hemodynamic load on cardiac mass. Circ Res 1991;68:1174–1178.
14 Gustafson T, Bahl J, Markham B, Roeske W, Morkin E: Hormonal regulation of myosin heavy chain and α-actin gene expression in cultured fetal rat heart myocytes. J Biol Chem 1987;262:13316–13322.

C. David Ianuzzo, PhD, Biology Department, York University,
Toronto, ON M3J 1P3 (Canada)

Sato Y, Poortmans J, Hashimoto I, Oshida Y (eds): Integration of Medical and Sports Sciences. Med Sport Sci. Basel, Karger, 1992, vol 37, pp 171–192

Decreased αB-Crystallin in Soleus Muscle Atrophy and Role of αB-Crystallin in Muscle

Yoriko Atomi

Department of Sports Sciences, College of Arts and Sciences, University of Tokyo, Komaba, Tokyo, Japan

What Is αB-Crystallin?

α-Crystallin in the vertebrate lens is a complex of two subunits, the A and B chains, which form large, stable aggregates. The B chain (αB-crystallin) shares strong sequence homology with the small heat shock proteins of *Drosophila* [1] and other species [2]. In addition, α-crystallin shows sequence homology with an egg antigen (p40) of the blood fluke *Schistosoma mansoni* [3] and binds to trypsin, inhibiting its proteolytic activity [4]. Iwaki et al. [5], Bhat and Nagineni [6], Durbin et al. [7] and we [8, 9] have shown that αB-crystallin is expressed in heart and skeletal muscle. This subunit transiently expresses as stress protein [5, 10] and is also seen in human [5] and sheep [11] brains in some pathological conditions. However, no functional or localization study has yet been reported for αB-crystallin in muscle. In this report, we describe that αB-crystallin is expressed in skeletal muscle, decreases in slow muscle atrophy, distributes in a fiber-type-specific manner and partitions at the Z band in myofibrils.

Soleus Muscle and Model of Atrophy by Hindlimb Suspension

Hindlimb plantar flexor muscles are composed of the gastrocnemius, soleus and some resting deep small muscles. The gastrocnemius muscle is a 'mixed' muscle, whereas the soleus muscle contains predominantly slow-twitch fibers [12] which have a high oxidative capacity, not only in mammalians but also in humans, according to Gollnick et al. [12]; the soleus greatly develops in humans, composing up to 40% of the total plantar flexor

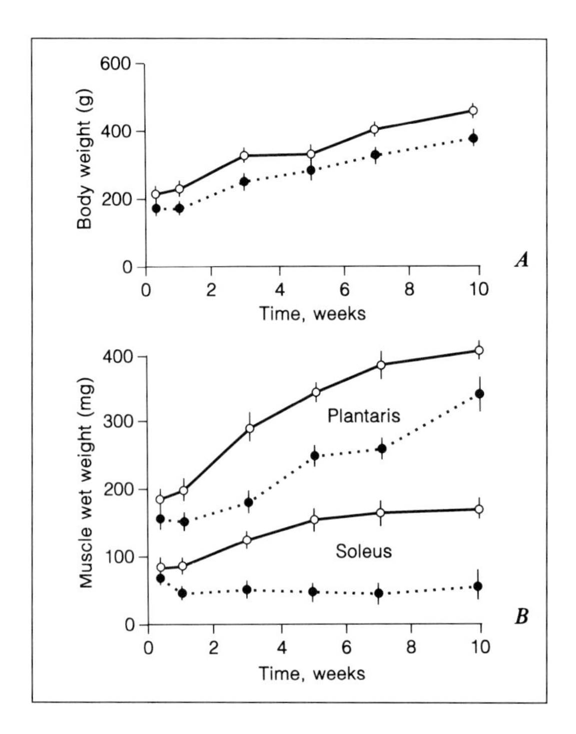

Fig. 1. Changes in mean (± SE) body weight (*A*) and weights of soleus and plantaris muscles (*B*) throughout a 10-week hindlimb suspension experiment. ○ = Control; ● = suspended.

to 56% of the triceps surae muscle in wet weight in cadavers [13]. In our data determined in 18 cadavers, the soleus comprised 50–68% of the triceps surae muscle [14]. The relative development of these leg muscles, especially in children, is related to a higher lactate threshold during running [15], whereas those in rats with four legs occupy only 0.6%. The soleus muscle plays an important role for the physical activity in standing posture in humans. The limb musculature in animals, especially slow muscles like the soleus muscle that functions to maintain posture and counteract the gravitational force on earth, undergoes a significant reduction in mass during space flight or during an unweighted condition, such as hindlimb suspension. Removing passive tension on the hindlimb plantar flexor muscle through the hindlimb suspension method (resulting in a shortening state) readily induces atrophy in the slow soleus but not in the fast muscles of the limb [see reviews in ref. 16 and 17]. In all experiments, adult Wistar rats were used. Muscle atrophy was induced by a standard hindlimb suspension method [18], modified by Atomi et al. [9, 19, 20], known to produce rapid and reproducible levels of muscle atrophy [16, 21]. In our first hindlimb suspension experiment for 10 weeks, the mean weight of the slow soleus muscle declined rapidly by 40% at the end of week 1 (fig. 1). By the 3rd week, muscle weight declined to 51% of control values and remained at this reduced level during the subsequent 10 weeks of the suspension procedure. The fast plantaris muscle of the atrophied limb showed a 27% decline during the 1st week, but soon recovered by 14% before week 3. This fast muscle then continued to grow throughout the 10-week experimental period so that at the end it showed an average 50% gain in weight. Total body weight of the animal did not show radical differences when compared to control (unsuspended) animals, but the former weighed roughly 10% less than controls after 10 weeks. According to Darr and Shultz [22], this decrease in muscle mass by hindlimb suspension, associated with a reduction in number and proliferation of satellite cells, persisted for 30 days in young rats from 20 to 50 days of age. These observations show that weight-supporting activity is an essential factor to maintain normal growth of soleus muscle.

A 22-kDa Specific Protein Component Is Decreased in Slow Muscle Atrophy

The relative disappearance of myofibrillar proteins during disuse atrophy has been described by many studies including the ones of Guba et al. [23], Templeton et al. [24] and Thomason et al. [17, 25] on slow myosin, troponin I and C, and tropomyosin in soleus muscle. Reports on changes in the muscle soluble protein compartment include that of Takacs et al. [26], which showed a decrease in 20- to 30-kDa proteins after 5 days immobilization of rabbit limbs. Howards et al. [27] reported a reduction of 15- to 25-kDa proteins using in vitro protein synthesis systems and mRNA preparations from atrophied muscles. Until now, there has been no study that attempted to describe the unique changes in muscle proteins or mechanisms which might help explain the early events leading to loss of muscle mass during disuse atrophy.

Experimental and control soleus muscles were taken at different times, and total muscle homogenates were subjected to SDS-PAGE [28] and densitometric analysis at the 1st week of hindlimb suspension (fig. 2). One of the first clear differences seen in this comparison was the decrease

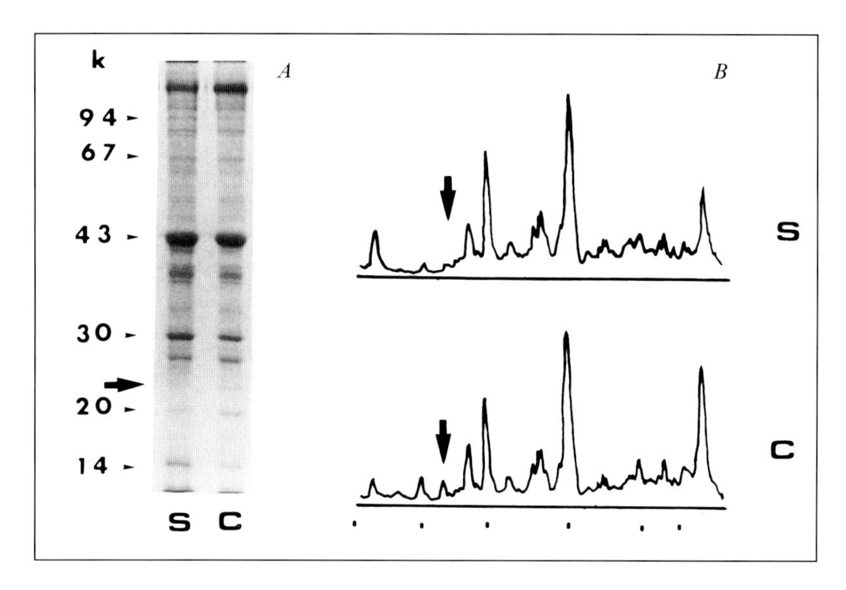

Fig. 2. SDS-PAGE (*A*) and densitometric analysis (*B*) of total muscle homogenate from atrophied rat soleus muscle from limbs suspended for 1 week (S) and from controls (C). Arrows point to the 22-kDa protein. Molecular-weight standards (Pharmacia) are given in kilodaltons (k).

in a 22-kDa protein band (arrow) in experimental muscle [29]. This 22-kDa protein was extracted into the soluble fraction of the muscle homogenate. This unique early decrease was detectable by day 3 and continued up to week 10 (fig. 3). No changes were observed in other components of this soluble protein compartment in the 1st week, although by week 3 changes in the 30- to 70-kDa range could be seen. In fast plantaris and extensor digitorum longus (EDL) muscle the 22-kDa protein was present, but barely detectable.

Passive Stretch Prevents Decrease of the 22-kDa Protein in Soleus Muscle and Induces Its Appearance in Denervated Fast Muscle with Hindlimb Suspension

Continuous mechanical stretching or lengthening of muscle is reported to increase muscle mass, while shortening of muscle length results in augmentation of muscle atrophy [30]. On the other hand, denervation of

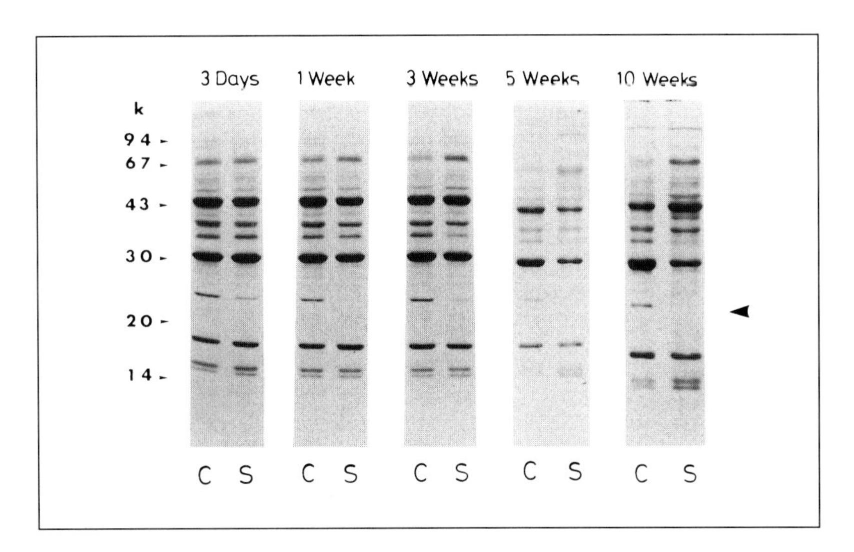

Fig. 3. SDS-PAGE of soluble proteins prepared from atrophied rat soleus muscle from limbs suspended for 10 weeks (S) and from controls (C). Coomassie blue staining of soluble fractions (30 μg) of total homogenate of rat soleus muscle is shown. Arrowhead points to the 22-kDa protein. Molecular-weight standards (Pharmacia) are given in kilodaltons (k).

skeletal muscle is known to aggravate muscle atrophy and cross-innervation between slow and fast muscle to reverse physiological and biochemical characteristics dependent on fiber type [31]. Therefore, we further examined the effects of continuous passive stretch and denervation of both slow and fast muscles under hindlimb suspension on the expression of the 22-kDa protein in skeletal muscle. Innervated or denervated animals under hindlimb suspension were restrained in a fully dorsiflexed (stretched state) or plantarflexed (shortened state) position with tape [9]. After 2 weeks experimentation, the soleus and plantaris muscles were removed from the suspended group and control group with no treatment. The mean muscle weights of each experimental group were compared with each other (fig. 4). Passive stretch showed significant protective effects on muscle atrophy in both muscles and under both suspended and denervated/suspended conditions. The soluble proteins separated from total homogenates of these muscles were analyzed by means of SDS-PAGE. In the suspended/innervated soleus muscles, the 22-kDa/αB-crystallin slightly increased in the innervated and stretched state compared with that in the

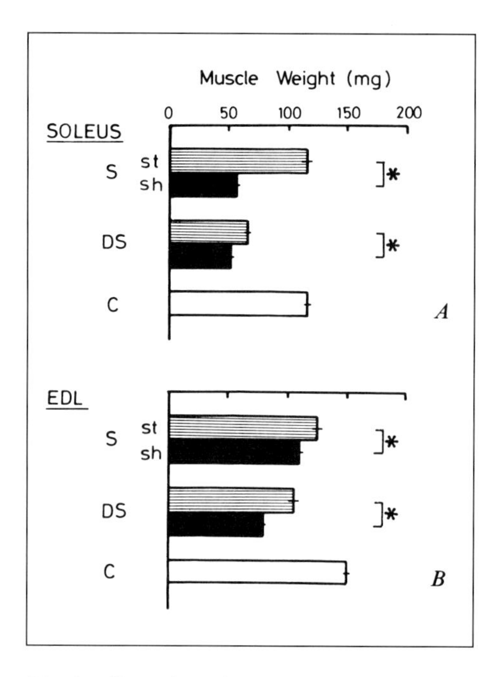

Fig. 4. Effect of passive stretching on mean muscle weight of suspended and suspended/denervated muscles. Soleus (*A*) and EDL (*B*) muscles were taken from control (C), suspended (S) and denervated/suspended (DS) animals. st = Stretched muscle; sh = shortened muscle. * $p < 0.05$.

control muscle, while it decreased in the shortened state and its content was the same as that of only suspended muscles, as shown in figure 5. Next, in the suspended/denervated soleus muscle, the 22-kDa protein disappeared completely in the shortened state, while it remained in the stretched state. In the suspended EDL muscle, in which the 22-kDa protein was scarcely expressed in the control group, the 22-kDa protein clearly appeared only in the denervated and stretched states. In the other conditions the 22-kDa protein was rarely expressed as in the control. These results suggest that passive stretching increases the expression of the 22-kDa protein in both slow and fast muscle, and especially in denervated fast skeletal muscles even under suspended conditions. It appears that the innervation in slow muscle facilitates the expression of the 22-kDa protein while the innervation in fast muscle inhibits the expression of the muscle 22-kDa protein because the 22-kDa protein appears only in the denervated state.

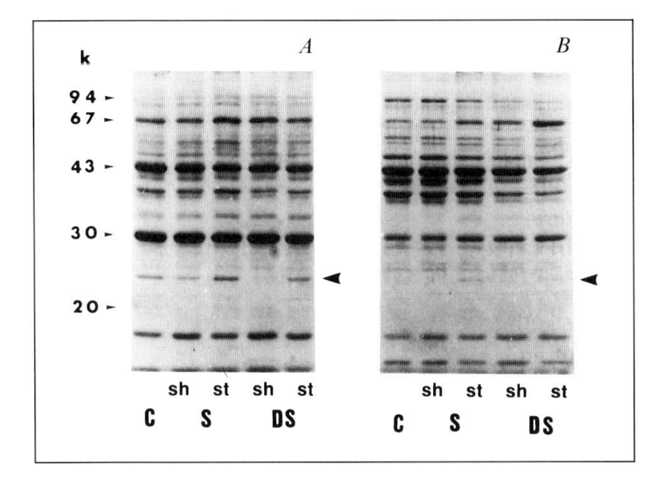

Fig. 5. Effects of passive stretching on the 22-kDa protein in suspended and suspended/denervated muscles. Coomassie brilliant blue staining of soluble proteins (30 μg) of total homogenates followed by immunoblotting of control (C), suspended (S) and denervated/suspended (DS) rat soleus (*A*) and EDL (*B*) muscles is shown. st = Stretched muscle; sh = shortened muscle. Arrowheads point to the 22-kDa protein. Molecular-weight standards (Pharmacia) are given in kilodaltons (k). [See ref.6.]

Taking this into consideration, since changes in the 22-kDa protein with hindlimb suspension well reflected soleus muscle atrophy and it dynamically responded to passive stretch, we continued our study of muscle atrophy by analyzing the 22-kDa protein, that is, by purification and identification of this protein in order to know what it is, and by preparation of an antibody to analyze its localization in muscle fiber.

Purification of the 22-kDa Protein and Preparation of the Specific Antibody

The 22-kDa protein was extracted from muscle homogenates by a low-ionic-strength solution and purified to homogeneity after three steps of gel filtration, DEAE-Sepharose and S-Sepharose column chromatography [20], as described in figure 6. Native 22-kDa protein forms large aggregates and was therefore equilibrated with 7 *M* urea and 0.1% Triton X-100 by dialysis. The amino acid composition of the purified 22-kDa protein was

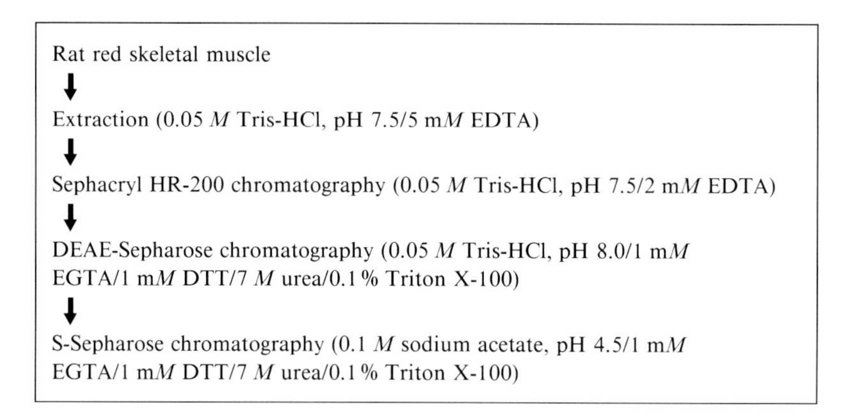

Rat red skeletal muscle

↓

Extraction (0.05 *M* Tris-HCl, pH 7.5/5 m*M* EDTA)

↓

Sephacryl HR-200 chromatography (0.05 *M* Tris-HCl, pH 7.5/2 m*M* EDTA)

↓

DEAE-Sepharose chromatography (0.05 *M* Tris-HCl, pH 8.0/1 m*M* EGTA/1 m*M* DTT/7 *M* urea/0.1% Triton X-100)

↓

S-Sepharose chromatography (0.1 *M* sodium acetate, pH 4.5/1 m*M* EGTA/1 m*M* DTT/7 *M* urea/0.1% Triton X-100)

Fig. 6. Procedure for purification of the 22-kDa protein. DTT = Dithiothreitol.

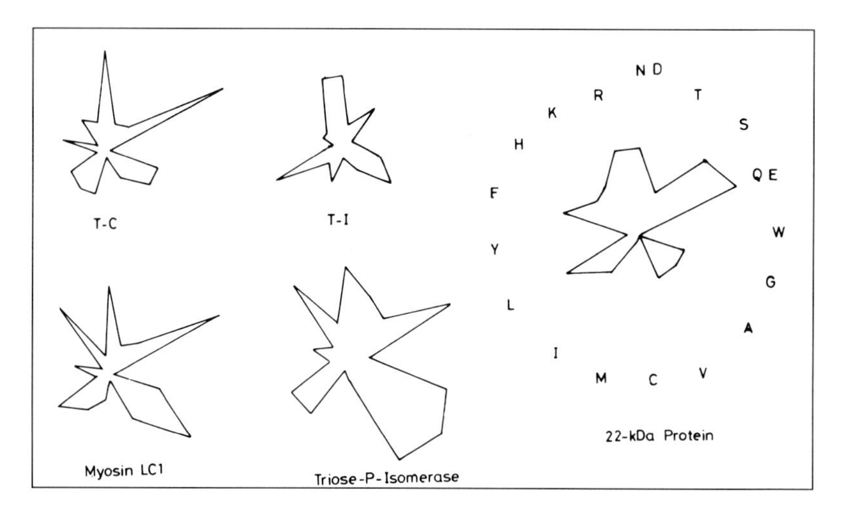

Fig. 7. Amino acid star diagram of the 22-kDa protein and other muscle proteins of the same molecular weight. T-C = Troponin C; T-I = troponin I; LC1 = light chain 1.

compared with that of other muscle proteins having similar molecular weight and being expressed in muscle, such as myosin light chains, triose-phosphate isomerase, troponin I and troponin C, by means of an amino acid star diagram (fig. 7). We could not identify the 22-kDa protein with any of them.

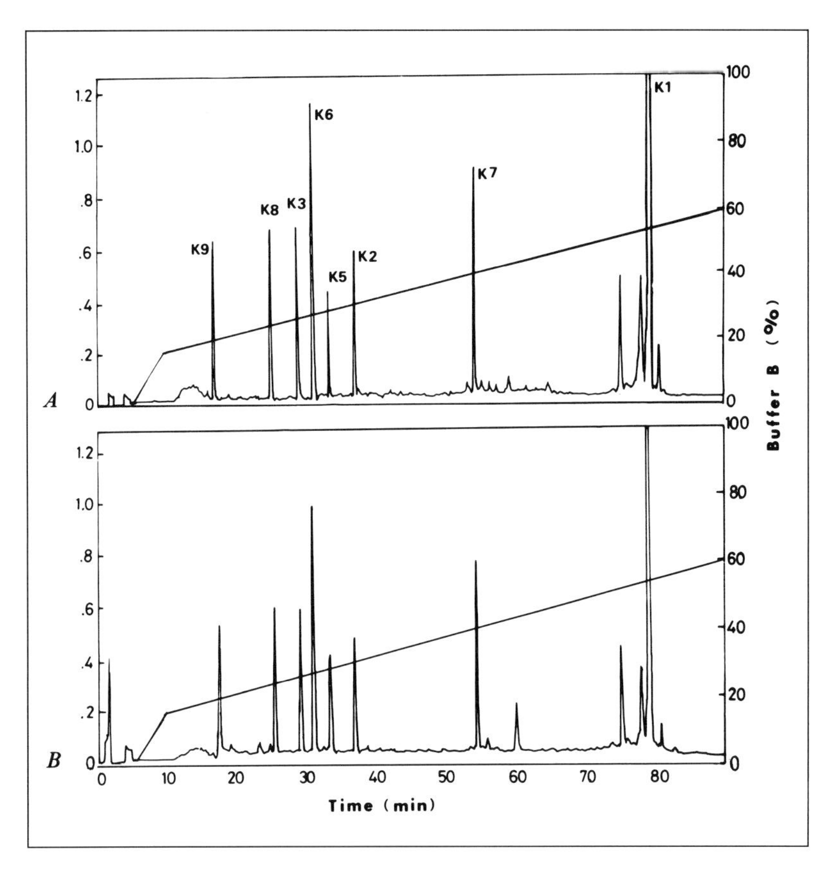

Fig. 8. HPLC elution patterns of purified rat muscle and lens 22-kDa protein cleaved by lysylendopeptidase. *A* 22-kDa protein purified from rat red muscle; *B* 22-kDa protein purified from rat lens. 10 peaks were obtained by reverse-phase column HPLC with 0.1% trifluoroacetic acid and two-step linear gradients of acetonitrile [20].

22-kDa Protein Is αB-Crystallin

We decided to determine the amino acid sequence of the 22-kDa protein. The N terminus was blocked and the protein was cleaved using lysylendopeptidase. Cleavage products were purified by reverse-phase HPLC. As shown in figure 8A, 10 peaks were obtained and sequencing was begun with the material in peak 7. 21 of 22 amino acids in this peptide showed a sequence identical to that found in bovine αB-crystallin [32] and in ham-

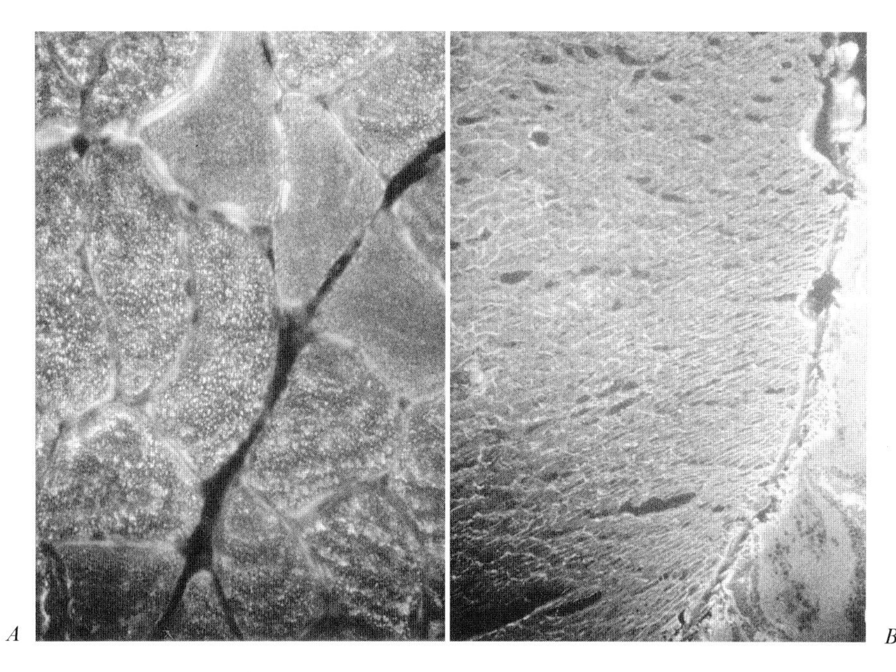

Fig. 9. Immunohistochemical staining of 22-kDa protein/αB-crystallin in cross-sections of rat soleus muscle and eye lens. Rat soleus muscle (*A*) and lens (*B*) were dissected out and immediately frozen in isopentane cooled with liquid nitrogen. These 10 μm cross-sections were stained by an indirect method using affinity-purified anti-muscle 22-kDa protein antibody and rabbit FITC-conjugated IgG.

ster lens αB-crystallin. The HPLC peak patterns of the 22-kDa protein and rat lens αB-crystallin were digested under the same conditions (fig. 8B) and were completely identical. A polyclonal antibody against muscle 22-kDa protein (staining specifically the 22-kDa protein band in SDS-PAGE as shown in fig. 5) stained both cross-sections of rat soleus muscle and eye lens (fig. 9). Of the total 115 amino acids sequenced from our 22-kDa slow muscle protein there was a homology of 95% compared with bovine and

---

Fig. 10. DNA sequence and deduced amino acid sequence of the rat αB-crystallin gene [33]. The fragments separated by lysylendopeptidase are indicated from k1 to k10 and those separated by aspartic proteinase from d1 to d5. The DNA sequence of rat αB-crystallin gene was compared with that of a hamster sequence [42]. Those amino acids determined by peptidase sequencing are underlined under the corresponding amino acid sequence.

```
                d1 | d2                                                                                                  | d3
Rat      GCCATC ATG |GAC ATA GCC ATC CAC CAC CCC TGG ATC CFF GGT CCC TTC TTT CCT TTC CAC TCC CCA AGC CGC CTC TTT |GAC CAG TTC TTC    90
                Met |Asp Ile Ala Ile His His Pro Thr Ile Arg Arg Pro Phe Phe Pro Phe His Ser Pro Ser Arg Leu Phe |Asp Gln Phe Phe    28
Hamster  ----C-  --- |--- --C --- --- --- --- --- --- --- --- --- --T --- --- --- --- --- --C --- --- --- --- --- |--- --- --- ---
                --- |--- --- --- --- --- --- --- --- --- --- --- --- --- --- --- --- --- --- --- --- --- --- --- |--- --- --- ---

                                    d4
         GGA GAG CAC CTG TTG GAG TCT |GAC CTC TTT TCT ACA GCC ACT TCC CTG AGC CCC TTC TAC CTT CGG CCA CCC TCC TTC CTG CGG GCA CCT   180
         Gly Glu His Leu Leu Glu Ser |Asp Leu Phe Ser Thr Ala Thr Ser Leu Ser Pro Phe Thy Leu Arg Pro Pro Ser Phe Leu Arg Ala Pro    58
         --- --- --- --- --- --- --- |--- --- --C --A --T --- --- --T --- --T --- --- --- --G --- --- --T --- --- --T --- --- --C
         --- --- --- --- --- --- --- |--- --- --- --- --- --- --- --- --- --- --- --- --- --- --- --- --- --- --- --- --- --- ---

                     | d5                                               k1 |                                       k2 |
         AGC TGG ATT |GAC ACT GGG CTC TCA GAG ATG CGT ATG GAG AAG |GAC AGG TTC TCT GTC AAC CTG GAC GTG AAG |CAC TTC TCT CCA GAG GAA   270
         Ser Trp Ile |Asp Thr Gly Leu Ser Glu Met Arg Met Glu Lys |Asp Arg Phe Ser Val Asn Leu Asp Val Lys |His Phe Ser Pro Glu Glu    88
         --- --- --- |--- --- --A --- --- --- --- --G --- --- --- |--- --A --- --C --C --- --- --T --- --- |--- --- --C --G --A --G
         --- --- --- |--- --- --- --- --- --- --- --- --- --- --- |--- --- --- --- --- --- --- --- --- --- |--- --- --- --- --- ---

               k3 |    k4 |                                       k5 |
         CTC AAA |GTC AAG |GTT CTG GGA GAC GTG ATT GAG GTG CAC GGC AAG |CAC GAA GAG CGC CAG GAC GAA CAT GGC TTC ATC TCC AGG GAG TTC   360
         Leu Lys |Val Lys |Val Leu Gly Asp Val Ile Glu Val His Gly Lys |His Glu Glu Arg Gln Asp Glu His Gly Phe Ile Ser Arg Glu Phe   118
         --G --- |--- --- |--G --- --G --- --- G-- --A --- --T --- --- |--- --- --- --- --- --- --- --C --- --- --- --T --- --- ---
         --- --- |--- --- |--- --- --- --- --- Val --- --- --- --- --- |--- --- --- --- --- --- --- --- --- --- --- --- --- --- ---

                   k6 |
         CAC AGG AAG |TAC CGG ATC CCA GCC GAC GTG GAT CCT CTC ACC ATT ACT TCT TCCCTG TCA TCG GAT GGA GTC CTC ACT GTG AAT GGA CCA  450
         His Arg Lys |Tyr Arg Ile Pro Ala Asp Val Asp Pro Leu Thr Ile Thr Ser Ser Leu Ser Ser Asp Gly Val Leu Thr Val Asn Gly Pro  148
         --T --- --- |--- --- --- --- --T --T --- --- --- --G --- --- --- --A --- --- --- --T --C --C --- --- --- --- --- --- ---
         --- --- --- |--- --- --- --- --- --- --- --- --- --- --- --- --- --- --- --- --- --- --- --- --- --- --- --- --- --- ---

               k7 |                                                    k8 |                                k9 |k10|
         AGG AAA |CAG GCC TCT GGC CCT GAG CGC ACC ATT CCC ATC ACC CGT GAA GAG AAG |CCT GCT GTC ACT GCA GCC CCT AAG |AAG|TAGATTCCCT T   542
         Arg Lys |Gln Ala Ser Gly Pro Glu Arg Thr Ile Pro Ile Thr Arg Glu Glu Lys |Pro Ala Val Thr Ala Ala Pro Lys |Lys|***            175
         --- --- |--- --- --- --- --C --- --T --- --- --- --- --- --- --- --- --- |--- --- --- --- --- --- --- --- |---|----C---- -
         --- --- |--- --- --- --- --- --- --- --- --- --- --- --- --- --- --- --- |--- --- --- --- --- --- --- --- |---|

         TC CTCGTTGC ATTTTTTAAG ACAAGGAAGT TTCCCATCAG CGAATGAACA TCTGTGACTA GTGCCGAAGC TTACTAATGC TAAGGGCTGG CCCAGATTAT TAAGCTAA  650
         --T---A---- ---------A -----A---- ---------- T---G---A- ---------- ----T----- ---T------ -------C-- ---------- --------

         TAA AAAATATCGT TCAGCAAC(Poly A)     671
         --- --T--CAT T -------TAGAT(Poly A) 674
```

99% with hamster αB-crystallin. We have also isolated a cDNA from a rat heart library which allowed us to deduce the entire amino acid sequence of rat muscle 22-kDa protein (see below). This entire sequence is illustrated in figure 10. We conclude on the basis of sequence data that the 22-kDa protein purified from rat slow muscle is an αB-crystallin molecule. This fact was also ascertained immunologically as described above.

The hydrophobic N-terminal region of αB-crystallin contains the phosphorylation sites of lens αB-crystallin [34] and is phosphorylated by cAMP [35]. α-Crystallin is a multimeric protein that can bind to trypsin causing trypsin inhibition [4]. The C-terminal region of αB-crystallin includes a sequence similar to that of *Drosophila* small heat shock proteins [1]. In fact, the promoter of the αB-crystallin gene fused to the bacterial chloramphenicol acetyltransferase gene was shown to confer heat inducibility on this reporter gene in transient assays with NIH 3T3 fibroblast cells [36]. The correlation between soleus muscle plasticity and heat inducibility or phosphorylation of αB-crystallin by cAMP is unknown. When we consider many of the changes inherent in skeletal and heart muscle plasticity [see ref. 37], it is obvious that this heat shock sequence and its regulation in muscle will require close attention in the future.

Protein Synthesis of αB-Crystallin Decreases in Atrophied Muscle

The disappearance of the 22-kDa αB-crystallin protein from atrophied soleus muscle, ascertained by immunoblotting (fig. 11A), was analyzed from the point of view of mechanism and whether or not transcriptional regulation was involved [33]. Control and experimental soleus muscles were removed from animals at specified times (after 24 h, and 1 and 3 weeks suspension with and without passive stretching), prepared for in situ labeling with [^{35}S]-methionine and analyzed by two-dimensional PAGE [39] and fluorography as described above. Accumulation and synthesis of the 22-kDa/αB-crystallin in atrophied muscles appear to be reduced to roughly half of control values by 1 week (fig. 11B), and we could detect

Fig. 11. Decreases in protein content as revealed by immunoblotting (*A*), protein synthesis (*B*) and mRNA level (*C*) of αB-crystallin in soleus muscle atrophy. *A* Protein content in homogenates from control soleus muscle (lane 1) and atrophied soleus muscle after 1 week of suspension (lane 2). *a* Coomassie blue staining of one-dimensional SDS-

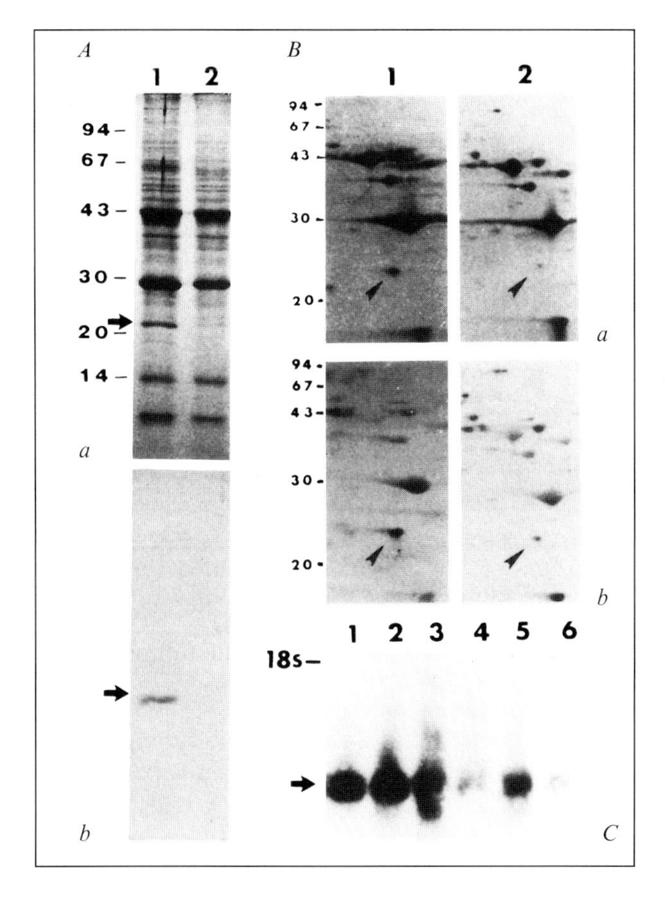

PAGE. 30 μg of soluble protein was applied for SDS-PAGE. *b* Immunoblotting with anti-22-kDa protein/αB-crystallin antibody. Arrows point to the 22-kDa protein/αB-crystallin. *B* Accumulation and protein synthesis in homogenates from control soleus muscle (lane 1) and atrophied soleus muscle after 1 week of suspension (lane 2). *a* Coomassie blue staining of two-dimensional SDS-PAGE. *b* Fluorography. About 20 mg muscle fiber bundles were dissected out and incubated in DMEM containing [^{35}S]-methionine for 2 h at 37 °C in a 5% CO_2 incubator. Supernatants of muscle homogenates were analyzed by two-dimensional PAGE and fluorography. Arrowheads indicate the αB-crystallin. Molecular-weight standards (Pharmacia) in *A* and *B* are given in kilodaltons (k). *C* Northern blotting of mRNA prepared from control and suspended muscles. Control (lanes 1, 3 and 5) and experimental (lanes 2, 4 and 6) muscles were examined after 24 h (lanes 1 and 2), 3 days (lanes 3 and 4) and 1 week (lanes 5 and 6) of suspension. 18s shows the position of ribosome RNA, and the arrow points to the position of αB-crystallin mRNA. Muscle total RNA was extracted by the AGCT method [38]. 20 μg total RNA from each sample was electrophoresed on agarose gel, transferred to Hybond N^+ and hybridized with labeled αB-crystallin cDNA probe.

changes in this direction as early as at 24 h of experimental treatment. Passive stretching under hindlimb suspension increased overall protein synthesis including αB-crystallin more than in control muscle, but the effect decreased gradually with experimental time (data not shown). These results demonstrated that at least a part of the decrease in αB-crystallin content in atrophied muscle was due to the decrease in protein synthesis.

cDNA Cloning of Rat Heart αB-Crystallin

In order to study whether protein synthesis of αB-crystallin in atrophied soleus was regulated at the mRNA level or not and to determine the complete amino acid sequence of the encoded protein, we obtained an αB-crystallin cDNA clone from a rat heart cDNA library. A 50-mer oligonucleotide probe derived from the C-terminal amino acid sequence of hamster lens αB-crystallin was synthesized and used to screen the library. 72,000 clones were examined [40], yielding 20 isolated clones that had hybridized to the oligonucleotide probe. One such clone containing a 0.7-kb insert was subcloned into pUC118. The DNA sequence of this insert was 671 bp. The deduced amino acid sequence from the cDNA contained an exact match to the 115-amino-acid sequence of the 22-kDa protein of rat skeletal muscle partially determined by us and shown in figure 10. This sequence also agreed with the amino acid sequence of rat lens αB-crystallin [41]. Alignment of the rat heart αB-crystallin cDNA with the genomic sequence of hamster αB-crystallin [42] revealed a high degree of homology (93%), with the coding region displaying a 92% homology while the 3′ noncoding region displayed an 88% DNA sequence homology.

Early Decrease of αB-Crystallin mRNA in Atrophied Soleus Muscle

αB-crystallin mRNA levels were examined in atrophied and control muscles at the times given above for measuring changes in protein accumulation and synthesis. RNA from the muscles was probed with our [^{32}P]-cDNA encoding αB-crystallin. No decrease in this RNA was detected at 24 h, but by 3 days there was a marked reduction in αB-crystallin mRNA and this continued til 1 week of suspension (fig. 11C). The disappearance of the 22-kDa αB-crystallin protein involves, therefore, an inhibition at the mRNA level. We have not yet measured changes that might also be occurring at the level of protein turnover.

Localization of αB-Crystallin in Myofibrils

When frozen muscles were homogenized in low-salt buffer and homogenates fractionated into soluble and pelleted material, the latter containing myofibrils, the 22-kDa protein appeared in the soluble fraction (fig. 3). Since αB-crystallin exists in the lens as a fibrous component, it is possible that the 22-kDa muscle αB-crystallin also exists as a part of the myofibrillar structure. In order to investigate this possibility, we prepared myofibrils in rigor buffer from glycerinated muscle [43].

When glycerinated muscle was prepared for myofibril isolation, we first homogenized the muscle and washed pelleted myofibrils. All supernatants from these washes were assayed by extraction for αB-crystallin. When homogenized in rigor buffer, the 22-kDa αB-crystallin was present in the initial supernatant (not shown) and was also detected associated with the myofibrillar pellet (fig. 12: lanes 1, 2). Furthermore, the 22-kDa protein was not coextracted with myosin-solubilizing buffers but remained with insoluble material which included disorganized networks of I-Z-I band material (fig. 12: lanes 7–10; fig. 13). 22-kDa αB-crystallin, while it does exist in a soluble pool, may also appear specifically associated with myofibrillar Z bands under defined conditions.

To further investigate this association, we carried out an immunocytochemical analysis using an affinity-purified αB-crystallin antibody and an immunogold assay at the electron-microscopic level [20]. In standard immunocytochemical studies using isolated myofibrils and fluorescent microscopy, the antibody showed a precise striated pattern associated with Z bands (fig. 14A, B).

Possible Role of αB-Crystallin in Muscle

We have now isolated and determined the amino acid sequence of the 22-kDa protein which specifically decreases in atrophied soleus, and have identified it as being equivalent to lens αB-crystallin by direct amino acid sequencing and DNA sequencing. In addition, we have shown that the expression of αB-crystallin mRNA is inhibited during muscle atrophy. The presence of αB-crystallin has recently been detected in striated muscle by Bhat and Nagineni [6], Iwaki et al. [5], Durbin et al. [7] and us [8]. However, none of these previous reports on αB-crystallin in skeletal muscle provides any further details regarding the structural or functional properties of this protein in muscle.

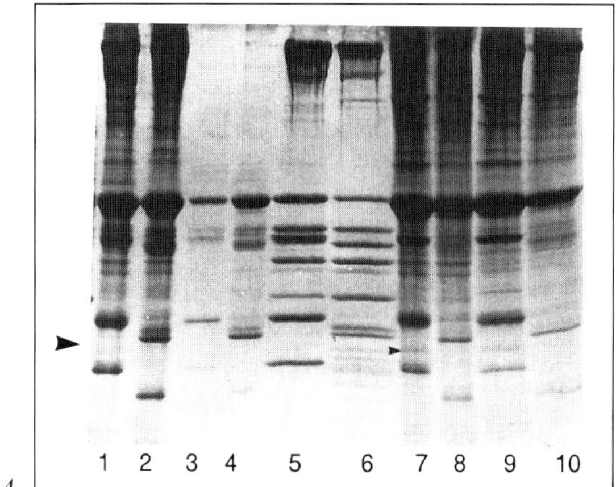

12A

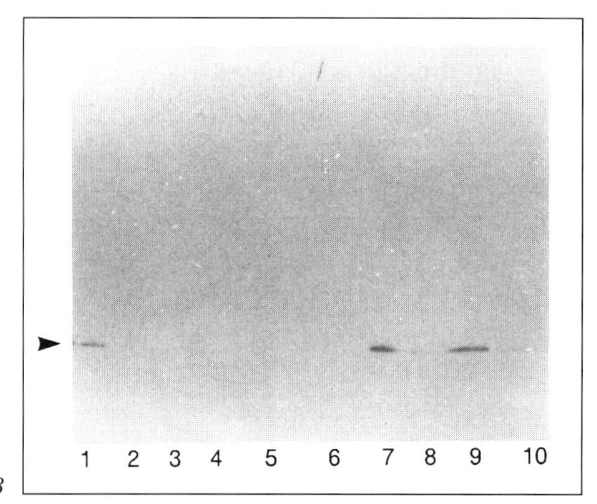

12B

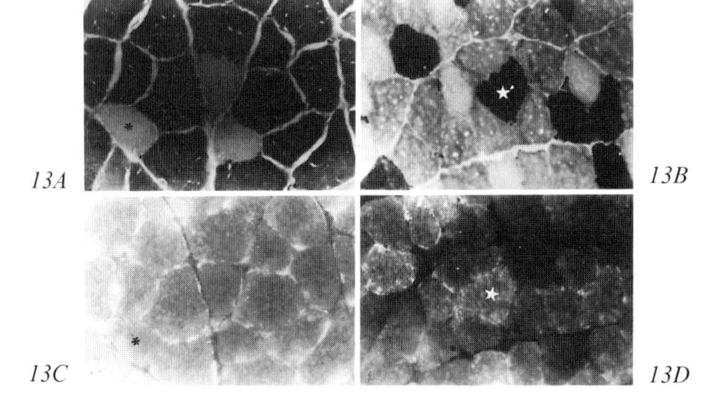

13A 13B 13C 13D

24 Templeton GH, Padalino M, Manton J, Glasberg M, Silver CJ, Silver P, DeMartino G, Leconey T, Klug G, Hagler H, Sutko JL: Influence of suspension hypokinesia during hindlimb suspension and recovery. J Appl Physiol 1984;63:130–137.
25 Thomason DB, Herrick RE, Surgyka D, Baldwin M: Time course of soleus muscle myosin expression during hindlimb suspension and recovery. J Appl Physiol 1987; 63:130–137.
26 Takacs O, Sohar II, Szilagyi T, Guba F: Experimental investigations on hypokinesis of skeletal muscles with different function. IV. Changes in the sarcoplasmic proteins. Acta Biol Sci Hung 1977;28:221–230.
27 Howards G, Steffen JM, Geoghegan TE: Transcriptional regulation of decreased protein synthesis during skeletal muscle unloading. J Appl Physiol 1989;66:1093–1098.
28 Laemmli UK: Cleavage of structural proteins during the assembly of the head of bacteriophage T4. Nature 1970;227:680–685.
29 Atomi Y, Yamada S, Hatta H, Yamamoto Y: Mechanism of muscle atrophy. IV. Changes of muscle protein contents and components with muscle atrophy induced by the tail suspension of rats. Proc Dept Sports Sci Univ Tokyo 1987;21:25–30.
30 Goldspink DF: Development and Specialization of Skeletal Muscle. London, Cambridge University Press, 1980, vol 7, pp 65–89.
31 Close RI: Dynamic properties of mammalian skeletal muscles. Physiol Rev 1972;52: 129–197.
32 Van der Ouderaa FJ, De Jong W, Hilderink A, Bloemendal H: The amino-acid sequence of the αB_2 chain of bovine α-crystallin. Eur J Biochem 1974;49:157–168.
33 Atomi Y, Yamada S, Nishida T: Early changes of αB-crystallin mRNA in rat skeletal muscle due to mechanical tension and denervation. Biochem Biophys Res Commun, in press.
34 Chiesa R, Gawinowicz-Kolks MA, Kleiman NJ, Spector A: Definition and comparison of the phosphorylation sites of the A and B chains of bovine α-crystallin. Exp Eye Res 1988;46:199–208.
35 Spector A, Chiesa R, Sredy J, Garner W: cAMP-dependent phosphorylation of bovine lens α-crystallin. Proc Natl Acad Sci USA 1985;82:4712–4716.
36 Klemenz R, Frohli E, Aoyama A, Hoffmann S, Simpson RJ, Moritz R, Schafer R: aB-crystallin accumulation is a specific response to Ha-ras and v-mos oncogene expression in mouse NIH 3T3 fibroblasts. Mol Cell Biol 1991;11:803–812.
37 Pette D (ed): The Dynamic State of Muscle Fibers. Berlin, de Gruyter, 1990.
38 Chomczynoki P, Sacci N: Single-step method of RNA isolation by acid guanidinium thiocyanate-phenol-chloroform extraction. Anal Biochem 1987;162:156–159.
39 O'Farrell PH: High resolution, two-dimensional electrophoresis of proteins. J Biol Chem 1975;250:4007–4021.
40 Benton WD, Davis RW: Screening λgt recombinant clones by hybridization to single plaques in situ. Science 1977;196:180–182.
41 Hendricks W, Weetink H, Voorter CEM, Sanders J, Bloemendal H, de Jong WW: The alternative splicing product αAins-crystallin is structurally equivalent to αA and aB subunits in the rat α-crystallin aggregate. Biochim Biophys Acta 1990;1037:58–65.

42 Quax-Jeuken Quax W, Van Rens G, Khan M, Bloemendal H: Complete structure of the αB-crystallin gene: Conservation of the exon-intron distribution in the two non-linked α-crystallin genes. Proc Natl Acad Sci USA 1985;82:5819–5823.
43 Knight P, Trinick JA: Preparation of myofibrils. Structural and contractile proteins. Part B. The contractile apparatus and cytoskeleton. Methods Enzymol 1985;85:9–12.
44 Schmalbruch H: Skeletal Muscle. Berlin, Springer, 1985.
45 Ellis J: Proteins as molecular chaperons. Nature 1987;328:378–379.
46 Masaki T, Endo M, Ebashi S: Localization of 6S component of α-actinin at Z-band. J Biochem, Tokyo 1967;68:630–633.
47 Ohashi K, Mikawa T, Maruyama K: Localization of Z-protein in isolated Z-disk sheets of chicken leg muscle. J Cell Biol 1982;95:85–90.
48 Chen WYJ, Dhoot GK, Perry SY: Characterization and fibre type distribution of a new myofibrillar protein of molecular weight 32-kDa. J Muscle Res Cell Motil 1986; 7:517–526.
49 Jakubiec-Puka AD, Kulesza-Lipka K, Krajewski K: Z-line in atrophying and regenerating skeletal muscle. Adv Physiol Sci 1981;24:333–339.
50 Hattori A, Takahashi KJ: Calcium-induced weakening of skeletal muscle Z-disks. J Biochem 1982;92:381–390.
51 Reddy JK, Etlinger JD, Rabinowitz J, Fishman DA, Zak R: Removal of Z-line and α-actinin from isolated myofibrils by a calcium-activated neutral protease. J Biol Chem 1975;259:4278–4284.
52 Ashley CA, Porter KR, Philpott DE, Hass GM: Observations by electron microscopy on contraction of skeletal myofibrils induced with adenosine triphosphate. J Exp Med 1951;94:9–19.
53 Ellis S, Nagainis PA: Activity of calcium activated protease in skeletal muscles and its changes in atrophy and stretch. Physiologist 1984;27(suppl):S74–S75.
54 Yoshida H, Murachi T, Tsukahara I: Limited proteolysis of bovine lens α-crystallin by calpain, a Ca^{2+}-dependent cysteine proteinase, isolated from the same tissue. Biochim Biophys Acta 1984;798:252–259.
55 Sohar I, Nagy I, Takacs O, Kovacs Zs, Guba F: Protein degradation processes in disused muscle. Adv Physiol Sci 1980;24:213–220.

Yoriko Atomi, MD, Department of Sports Sciences, College of Arts and Sciences, University of Tokyo, 3-8-1, Komaba, Meguroku, Tokyo 153 (Japan)

Sato Y, Poortmans J, Hashimoto I, Oshida Y (eds): Integration of Medical and Sports Sciences. Med Sport Sci. Basel, Karger, 1992, vol 37, pp 193–200

Biochemical Determination of Training Effects Using Insulin Clamp and Microdialysis Techniques

Yuzo Sato[a], *Yoshiharu Oshida*[a], *Isao Ohsawa*[a], *Juichi Sato*[b], *Kunio Yamanouchi*[c]

[a] Research Center of Health, Physical Fitness and Sports, Nagoya University, Nagoya; [b] Third Department of Internal Medicine, Nagoya University School of Medicine, Nagoya; [c] First Department of Internal Medicine, Aichi Medical University, Aichi, Japan

Regular physical exercise has been known to be beneficial in the treatment of so-called hypokinetic diseases or syndrome X as named by Reaven [1], including diabetes [2], hypertension and coronary artery disease. Continued physical exercise improves reduced peripheral tissue sensitivity to insulin in non-insulin-dependent diabetes mellitus (NIDDM) [3] and obesity [4]. On the other hand, metabolic changes during and after physical exercise are modified by the degree of training of the subjects [2]. Therefore, the necessity for the monitoring of the training effect in the treatment of diabetic patients and obese subjects becomes increasingly mandatory [2, 5]. However, it is difficult to estimate the physical training effects separately from the metabolic changes derived from dietary restriction [2, 3]. Consequently, we have evaluated the effects of physical training in terms of in vivo action of insulin by the multiple euglycemic clamp technique [6] and by the microdialysis procedure [7].

Materials and Methods

Human Study

Eighteen obese NIDDM patients, 22 simply obese subjects, 28 long-distance runners and 34 untrained healthy controls were studied after an overnight fast. Twenty-one of the obese diabetic patients and 6 of the simply obese subjects started and continued a physical exercise program for 1–2 months. They were studied before and after the training program. Each subject gave informed consent. In order to evaluate the effects of physical training in terms of in vivo action of insulin, the multiple euglycemic clamp technique [6] was performed as described previously [5, 8] with a regular dose of 40 $mU/m^2 \cdot min$ (insulin sensitivity – binding defect) or a high dose of 400 $mU/m^2 \cdot min$ (insulin responsiveness – postbinding defect) [9]. Both glucose metabolism (GM; glucose infusion rate) and glucose metabolic clearance rate (MCR) provide a quantitative estimate of insulin action at the whole body or lean body mass level [10].

Animal Study

The effect of physical exercise on peripheral insulin action (sensitivity/responsiveness) was studied in 7 anesthetized male Wistar trained rats and 8 untrained controls using a two-step sequential euglycemic clamp procedure (insulin infusion rate: 6.0 and 30.0 mU/kg·min) in combination with a microdialysis technique [7] in skeletal muscle and adipose tissue [11]. The glucose infusion rate during the euglycemic clamp period and the lactate concentration in dialysate are a quantitative estimate of in vivo insulin action in the whole body or within individual tissues [7, 10, 11].

Data are presented as means ± SE. Statistical analyses were performed with paired and unpaired Student's t test. $p < 0.05$ was considered statistically significant.

Results

Human Study

Under regular-dose conditions, GM (in mg/kg·min; $p < 0.001$; fig. 1) and MCR (in ml/kg·min: $p < 0.001$) directly correlated with maximal oxygen uptake ($\dot{V}O_2max$). MCR in 18 obese diabetic patients (3.1 ± 0.4) and 16 simply obese subjects (4.5 ± 0.2) were significantly ($p < 0.001$) lower than those in 34 healthy controls (9.1 ± 0.6), whose values were, on the other hand, significantly ($p < 0.001$) lower than those in 19 athletes (12.9 ± 0.9). Inverse correlations existed between body mass index and GM ($p < 0.001$), and between body mass index and MCR ($p < 0.001$).

MCR was decreased after rapid weight reduction by fasting for 72 h. After dietary restriction and physical training, body weight was significantly ($p < 0.001$) decreased, while GM (fig. 2) and MCR were significantly ($p < 0.001$) increased. Thus, combined treatment with dietary restriction and physical exercise is effective for both weight reduction and increased insulin sensitivity in decreased insulin action in simply obese and obese diabetic patients [3, 4].

However, if individual data are considered, GM of some patients was improved in comparison with the degree of weight reduction, while GM of other patients was not improved so much after dominant weight reduction

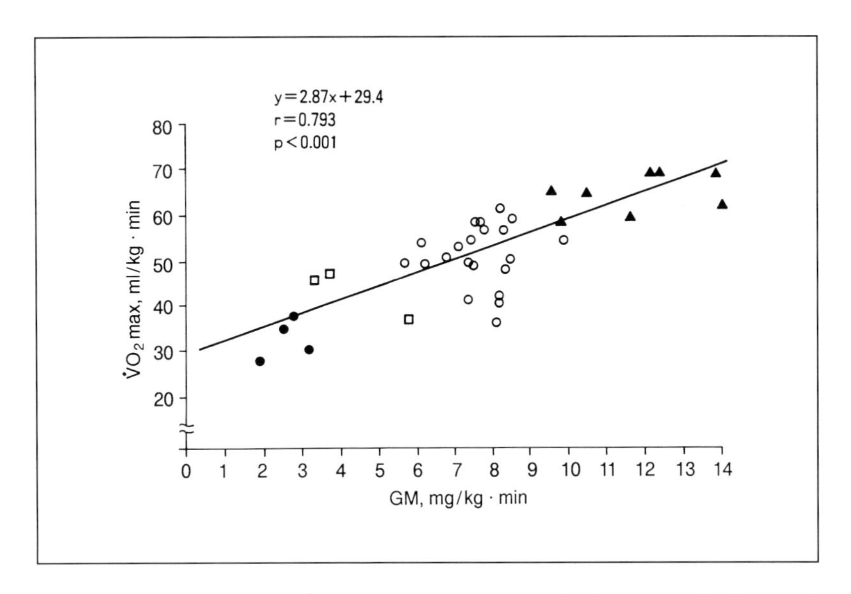

Fig. 1. Correlation between $\dot{V}O_2$max and GM. ▲ = Trained; ○ = untrained normal; □ = overweight; ● = obese.

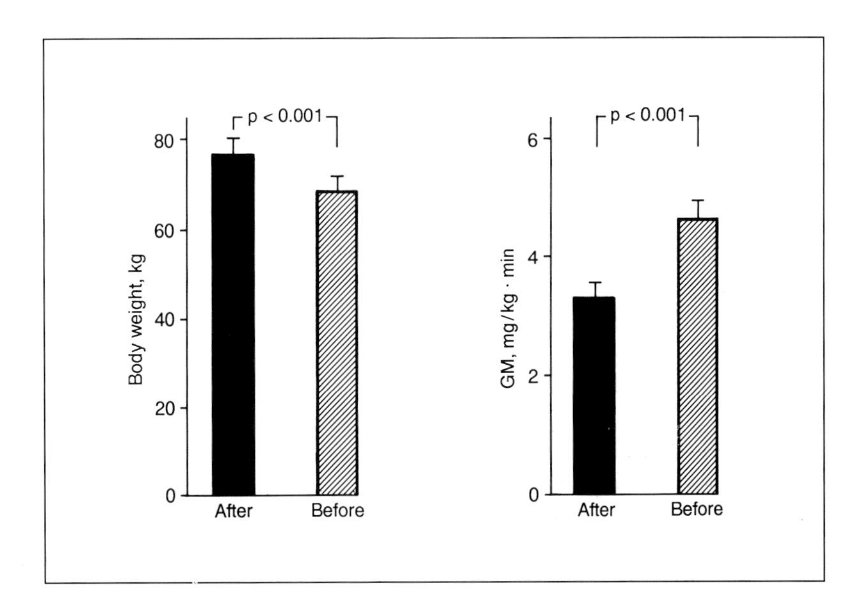

Fig. 2. Changes in GM before and after weight reduction.

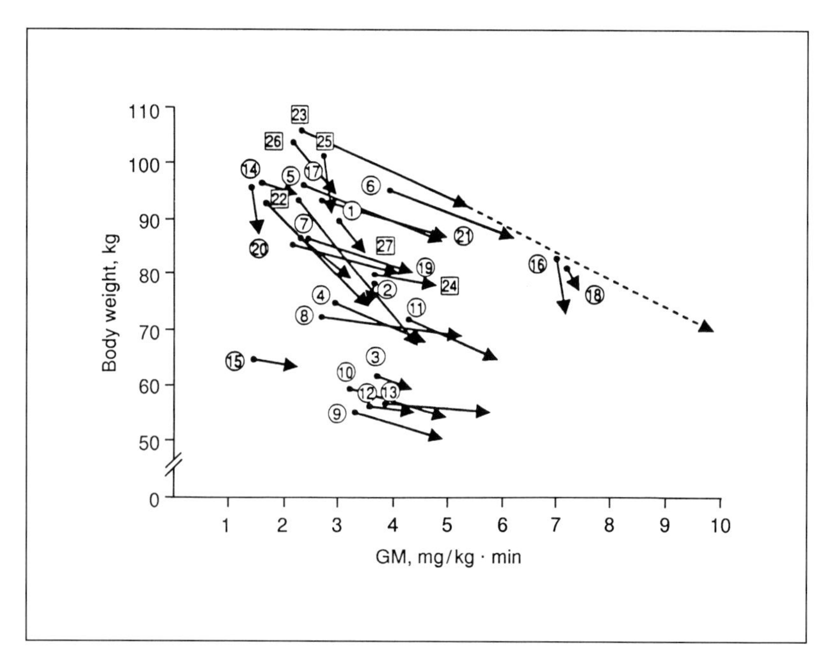

Fig. 3. Individual data of changes in body weight/GM. ○ = Obese diabetics; □ = simply obese subjects.

(fig. 3). Therefore, the improvement in GM was not correlated with the degree of weight reduction in each case. We examined each patient's record carefully. It became clear that the latter patients kept only dietary restriction and did not practise physical exercise. Then, we classified these patients into two groups according to daily walking steps. Insulin sensitivity was significantly increased only in the group with exercise score 2 (10,000–20,000 paces/day), but not in the group with exercise score 1 (less than 10.000 paces/day; fig. 4).

These results suggest that physical exercise is essential to improve insulin action in obese NIDDM and simply obese subjects, and that by using the euglycemic clamp technique physical training effects can be determined separately from dietary factors. Further physical training in addition to dietary restriction provides a reasonable therapeutic means for obese NIDDM and simply obese patients from pathophysiologic viewpoints.

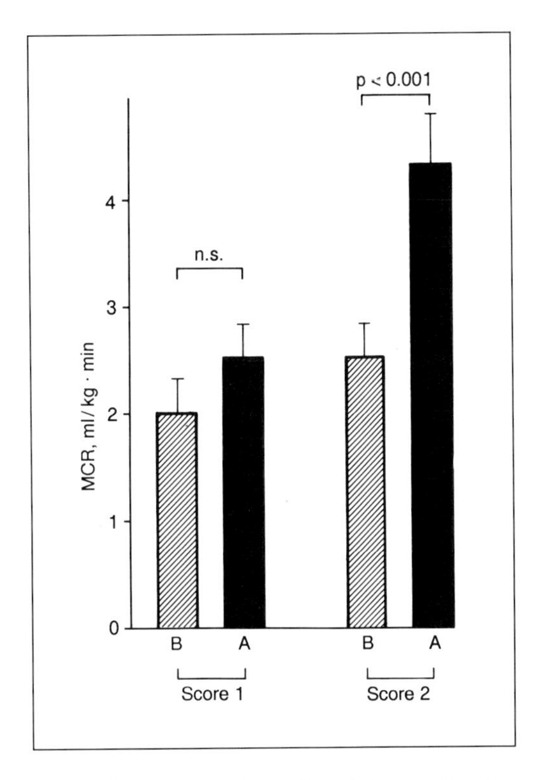

Fig. 4. Changes in MCR before (B) and after (A) weight reduction, classified by exercise score. Score 1 = 5,000–9,999 walking paces/day; score 2 = 10,000–20,000 walking paces/day.

Under high-dose conditions, MCR in simply obese subjects (7.4 ± 0.7) were significantly lower than those in controls, whose values were not lower than those in athletes (16.4 ± 0.6). After physical training and dietary restriction for 1–2 months, MCR was significantly ($p < 0.001$) increased.

Animal Study

At an insulin infusion rate of 6.0 mU/kg·min the glucose infusion rate was significantly higher in trained rats than in untrained controls. However, significant differences between trained and untrained controls disappeared at an inuslin infusion rate of 30 mU/kg·min (fig. 5).

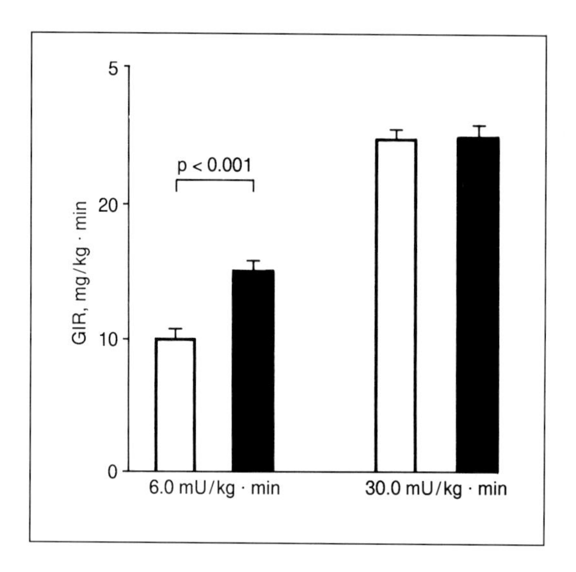

Fig. 5. Glucose infusion rate (GIR) during sequential euglycemic clamp in rats. ■ = Trained; □ = control.

With the microdialysis technique, a leftward shift of the insulin dose-response curve was observed only in skeletal muscle of trained rats, as compared with controls, when the lactate concentration in dialysate was plotted against the logarithm of the mean plasma insulin concentration.

Discussion

Resistance to insulin-stimulated glucose uptake is a common phenomenon and plays a principal role in the pathogenesis and clinical course of several important human diseases, such as NIDDM, hypertension and coronary artery disease, which were named syndrome X by Reaven [1]. He noticed that variations in life-style, in particular avoiding obesity and remaining physically active, provide an approach to minimize the risk factors for coronary artery disease associated with resistance to insulin-stimulated glucose uptake.

In addition to the present results, previous studies from our laboratory showed that long-term mild jogging increases insulin action despite no influence on body mass index or $\dot{V}O_2max$ [12]. Thus, long-term physical training improves insulin action under physiological hyperinsulinemic conditions (70–100 μU/ml). However, the determination of insulin action at a single submaximal insulin concentration does not differentiate increased insulin sensitivity and increased insulin responsiveness [6]. Using the multiple euglycemic technique, several studies concerning insulin action sites [6, 13, 14] have been performed: however, these reports were controversial.

In the present study, a two-step sequential euglycemic procedure in combination with a microdialysis technique suggested that an enhanced insulin action by physical training is due to an increase in insulin sensitivity. However, MCR in simply obese subjects was increased after physical training under both regular-dose and high-dose conditions. Further investigation is required including studies of the glucose transport system [15, 16].

In conclusion, the present study suggests that the multiple euglycemic clamp technique and the microdialysis procedure provide reliable estimates of training effects. Physical training could increase insulin action at the level of binding and postbinding mechanisms in skeletal muscles.

Acknowledgments

The authors would like to thank their colleagues of the Diabetes Research Group of the First Department of Internal Medicine, Aichi Medical University, for assisting in this study, and Miss Y. Hattori and Mrs. Y. Uno for typing the manuscript.

References

1 Reaven GM: Role of insulin resistance in human disease. Diabetes 1988;37:1595–1607.
2 Sato Y: Practical method of physical exercise for diabetic patients. Asian Med J 1988;31:438–444.
3 Horton ED: Exercise and diabetes mellitus. Med Clin North Am 1988;72:1301–1321.
4 Ishiguro T, Sato Y, Oshida Y, Yamanouchi K, Okuyama M, Sakamoto N: The relationship between insulin sensitivity and weight reduction in simple obese and obese diabetic patients. Nagoya J Med Sci 1987;49:61–69.
5 Sato Y, Iguchi A, Sakamoto N: Biochemical determination of training effects using insulin clamp technique. Horm Metab Res 1984;16:483–486.

6 King DS, Dalsky GP, Staten MA, Clutter WE, Van Houton DR, Holloszy JO: Insulin action and secretion in endurance-trained and untrained humans. J Appl Physiol 1987;63:2247–2252.
7 Ungerstedt U: Microdialysis – A new bioanalytical sampling technique. Curr Separations 1986;7:43–46.
8 Sato Y, Hayamizu S, Yamamoto C, Ohkuwa Y, Yamanouchi K, Sakamoto N: Improved insulin sensitivity in carbohydrate and lipid metabolism after physical training. Int J Sports Med 1986;7:307–310.
9 Olefsky JM, Molina JM: Insulin resistance in man; in Rifkin H, Porte D (eds): Diabetes mellitus. Theory and Practice. Amsterdam, Elsevier, 1990, pp 121–153.
10 DeFronzo RA, Tobin JD, Andres R: Glucose clamp technique: A method for quantifying insulin secretion and resistance. Am J Physiol 1979;237:E214–E223.
11 Oshida Y, Ohsawa I, Sato J, Sato Y: In vivo insulin-stimulated glycolysis in skeletal muscle and adipose tissue. Endocrinol Jpn, submitted.
12 Oshida Y, Yamanouchi K, Hayamizu S, Sato Y: Long-term mild jogging increases insulin action despite no influence on body mass index or $\dot{V}O_{2max}$. J Appl Physiol 1989;66:2206–2210.
13 Mikines KJ, Sonne B, Farrell PA, Tronier B, Galbo H: Effect of physical exercise on sensitivity and responsiveness to insulin in humans. Am J Physiol 1988;254:E248–E259.
14 Bourey RE, Coggan AR, Kohrt WM, Kirwan JP, King DS, Holloszy JO: Effects of exercise on glucose disposal: Response to a maximal insulin stimulus. J Appl Physiol 1990;69:1689–1694.
15 Rodnick KJ, Holloszy JO, Mondon CE, James DE: Effects of exercise training on insulin-regulatable glucose-transporter protein levels in rat skeletal muscle. Diabetes 1990;39:1425–1429.
16 Youn JH, Gulve EA, Holloszy JO: Calcium stimulates glucose transport in skeletal muscle by a pathway independent of contraction. Am J Physiol 1991;260:C555–C561.

Yuzo Sato, MD, Research Center of Health, Physical Fitness and Sports,
Nagoya University, Furo-cho, Chikusa-ku, Nagoya 464–01 (Japan)

Sato Y, Poortmans J, Hashimoto I, Oshida Y (eds): Integration of Medical and Sports Sciences. Med Sport Sci. Basel, Karger, 1992, vol 37, pp 201–215

The Glucose Transport System in Skeletal Muscle: Effects of Exercise and Insulin

Laurie J. Goodyear[a], *Michael F. Hirshman*[b], *Edward S. Horton*[b]

[a]Harvard University Medical School, Joslin Research Laboratory, Boston, Mass.; [b]Division of Endocrinology, Metabolism and Nutrition, Department of Medicine, University of Vermont College of Medicine, Burlington, Vt., USA

Introduction

Under normal physiological conditions, glucose transport is the rate limiting step in glucose utilization [1], and thus an important factor in regulating an organism's overall glucose homeostasis. In skeletal muscle, exercise and insulin are the two most potent and physiologically important activators of glucose transport. This paper will review current research that has investigated potential mechanisms by which the glucose transport system is activated by these stimuli.

Skeletal Muscle Glucose Transport

Glucose transport occurs primarily by facilitated diffusion, an energy independent process that uses a carrier protein for transport of a substrate across a membrane [2]. Glucose transport follows saturation kinetics [3–6] and can be inhibited by cytochalasin B [6]. Recent studies have demonstrated that the glucose transporters in mammalian tissues are a family of structurally related proteins that are encoded by distinct genes and are expressed in a tissue-specific manner [7]. They are transmembrane glycoproteins with an average molecular size of approximately 45–55 kDa. To date, 2 transporter isoforms, designated GLUT-1 and GLUT-4, have been shown to be significantly expressed in skeletal muscle tissue [8]. The

GLUT-4 isoform is thought to be the major isoform present in skeletal muscle and is probably responsible for insulin-stimulated glucose transport. The GLUT-1 isoform is expressed at a much lower abundance than GLUT-4 and may be present largely in non-muscle cells.

Both insulin [4–6, 9] and exercise [3, 4, 6, 10, 11] can greatly increase the rate of glucose transport in skeletal muscle. In addition to these major mediators of transport, numerous other factors including catecholamines, growth factors, hypoxia, corticosteroids, thyroid and growth hormones, and glucose can alter skeletal muscle glucose transport. The increase in glucose transport by insulin is dose dependent while exercise increases transport proportional to the frequency of contraction. The effect of exercise on glucose transport can persist for several hours following the cessation of exercise [12, 13], whereas following the removal of an acute insulin stimulation, glucose transport rapidly returns to basal rates [14]. Rates of basal, insulin-stimulated, and contraction-stimulated glucose transport generally are greater in muscle composed of predominately red, oxidative fibers, as compared to white, more glycolytic fibers [15, 16].

Mechanism of Insulin-Mediated Glucose Transport

Insulin action is initiated by the binding to specific cell-surface membrane receptors. Upon insulin binding, there is autophosphorylation of the receptor and other endogenous proteins. These phosphorylation events may then modulate several metabolic and growth processes including the stimulation of glucose transport. Although it is known that insulin acts through its receptor, the initial signalling event that eventually leads to the stimulation of glucose transport by muscle contractions is not known. However, it is clear that at least the initial signalling mechanisms for contraction and insulin-stimulated transport are distinct because exercise-stimulated glucose transport in vitro occurs independent of insulin [17–19], and exercise does not alter the structure or function of the insulin receptor [20].

In adipose cells, insulin stimulates glucose transport by the translocation of glucose transporters from an intracellular microsomal membrane pool to the plasma membrane [21, 22]. Karnieli et al. [14] have shown that the stimulation and reversal of transporter translocation are closely correlated with glucose transport activity. Limited studies suggest a similar phenomenon exists in heart muscle and diaphragm [23, 24].

Methods to Study Glucose Transporter Proteins in Skeletal Muscle

Studies of glucose transporter distribution and function were not initially done in skeletal muscle because the complex ultrastructure and great abundance of contractile proteins in muscle made this tissue difficult to study. Recently, however, new techniques have been developed and successfully used to study glucose transporter proteins in rat skeletal muscle [25–28]. These techniques involve the isolation of partially purified plasma and intracellular microsomal membranes and the quantification of glucose transporters in membrane fractions by the cytochalasin B binding assay [29] or by the identification of different glucose transporter isoforms using specific antibody preparations. In addition, isolated plasma membranes can be used to measure rates of vesicle glucose transport and to calculate carrier turnover number, a measure of the average intrinsic activity of the transporter proteins [30].

We have developed a membrane fractionation technique [27], a modification of the techniques of Grimditch et al. [26] and Bers [25], which involves the use of differential and density-gradient centrifugation, and a DNAase incubation that improves plasma membrane yield and purity. The procedure results in a microsomal fraction enriched by approximately 35-fold in galactosyl transferase activity, an enzyme that is associated with the trans region of the Golgi apparatus, and a plasma membrane fraction enriched by approximately 25- to 35-fold in the plasma membrane marker enzymes, potassium stimulated p-nitrophenol phosphatase (Kpnppase) and 5′nucleotidase [27]. In addition, the membrane fractions are virtually deficient of enzymes that are markers of sarcoplasmic reticulum and mitochondrial membranes [27, 31, unpubl. observation]. Using this technique about 15–20% of the plasma membranes are recovered. A major advantage of our method over that of another described in the literature [32], is that only about 5–6 g of muscle are needed to isolate both plasma and microsomal membranes and 1.0 g of muscle to isolate only plasma membranes.

Exercise and Insulin Stimulation of Glucose Transporter Translocation

Hindlimb skeletal muscles of male Sprague-Dawley rats were perfused in either the absence of insulin or in the presence of a supramaximal dose of insulin and the number and subcellular distribution of glucose trans-

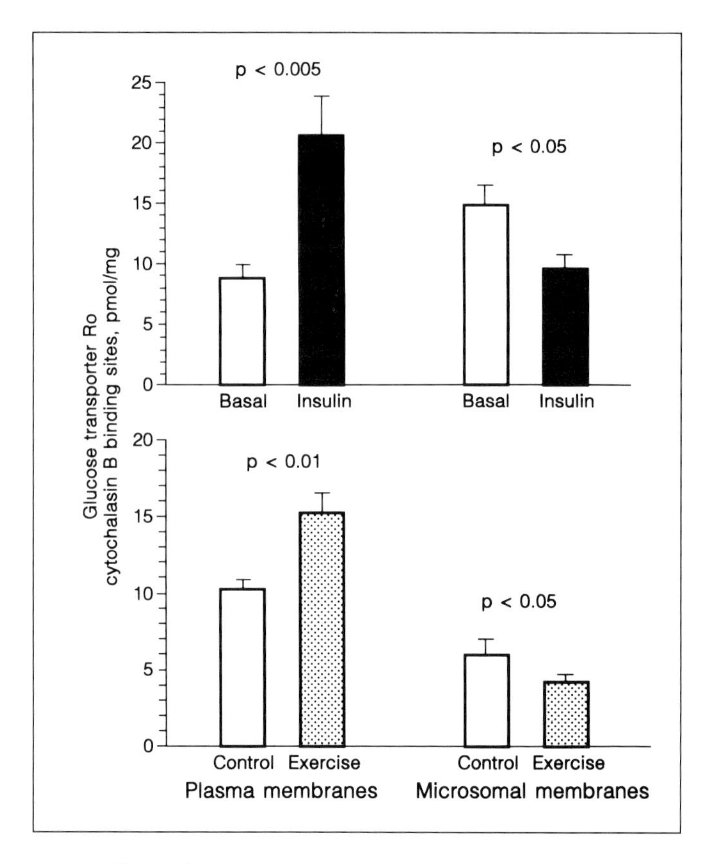

Fig. 1. Effect of insulin (top panel) and exercise (bottom panel) on the number and subcellular distribution of glucose transporter proteins as determined by cytochalasin B binding. Results are the mean ± SE of 8 observations (insulin study) or 9–10 observations (exercise study). Adapted from Hirshman et al. [27], with permission from the *Journal of Biological Chemistry,* and from Goodyear et al. [39], with permission from the *American Journal of Physiology.*

porters was studied [27]. In the insulin-stimulated muscle (insulin increased hindlimb glucose uptake by > 3-fold), plasma membrane glucose transporter number, measured by cytochalasin B binding, increased by 2-fold. Concomitant to this increase in the plasma membrane fraction was a 40% decrease in microsomal membrane glucose transporter number. These findings are similar to those of Klip et al. [28], who also studied perfused hindlimb muscle using a different fractionation procedure, and by our own studies in resting, postprandial rats where maximal insulin stimulation was achieved by intraperitoneal injection [33–35]. These findings suggest that the increase in glucose uptake associated with insulin stimulation in skeletal muscle is due, at least in part, to the translocation of glucose transporter proteins from an intracellular microsomal pool to the plasma membrane.

The effect of exercise on the subcellular distribution of glucose transporters has also been studied in rat skeletal muscle. Hirshman et al. [36] demonstrated that 1 h of treadmill exercise causes an increase in the number of glucose transporters present in the plasma membrane of rat gastrocnemius skeletal muscle. This finding has been confirmed in studies of both exercise in vivo [30, 35, 37, 38] and contractile activity of hindlimb muscles, produced by electrical stimulation of the sciatic nerve [31]. In addition, similar to insulin, the exercise-induced increase in plasma membrane glucose transporter number has been shown to be associated with a significant decrease in glucose transporter number in an intracellular microsomal pool [35, 39]. Thus, both insulin and exercise cause the translocation of glucose transporters from a common intracellular membrane pool to the plasma membrane (fig. 1).

Subcellular Distribution of GLUT-1 and GLUT-4 after Insulin and Exercise

In the isolated rat adipose cell, the relative abundance of both the GLUT-1 and GLUT-4 transporter isoforms is increased in the plasma membrane in response to insulin stimulation with GLUT-4 showing a greater response than GLUT-1 [40, 41]. Using specific antibodies against the individual glucose transporter isoforms, we determined if exercise and insulin cause the translocation of GLUT-1 and GLUT-4 proteins in skeletal muscle. Figure 2 shows an autoradiograph of Western blots where the GLUT-4 proteins were measured in skeletal muscle plasma and microsomal membrane fractions [27]. The left four lanes show the effects of a maximal insulin stimulus (hindlimb perfusion) and the right four lanes the effects of treadmill exercise. Both insulin and exercise increased the relative abundance of GLUT-4 in the plasma membrane fraction and decreased GLUT-4 in the microsomal membrane fraction. This translocation of GLUT-4 in response to exercise and insulin has been confirmed in additional studies [33, 39, 42]. On the other hand, neither stimulus had a

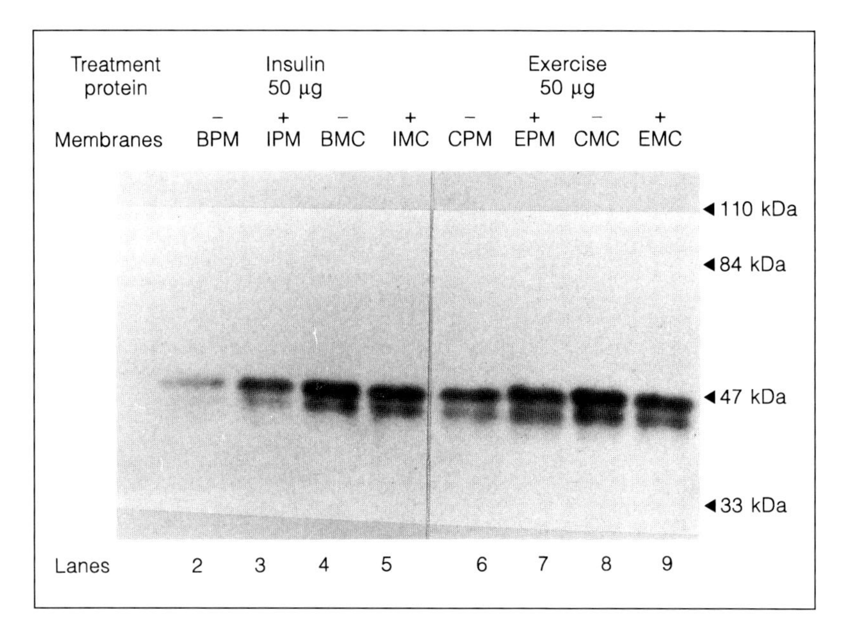

Fig. 2. Autoradiograph of a Western blot in which mAb 1F8 was used to identify GLUT-4 proteins. Lanes 2–5 are plasma and microsomal membranes from skeletal muscle perfused in the absence or presence of a maximal concentration of insulin (20 mU/ml). Lanes 6–9 are plasma and microsomal membranes from control and treadmill exercised (1 h, 20 m/min, 10% grade) skeletal muscle. Lanes 2–3 and 6–7 show that GLUT-4 density increases in plasma membranes in response to insulin and exercise, respectively. Lanes 4–5 and 8–9 show that GLUT-4 density decreases in microsomal membranes in response to insulin and exercise, respectively. PM = Plasma membrane; MC = microsomal membrane; B = basal perfused; I = insulin perfused; C = control; E = exercise. Reproduced from Goodyear et al. [39] with permission from the *American Journal of Physiology.*

significant effect on the redistribution of the GLUT-1 protein [33, 39, 42]. These findings support the hypothesis that GLUT-4, but not GLUT-1 is an 'insulin and exercise sensitive' transporter.

Exercise and Insulin Increase Glucose Transporter Intrinsic Activity

Exercise and insulin increase glucose transport in isolated skeletal muscle preparations through an increase in the maximal velocity of glucose transport (V_{max}) with no change in the substrate concentration at which

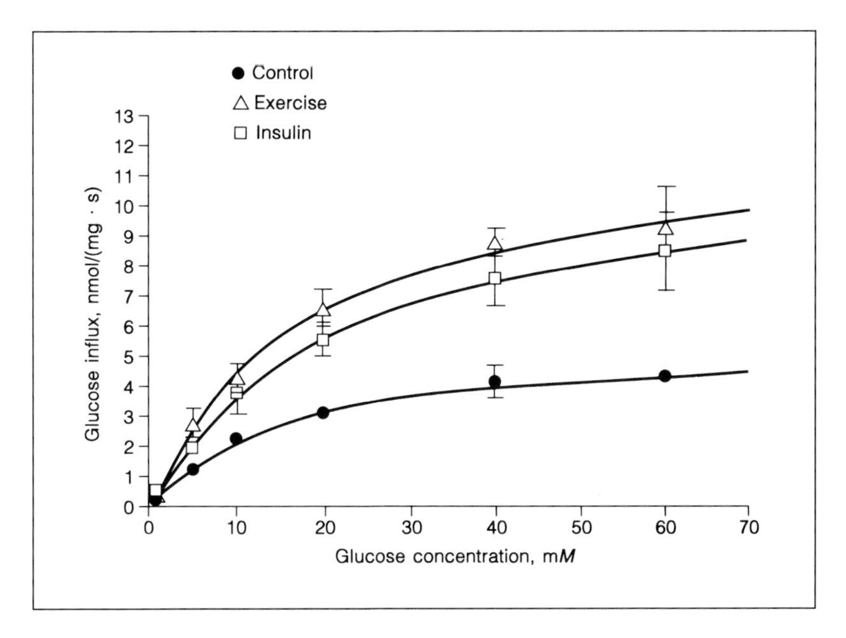

Fig. 3. Relationship between glucose influx and glucose concentration in plasma membrane vesicles from red gastrocnemius muscle. Effects of insulin and exercise on glucose influx are shown. Each point is the mean ± SE of 4–6 preparations and represents facilitated *D*-glucose influx rate (*D*-glucose uptake minus *L*-glucose uptake) measured at 1, 5, 10, 20, 40, or 60 m*M* glucose. Lines are calculated using nonlinear least-squares fit regression. Adapted from Goodyear et al. [33], with permission from the *American Journal of Physiology.*

glucose transport is half maximal ($K_{1/2}$) [3–6, 30]. An increase in V_{max} can occur by an increase in the number of carrier proteins present in the plasma membrane, an increase in the rate that each carrier transports glucose, or both. Techniques to measure glucose transport in membrane vesicles using the same plasma membrane fraction that are used to measure glucose transporter number by the cytochalasin B binding technique have been described [30, 43]. Using these procedures, the carrier turnover number is calculated by dividing the V_{max} of glucose transport by the total glucose transporter number (R_o), determined by Scatchard analysis of cytochalasin B binding data. Thus, the turnover number represents the rate of glucose transported per glucose transporter, and is an indication of the average intrinsic activity of the glucose transporters that are present in the plasma membrane.

Plasma membrane vesicle glucose transport data from red gastrocnemius muscles are shown in figure 3. The mean rate of glucose influx was measured at 6 glucose concentrations with the non linear least squares fit of the data determined. The mean calculated V_{max} from each group is shown in figure 4 (lower left). Exercise and insulin increased glucose influx throughout the range of gucose concentrations resulting in an approximately 2-fold increase in the transport V_{max}. The calculated $K_{½}$ values were between 15 and 20 mM glucose and were not altered by exercise or insulin treatment (data not shown). Figure 4 also shows glucose transporter turnover number (V_{max}/R_o) (upper panel) and glucose transporter number (R_o) data (lower right) from the same plasma membrane preparations. While exercise and insulin increased glucose transport V_{max} by approximately 2-fold, the R_o was increased by approximately 40%. Thus, the glucose transporter turnover number (fig. 4, top panel) was increased by approximately 2-fold by both insulin and exercise. Thus, the mechanism of increased insulin and exercise-stimulated skeletal muscle glucose uptake involves both a translocation of glucose transporters from an intracellular pool to the plasma membrane and an increase in the average intrinsic activity of the transporters that are present in the plasma membrane.

The mechanism by which exercise and insulin increase transporter turnover number is unknown, but may involve the recruitment of a more active transporter to the plasma membrane, the direct activation of transporter proteins in the plasma membrane, or a combination of both. Alternatively, changes in transporter intrinsic activity could reflect cross-contamination of membrane fractions, the incomplete recovery of plasma and intracellular membranes, the presence of an unidentified transporter species that does not bind cytochalasin B, or a change in the binding affinity of cytochalasin B in plasma membranes from stimulated muscles.

In the studies discussed in this review, the carrier turnover numbers were calculated from V_{max} and R_o data generated from the same membrane preparations and the protein and marker enzyme profiles of these membranes and the transporter binding affinities were not different among treatment groups. Furthermore, we have evidence to suggest that, at least for the exercise-induced increase in plasma membrane glucose transporter number and intrinsic activity, these processes can be dissociated [38]. Hindlimb skeletal muscle was obtained from male rats either immediately, 30 min or 2 h after treadmill exercise. Immediately after exercise plasma membrane glucose transporter number and glucose transport were increased by 2- and 4-fold, respectively, while at 30 min after

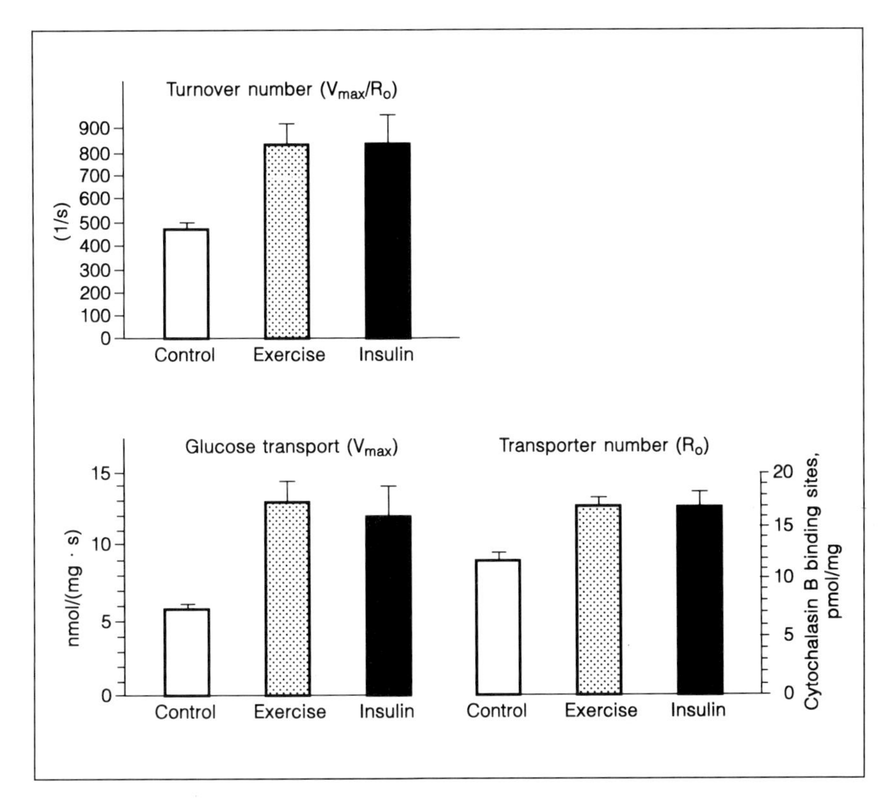

Fig. 4. Effects of exercise and insulin on red gastrocnemius skeletal muscle plasma membrane glucose transport maximal velocity (V_{max}) calculated from the nonlinear least-squares fit of the data from figure 3 (lower left panel), glucose transporter number (R_o) determined from Scatchard analysis of cytochalasin B binding data (lower right panel), and glucose transporter turnover number calculated by dividing the V_{max} by the R_o (top center panel). Each bar represents the mean ± SE of 4–6 preparations. Adapted from Goodyear et al. [33], with permission from the *American Journal of Physiology.*

exercise, transporter number remained elevated (2-fold over controls) but glucose transport was decreased to only 2-fold above baseline. Therefore, immediately after exercise both transporter number and intrinsic activity were elevated whereas at 30 min postexercise, transporter number remained elevated but transporter intrinsic activity was not. This dissociation between transporter translocation and activation suggest that these processes are distinct.

Interaction of Insulin and Exercise on Glucose Transport

Contractile activity in vitro can increase rates of skeletal muscle glucose uptake in the total absence of insulin [17–19]. This finding has important implications in terms of our understanding the signalling mechanism(s) of insulin- and exercise-stimulated glucose transport. We have recently determined that the effect of insulin-independent contractile activity to increase glucose transport occurs through a change in the distribution and/or function of glucose transporter proteins. Plasma membranes were prepared from rat hindlimb skeletal muscle perfused in the absence of insulin with or without electrical stimulation to produce muscle contractions [31]. Muscle contractions alone significantly increased plasma membrane vesicle glucose transport, transporter number, and the ratio of transport to transporter number, demonstrating that insulin-independent contractile activity increases skeletal muscle glucose transport by an increase in both the number and average intrinsic activity of glucose transporters in the plasma membrane. These results [44] and the finding that exercise does not alter insulin receptor function [45], demonstrate that the signal transduction process that results in the movement (and possibly activation) of glucose transporters is not initiated by the phosphorylation events associated with the insulin receptor.

The stimulation of glucose transport in mammalian skeletal muscle by contractile activity and maximal insulin stimulation has been reported to be additive or partially additive [6, 12, 46, 47]. Different mechanisms for contraction stimulated and insulin stimulated glucose transport suggest that there are two separate cellular signalling systems leading to activation of the transport system, that there are distinct transporter isoforms for each stimulus, or that the two stimuli activate two independent 'pools' of glucose transporter proteins. While there is substantial evidence to support the concept of different intracellular signalling mechanisms for insulin- and exercise-induced glucose transport [48], to date, there is no evidence to suggest the existence of distinct glucose transporter isoforms, one that is insulin-sensitive and one that is exercise-sensitive [39, 42]. The issue of separate pools of glucose transporters is not well understood. Douen et al. [42] have reported that exercise does not decrease glucose transporter number in an insulin-sensitive intracellular membrane fraction while our group [39] and Fushiki et al. [35] have demonstrated that both exercise and insulin decrease intracellular glucose transporter number from a common intracellular pool. The case for a common pool of glucose transporters is strengthened by the finding that the increase in plasma membrane glucose transporter number [31, 49] or GLUT-4 protein [49] in response to the combination of maximal insulin and exercise stimulation is not additive. If stimuli-specific isoforms or separate pools of glucose transporters are not present in muscle, then the additive effect of exercise and insulin on glucose transport in vivo might be explained by other mechanisms such as changes in tissue blood flow or interaction of submaximal stimuli. At this time, there is no clear evidence that the combination of maximal insulin and exercise treatments results in a greater increase in the number or intrinsic activity of glucose transporter proteins compared to either stimulus alone [31]. Future studies in this area are needed to reconcile the differences between physiological measurements showing additivity of insulin and exercise stimulated glucose uptake in vivo and the lack of additivity in measurements of glucose transporter number and function in skeletal muscle membrane preparations. Further studies of the skeletal muscle glucose transport system are also needed to determine whether insulin and exercise at submaximal doses have additive effects.

Glucose Transporters in Red and White Skeletal Muscle

Rates of skeletal muscle glucose uptake are dependent on the fiber type composition of a muscle [50]. In order to determine if differences in basal, insulin- and exercise-stimulated rates of glucose uptake among muscles composed of different fiber types correspond to differences in plasma membrane glucose transporter number and function, portions of red and white gastrocnemius muscle were obtained from insulin injected, exercised, and control rats [33]. In control rats, both glucose transport V_{max} and glucose transporter number were approximately 2-fold greater in plasma membranes prepared from red as compared to white gastrocnemius muscle. Therefore, there was no difference in the carrier turnover number between the red and white muscle preparations indicating no difference in the intrinsic activity of glucose transporter proteins present in different fiber types. Both insulin and exercise increased plasma membrane glucose transporter number, transport V_{max}, and carrier turnover number. While the fold increase in transport V_{max} and transporter number was greater in white muscle, the absolute change (or the delta value) in these parameters was not different between the red and white muscle preparations. These data suggest that differences in basal, exercise- and insulin-stimulated rates of glucose transport between red and white muscle fibers are due to differences in the number of glucose transporters that are present in the plasma membrane in the basal state, and not due to differences in the translocation or the intrinsic activity of transporters. The increase in glucose transporter number in red muscle was shown to be due to an increase in both the GLUT-1 and GLUT-4 isoforms [33].

Conclusion

In conclusion, considerable evidence has been presented to suggest that exercise and insulin increase skeletal muscle glucose uptake through both the translocation of glucose transporters from an intracellular pool to the plasma membrane and through an increase in the average intrinsic activity of glucose transporter proteins in the plasma membrane. Although the effects of skeletal muscle contractile activity on the glucose transport system can occur independent of insulin, both insulin and contraction cause the translocation of the same glucose transporter protein isoform (GLUT-4). The other glucose transporter isoform thought to be present in skeletal muscle tissue, GLUT-1, does not appear to be translocated to the plasma membrane following either insulin stimulation or exercise. Additional studies in exercise-stimulated skeletal muscle have provided evidence that the glucose transporter translocation and activation processes are distinct phenomena. Further research is necessary to fully understand the mechanism(s) regulating the activation of the glucose transport system by insulin and exercise in skeletal muscle. Greater knowledge of the glucose transport system in this tissue will provide new insights into our understanding of glucose homeostatic mechanisms.

Acknowledgements

This work was supported by the National Institutes of Diabetes and Digestive and Kidney Diseases (NIDDK) Research Grant R01 DK-26317 (E.S. Horton) and a grant from the Juvenile Diabetes Foundation (E.S. Horton). L.J. Goodyear was supported by NIDDK Postdoctoral Research Fellowships T32-DK-07523 (University of Vermont) and T32-DK-07260-15 (Joslin Research Laboratory).

References

1 Kubo K, Foley JE: Rate-limiting steps for insulin-mediated glucose uptake into perfused rat hindlimb. Am J Physiol 1986;250:E100–E102.
2 Guyton AC: Textbook of Medical Physiology. Philadelphia, Saunders, 1976.
3 Holloszy JO, Narahara HT: Changes in permeability to 3-methylglucose associated with contraction of isolated frog muscle. J Biol Chem 1965;240:3493–3500.
4 Narahara HT, Ozand P: Studies of tissue permeability. IX. The effect of insulin on the penetration of 3-methylglucose-H3 in frog muscle. J Biol Chem 1963;238:40–49.
5 Narahara HT, Ozand P, Cori CF: Studies of tissue permeability. VII. The effect of insulin on glucose penetration and phosphorylation in frog muscle. J Biol Chem 1960;235:3370–3378.
6 Nesher R, Karl IE, Kipnis DM: Dissociation of effects of insulin and contraction on glucose transport in rat epitrochlearis muscle. Am J Physiol 1985;249:C226–C232.
7 Bell GI, Kayano T, Buse JB, Burant CF, Takeda J, Lin D, Fukumoto H, Seino S: Molecular biology of mammalian glucose transporters. Diabetes Care 1990;13:198–208.
8 Klip A, Paquet MR: Glucose transport and glucose transporters in muscle and their metabolic regulation. Diabetes Care 1990;13:228–242.
9 Sternlicht E, Barnard RJ, Grimditch GK: Exercise and insulin stimulate skeletal muscle glucose transport through different mechanisms. Am J Physiol 1989;256: E227–E230.
10 Garthwaite SM, Holloszy JO: Increased permeability to sugar following muscle contraction: Inhibitors of protein synthesis prevent reversal of the increase in 3-methylglucose transport rate. J Biol Chem 1965;257:5008–5012.
11 Instrom JP, Rennie MJ, Schersten T, Bylund-Fellenius A-C: Membrane transport in relation to net uptake of glucose in the perfused rat hindlimb: Stimulatory effect of insulin, hypoxia and contractile activity. Biochem J 1986;233:131–137.
12 Garetto LP, Richter EA, Goodman MN, Ruderman NB: Enhanced muscle glucose metabolism after exercise in the rat: The two phases. Am J Physiol 1984;246:E471–E475.
13 Elbrink J, Phipps BA: Studies on the persistence of enhanced monosaccharide transport in rat skeletal muscle following the cessation of the initial stimulus. Cell Calcium 1980;1:349–358.
14 Karnieli E, Hissin PJ, Simpson IA, Salans LB, Cushman SW: A possible mechanism of insulin resistance in the rat adipose cell in streptozotocin-induced diabetes mellitus: Depletion of intracellular glucose transport systems. J Clin Invest 1981;68: 811–814.
15 James DE, Jenkins AB, Kraegen EW: Heterogeneity of insulin action in individual muscles in vivo: Euglycemic clamp studies in rats Am J Physiol 1985;248:E567–E574.
16 Richter EA, Garreto LP, Goodman MN, Ruderman NB: Enhanced muscle glucose metabolism after exercise: Modulation by local factors. Am J Physiol 1984;246: E476–E482.
17 Ploug T, Galbo H, Richter EA: Increased muscle glucose uptake during contractions: no need for insulin. Am J Physiol 1984;247:E726–E731.

18 Wallberg-Henriksson H, Holloszy JO: Contractile activity increases glucose uptake by muscle in severely diabetic rats. J Appl Physiol 1984;57:1045–1049.
19 Wallberg-Henriksson H, Holloszy JO: Activation of glucose transport in diabetic muscle: Responses to contraction and insulin. Am J Physiol 1985;249:C233–C237.
20 Treadway JL, James DE, Burcel E, Ruderman NB: Effect of exercise on insulin receptor binding and kinase activity in skeletal muscle. Am J Physiol 1989;256: E138–E144.
21 Cushman SW, Wardzala LJ: Potential mechanism of insulin action on glucose transport in the isolated rat adipose cell: Apparent translocation of intracellular transport systems to the plasma membrane. J Biol Chem 1980;255:4758–4762.
22 Suzuki K, Kono T: Evidence that insulin causes translocation of glucose transport activity to the plasma membrane from an intracellular storage site. Proc Natl Acad Sci USA 1980;77:2542–2545.
23 Wardzala LJ, Jeanrenaud B: Potential mechanism of insulin action on glucose transport in the isolated rat diaphragm – apparent translocation of intracellular transport units to the plasma membrane. J Biol Chem 1981;256:7090–7093.
24 Watanabe T, Smith MM, Robinson FW, Kono T: Insulin action on glucose transport in cardiac muscle. J Biol Chem 1984;260:13117–13122.
25 Bers DM: Isolation and characterization of cardiac sarcolemma. Biochim Biophys Acta 1979;555:131–146.
26 Grimditch G, Barnard J, Kaplan S, Sternlicht E: Insulin binding and glucose transport in rat skeletal muscle sarcolemmal vesicles. Am J Physiol 1985;249:E398–E408.
27 Hirshman MF, Goodyear LJ, Wardzala LJ, Horton ED, Horton ES: Identification of an intracellular pool of glucose transporters from basal and insulin-stimulated rat skeletal muscle. J Biol Chem 1990;265:987–991.
28 Klip A, Ramlal T, Young DA, Holloszy JO: Insulin induced translocation of glucose transporters in rat hindlimb muscles. FEBS Lett 1988;244:224–230.
29 Wardzala LJ, Cushman SW, Salans LB: Mechanism of insulin action on glucose transport in the isolated rat adipose cell. J Biol Chem 1978;260:2197–2201.
30 King PA, Hirshman MF, Horton ED, Horton ES: Glucose transport in skeletal muscle membrane vesicles from control and exercised rats. Am J Physiol 1989;257: C1128–C1134.
31 Goodyear LJ, King PA, Hirshman MF, Thompson CM, Horton ED, Horton ES: Contractile activity increases plasma membrane glucose transporters in absence of insulin. Am J Physiol 1990;258:E667–E672.
32 Klip A, Ramlal T, Young DA, Holloszy JO: Insulin-induced translocation of glucose transporters in rat hindlimb muscles. FEBS Lett 1988;224:224–230.
33 Goodyear LJ, Hirshman MF, Smith RJ, Horton ES: Glucose transporter number, activity and isoform content in plasma membranes of red and white skeletal muscle. Am J Physiol 1991;261:E556–E561.
34 Sternlicht E, Barnard RJ, Grimditch GK: Mechanism of insulin action on glucose transport in rat skeletal muscle. Am J Physiol Endocrinol Metab 1988;254:E633–E638.
35 Fushiki T, Wells JA, Tapscott EB, Dohm GL: Changes in glucose transporters in muscle in response to exercise. Am J Physiol 1989;256:E580–E587.

36 Hirshman MF, Wallberg-Henriksson H, Wardzala LJ, Horton ED, Horton ES: Acute exercise increases the number of plasma membrane glucose transporters in rat skeletal muscle. FEBS Lett 1988;238:235–239.
37 Douen AG, Ramlal T, Klip A, Young DA, Cartee GD, Holloszy JO: Exercise-induced increase in glucose transporters in plasma membranes of rat skeletal muscle. Endocrinology 1989;124:449–454.
38 Goodyear LJ, Hirshman MF, King PA, Thompson CM, Horton ED, Horton ES: Skeletal muscle plasma membrane glucose transport and glucose transporters after exercise. J Appl Physiol 1990;68:193–198.
39 Goodyear LJ, Hirshman MF, Horton ES: Exercise induced translocation of skeletal muscle glucose transporters. Am J Physiol 1991;261:E795–E799.
40 Holman GD, Kozka IJ, Clark AE, Flower CJ, Saltis J, Habberfield AD, Simpson IA, Cushman SW: Cell surface labeling of glucose transporter isoform GLUT-4 by bis-mannose photolabel. Correlation with stimulation of glucose transport in rat adipose cells by insulin and phorbol ester. J Biol Chem 1990;265:18172–18179.
41 James DE, Brown R, Navarro J, Pilch P: Insulin-regulatable tissues express a unique insulin-sensitive glucose transport protein. Nature 1988;333:183–185.
42 Douen AG, Ramlal T, Rastogi S, Bilan PJ, Cartee GD, Vranic M, Holloszy JO, Klip A: Exercise induces recruitment of the 'insulin-responsive glucose transporter'. Evidence for distinct intracellular insulin- and exercise-recruitable transporter pools in skeletal muscle. J Biol Chem 1990;265:13427–13430.
43 Klip A, Ramlal T, Douen AG, Burdett E, Young D, Cartee GD, Holloszy JO: Insulin-induced decrease in 5′-nucleotidase activity in skeletal muscle membranes. FEBS Lett 1988;238:419–423.
44 Richter EA, Ploug T, Glabo H: Increased muscle glucose uptake after exercise: no need for insulin during exercise. Diabetes 1985;34:1041–1048.
45 Coppack SW, Frayn KN, Whyte PL, Humphreys SM: Carbohydrate metabolism in human adipose tissue in vivo. Biochem Soc Trans 1989;17:145–146.
46 Wallberg-Henriksson H, Constable SH, Young DA, Holloszy JO: Glucose transport into rat skeletal muscle: Interaction between exercise and insulin. J Appl Physiol 1988;65:909–913.
47 Zorzano A, Balon TW, Goodman MN, Ruderman NB: Additive effects of prior exercise and insulin on glucose and AIB uptake by rat muscle. Am J Physiol 1986; 251 14:E21–E26.
48 Holloszy JO, Constable SH, Young DA: Activation of glucose transport in muscle by exercise. Diabetes Metab Rev 1986;1:409–424.
49 Douen AG, Ramlal T, Cartee GD, Klip A: Exercise modulates the insulin-induced translocation of glucose transporters in rat skeletal muscle. FEBS Lett 1990;261: 256–260.
50 Henriksen EJ, Bourey RE, Rodnick KJ, Koranyi L, Permutt MA, Holloszy JO: Glucose transporter protein content and glucose transport capacity in rat skeletal muscles. Am J Physiol Endocrinol Metab 1990;259:E593–E598.

Dr. Edward S. Horton, Division of Endocrinology, Metabolism and Nutrition, Department of Medicine, C-350 Given Bldg., University of Vermont College of Medicine, Burlington, VT 05405 (USA)

Sato Y, Poortmans J, Hashimoto I, Oshida Y (eds): Integration of Medical and Sports Sciences. Med Sport Sci. Basel, Karger, 1992, vol 37, pp 216–226

Acute Metabolic Effects of Exercise on Glucose Fluxes in Splanchnic and Peripheral Tissues in Diabetics, Determined with an Innovative Approach

Ryuzo Kawamori, Minoru Kubota, Masahiko Ikeda, Munehide Matsuhisa, Masashi Kubota, Toyohiko Morishima, Takenobu Kamada

First Department of Medicine, Osaka University Medical School, Fukushima, Osaka, Japan

The liver plays an important role in glucose homeostasis. However, the factors promoting hepatic glucose uptake still remain unclear. Clinically, deranged hepatic glucose handling during and after meal might be the cause of the exaggerated postprandial glucose rise seen in diabetics. Therefore, by clarifying the mechanism of regulation of hepatic glucose uptake, hepatic glucose handling could be normalized with therapeutic modalities.

In this series of experiments, firstly, we tried to investigate factors promoting hepatic glucose uptake in animal studies. Secondly, we examined the effect of strict glycemic control or acute exercise on glucose handling by splanchnic and peripheral tissues in subjects with non-insulin-dependent diabetes mellitus (NIDDM).

Factors Promoting Hepatic Glucose Uptake

As regards factors regulating hepatic glucose uptake, we have firstly reported that portally administered insulin enhances much more hepatic glucose uptake as compared with the same amount of insulin administered peripherally [1]. Secondly, we have shown that the amount of glucose taken up by the liver after glucose loading depends on the route of glucose administration and that hepatic removal of glucose is greater after portal glucose infusion compared with peripheral glucose delivery. Even though the exact mechanism of this phenomenon is not clearly understood, the arterial-portal venous glucose gradient has been speculated to be an important signal for augmenting hepatic glucose disposal. We have also demonstrated that the portal and hepatic arterial glucose gradient cannot explain this phenomenon [2]. Therefore, we investigated the significance of the glucose gradient between the hepatoportal system and the central nervous system (CNS) in promoting hepatic glucose disposal in normal dogs.

Significance of the Glucose Gradient between the Central Nervous and Hepatoportal Systems in Hepatic Glucose Disposal

The significance of the glucose gradient between the CNS and hepatoportal system in hepatic glucose uptake was investigated in 5 normal dogs with chronically implanted pulsed Doppler flow probes on the portal vein (PV) and hepatic artery (HA) and cannulae in the hepatic vein (HV), superior mesenteric vein (SMV) and PV. Intravenous glucose load was performed for 90 min according to the following 4 protocols: (1) PE: peripheral infusion (7 mg/kg/min); (2) PO: intraportal infusion via the SMV (same amount as PE); (3) PO + CNS: portal infusion (same as PO) plus additional glucose infusion into the unilateral carotid (CA) and vertebral (VA) arteries to abolish the glucose gradient between the CNS and hepatoportal system (glucose amounts infused into the CA and VA were calculated based on the plasma flow ratio of PV to CA and VA), and (4) PO + PE: the same amount of additional glucose was infused into the peripheral vein instead of the CNS. Net hepatic glucose balance (NHGB) and hepatic extraction of glucose (HERG) were calculated from plasma flows and plasma glucose levels in PV, HA and HV.

As results, during PE and PO + CNS glucose infusion, the hepatoportal-CNS glucose gradient was negative and almost zero, respectively. During PO and PO + PE glucose infusion, on the other hand, the hepatoportal-CNS glucose gradient was positive. PO glucose infusion caused a significant increase in both NHGB and HERG as compared to PE under comparable levels of plasma glucose and IRI. This increase in hepatic glucose uptake significantly decreased during PO + CNS, but returned to a value similar to that of PO during PO + PE (table 1).

Table 1. Arterial plasma glucose level (A-PG), arterial IRI level (A-IRI), glucose gradient between hepatoportal system and CNS ($\Delta PG_{HPS\text{-}CNS}$), NHGB and HERG during PE, PO, PO + CNS and PO + PE glucose infusion

Glucose infusion	A-PG mg/dl	A-IRI μU/ml	$\Delta PG_{HPS\text{-}CNS}$ mg/dl	NHGB mg/kg/min	HERG %
PE	163 ± 50	38 ± 16	−5 ± 3	2.1 ± 2.3	5.6 ± 3.8
PO	169 ± 56	41 ± 15	28 ± 5	5.1 ± 1.8*	10.3 ± 4.8*
PO + CNS	188 ± 43	49 ± 19	0	3.7 ± 2.1**	5.9 ± 3.5**
PO + PE	201 ± 49	53 ± 21	27 ± 2	5.7 ± 3.1	9.3 ± 5.3

* $p < 0.05$: PO vs. PE; ** $p < 0.05$: PO + CNS vs. PO.

Table 2. Arterial plasma glucose level (A-PG), arterial IRI level (A-IRI), NHGB and HERG during euglycemic and hyperglycemic clamp with portal glucose infusion (7 mg/kg/min)

Clamp	A-PG mg/dl	A-IRI μU/ml	NHGB mg/kg/min	HERG %
Euglycemic	98 ± 8	120 ± 10	4.0 ± 0.5	11.3 ± 2.1
Hyperglycemic	220 ± 7	122 ± 7	6.1 ± 1.0	8.1 ± 1.2
Difference	$p < 0.001$	n.s.	$p < 0.05$	$p < 0.05$

These results suggest that the glucose gradient between the CNS and hepatoportal system plays a crucial role in regulating hepatic glucose disposal.

Hepatic Glucose Handling during Euglycemic or Hyperglycemic Clamp with the Same Arterial-Portal Venous Glucose Gradient

In order to examine the effect of prevailing plasma glucose levels on hepatic glucose handling, hepatic glucose uptake during intraportal glucose infusion (7 mg/kg/min) was measured under euglycemic and hyperglycemic clamp in 7 normal dogs. Insulin was administered into the PV at a rate of 20 × B (B = 225 μU/kg·min) and somatostatin (0.5 μg/kg·min) was infused to suppress endogenous insulin secretion. Pulsed Doppler flow

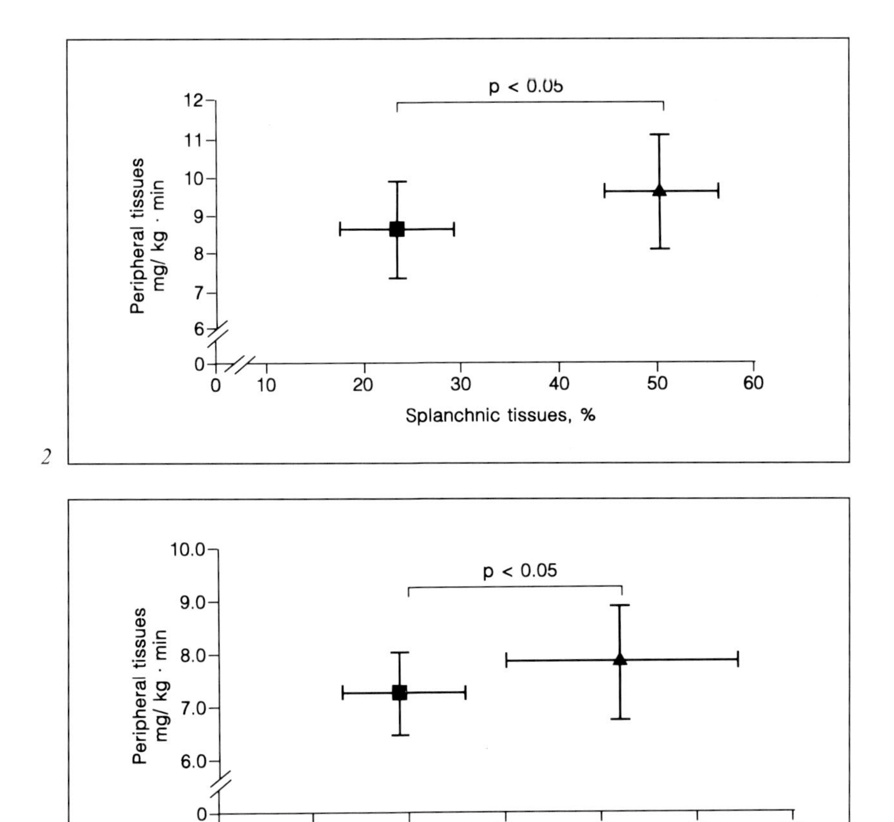

Fig. 2. Acute effect of exercise on glucose handling by peripheral and splanchnic tissues in NIDDM patients (mean ± SEM). ■: before exercise (sedentary); ▲: after exercise.

Fig. 3. Effect of strict glycemic control by intensified insulin therapy on glucose handling by peripheral and splanchnic tissues in NIDDM patients (mean ± SEM). ■: before strict glycemic control; ▲: after strict glycemic control.

take. Through these mechanisms, normalization of postmeal glycemia will be realized. Sufficient insulin supplements before each meal prevent beta cell exhaustion followed by restoration of endogenous basal insulin secretion. If secreted endogenous insulin is sufficient to normalize midnight and early-morning glycemia, this therapy is most suitable for NIDDM.

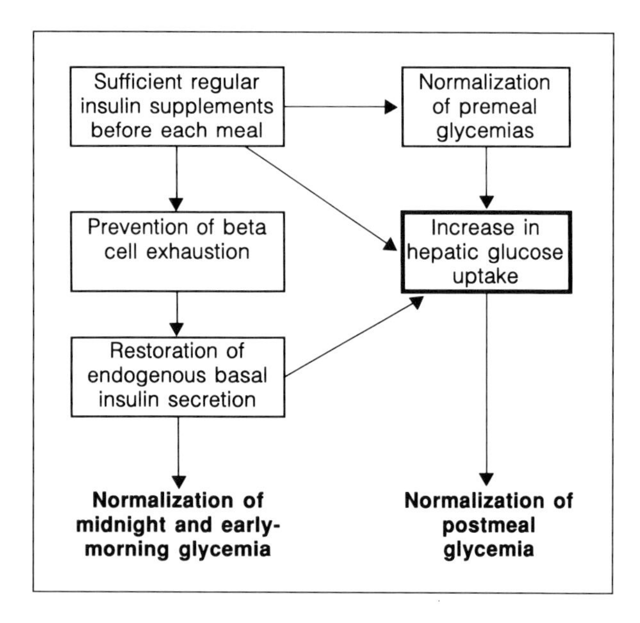

Fig. 4. Working hypothesis for the treatment of nonobese NIDDM patients with secondary failure on sulfonylureas.

This hypothesis was tested in NIDDM subjects with secondary failure on sulfonylureas. Patients were admitted to switch to insulin therapy. In these patients, the plasma profile of insulin stayed in lower levels, even though endogenous insulin secretion was stimulated with maximal amounts of glibenclamide and hyperglycemia. Patients were given regular insulin 30 min preprandially for 3 meals. Insulin injections were initiated at doses of 10, 8 and 6 U for breakfast, lunch and dinner, respectively. Then, according to the daily profile of plasma glucose taken every 3–4 days, insulin doses were adjusted to obtain normal preprandial and 2-hour postprandial glycemia. At 4–5 weeks after the initiation of insulin therapy, in 56 out of 77 patients perfect normalization of glycemia was established with the mean dose of 10, 7 and 7 U for breakfast, lunch and dinner, respectively (fig. 5). Mean urinary excretion of CPR at night from 11 p.m. to 7 a.m. at 4 weeks, was 1.85 μg/h, and UCPR divided by plasma glucose was increased 2-fold. These amounts seemed to be sufficient to suppress hepatic glucose production and to increase peripheral glucose uptake.

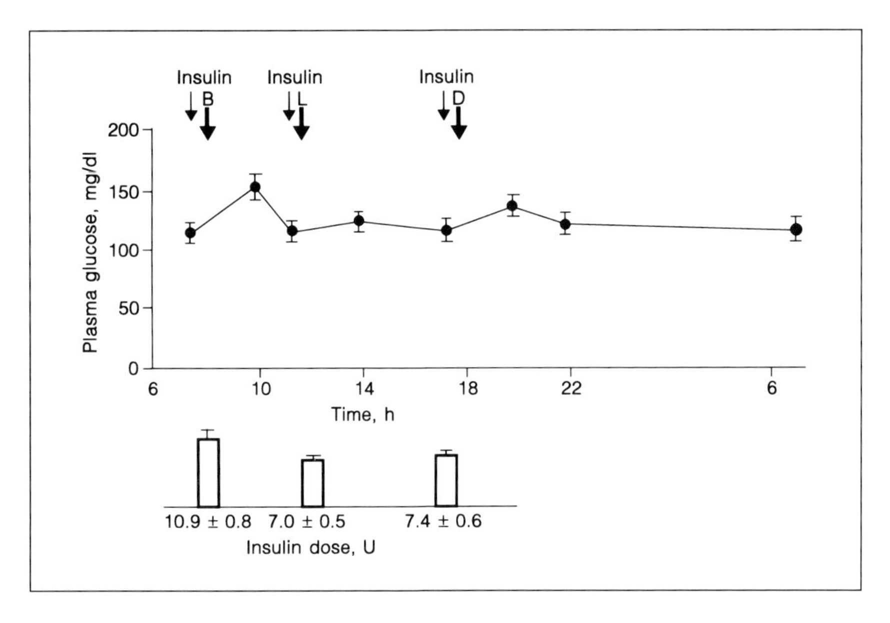

Fig. 5. Mean plasma glucose profile in response to regular prandial subcutaneous insulin injections in 33 nonobese NIDDM patients with secondary failure on sulfonylureas. B: Breakfast, L: lunch, D: dinner.

Thus, it was clearly shown that in NIDDM the insulin-secretory ability, especially the basal insulin secretion, is dynamic and reversible. Hemoglobin A_{1c} decreased markedly, even though insulin doses were reduced. This phenomenon may be caused by increased hepatic glucose uptake after each meal, showing that insulin sensitivity of the patients increased gradually.

Conclusions

(1) To realize the maximal metabolic effect of exercise in diabetic patients, exercise should be performed in a normoglycemic state. Insulin-treated patients who are willing to do heavy exercise should not take carbohydrates before exercise, but should take them during and after exercise.

(2) Both strict glycemic regulation with sufficient regular insulin supplements before each meal and exercise therapy are effective in improving glucose handling by splanchnic tissues in patients with NIDDM.

References

1 Morishima T, Kubota M, Saito Y, Kawamori R, Shichiri M: Physiological significance of intraportal delivery of insulin on glucose regulation; in Brunetti P, Waldhaus WK (eds): Advanced Models for the Therapy of Insulin-Dependent Diabetes. New York, Raven Press, 1987, pp 35–39.
2 Kubota M, Morishima T, Sekiya M, Kawamori R, Shichiri M, Kamada M: The significance of arterial-portal vein glucose gradient in the hepatic glucose uptake. Diabetes 1987;36(suppl):37A.
3 Kawamori R, Shichiri M, Murata T, Nomura M, Shigeta Y, Abe H: Study of glucose tolerance and the dynamic property of insulin secretion: Analysis of intravenous glucose tolerance test with the aid of control theory. Acta Endocrinol (Copenh) 1979; 90:283–294.
4 Kawamori R, Bando K, Yamasaki Y, Kubota M, Watarai T, Iwama N, Shichiri M, Kamada T: Fasting plus prandial insulin supplements improve insulin secretory ability in non-insulin dependent diabetics. Diabetes Care 1989;12:680–685.

Ryuzo Kawamori, MD, First Department of Medicine, Osaka University Medical School, 1-1-50, Fukushima, Fukushima-ku, Osaka (Japan)

Sato Y, Poortmans J, Hashimoto I, Oshida Y (eds): Integration of Medical and Sports Sciences. Med Sport Sci. Basel, Karger, 1992, vol 37, pp 227–236

Exercise and Diabetes

Henrik Galbo, Michael von Linstow, Flemming Dela, Michael Kjaer, Kari Mikines

Department of Medical Physiology B, The Panum Institute, and Department of Internal Medicine TTA, University Hospital, University of Copenhagen, Denmark

The literature on diabetes and exercise has recently been reviewed in detail [1]. So, in the present paper only a brief description of some essential features of exercise in type I diabetics will be given. Then, recent studies showing that exercise is associated with similar problems in type II and type I diabetics will be mentioned. Finally, data from ongoing studies of the effect of training on secretion and action of insulin in type II diabetics are discussed.

In type I diabetics the major problem at rest is that the plasma insulin is not spontaneously adjusted to the needs. They either take too much or too little insulin and accordingly tend to become either hypo- or hyperglycemic. In relation to exercise the diabetics face exactly the same problem. However, it is aggravated and the underlying mechanisms have become apparent within the last decade [1]. In healthy subjects exercise elicits an increased metabolic rate and in turn glucose uptake in muscle, which tends to decrease the plasma glucose concentration. The contraction-induced glucose uptake does not depend on insulin [1]. In fact, the effects of insulin and contractions on glucose transport are additive in isolated muscle [1]. In vivo the effects on glucose uptake in muscle are even synergistic [2], probably reflecting that during exercise the direct effect of insulin may be enhanced when blood flow and in turn glucose delivery are increased, and also the indirect effect via inhibition of fat cell lipolysis may be augmented because lipolysis is accelerated. However, exercise also triggers a decrease in plasma insulin and an increase in counterregulatory hormones, an endocrine response enhancing glucose production and inhibiting glucose uptake, and accordingly tending to increase the plasma glucose concentration. Normally, these opposing forces are well matched and the plasma glucose concentration is essentially constant during exercise. However, in diabetics plasma insulin does not decrease in response to exercise. On the contrary, the plasma insulin concentration may even increase due to enhanced absorption from depots when subcutaneous blood flow increases [3]. Thus, if a diabetic takes his usual dose of insulin, his blood glucose decreases in case he exercises [4, 5]. If, on the other hand, he reduces the dose of insulin to such an extent that he becomes ketotic the response of counterregulatory hormones to exercise becomes exaggerated. The exaggerated hormonal response intensifies hepatic glucose production, an effect which is enhanced by the coexisting low insulin level [6, 7]. Studies in exercising diabetic dogs have shown that counterregulatory hormones also favor hyperglycemia by decreasing glucose clearance, an effect which is predominantly due to inhibition of lipolysis and glycogenolysis [8].

Thus, the diabetic's usual metabolic problems are aggravated by exercise. If he takes too much insulin the ensuing decrease in plasma glucose is accelerated by contraction-induced glucose uptake and by diminished exercise-induced increase in hepatic glucose production. If he takes too little insulin the ensuing increase in plasma glucose is accelerated by the exercise-induced increase in counterregulatory hormones. That the balance is very delicate appears from a study of diabetics with a fasting plasma glucose concentration of 13 m*M*, in whom glucose decreased by 5 m*M* during exercise if 4 U of regular insulin was injected 1 h before exercise, whereas glucose increased by 2 m*M* if the insulin was injected 3 h before exercise [9]. If exercise only requires little energy no precautions may be necessary in the well-regulated diabetic. If exercise is more demanding he should mimic normal physiology by reducing the insulin dose prior to exercise. This means that before exercise glucose will rise to levels higher than usual but during exercise hypoglycemia is avoided [4, 5]. The advices to the physically active diabetic are that he should: be educated in both exercise physiology and diabetes; use frequent insulin injections or continuous subcutaneous delivery by a pump – the two regimens allow similar coping with exercise [4, 5]; reduce insulin dose prior to prolonged exercise; monitor blood glucose; take extra slowly digested carbohydrate; always carry readily digested carbohydrate – in case of hypoglycemia; live a regular life; have regular medical examinations – besides acute metabolic complications exercise interacts with diabetic long-term complications [1, 10]. The benefits diabetics can obtain from exercise are physiological, medical, psychological and social [1, 11]. However, they should not expect improved metabolic control in terms of lower average blood glucose and hemoglobin A_{1c} levels [1]. Usually, total insulin requirements are also unaltered in physically active compared to sedentary diabetics, probably because in the former the insulin-saving effect of exercise is outweighed by a need for extra food intake. On nonexercise days, in order to avoid weight gain the diabetic athlete has to reduce calorie intake rather than increasing the insulin dose.

Exercise has been thought to be less problematic in type II than in type I diabetics. Type II diabetics have some endogenous insulin secretion, which responds to needs in principle as it does in healthy subjects. During submaximum exercise the plasma glucose concentration often decreases markedly in type II diabetics [2], probably because contraction-induced muscular glucose uptake is favored by the high plasma glucose level mostly prevailing at onset of exercise. However, the risk of hypoglycemia is less than in type I diabetics, because plasma insulin spontaneously decreases with glucose levels [2]. Nevertheless, type II diabetics who are treated with oral antidiabetic drugs may experience dangerous hypoglycemia in response to exercise. This fact probably reflects that insulin secretion is artificially stimulated [12]. In these type II diabetics the insulin level is inappropriately high during exercise just as it is in type I diabetics who develop hypoglycemia during exercise. By reducing antidiabetic tablet dose they have to reduce the insulin level prior to exercise just as type I diabetics. Interestingly, and again analogous with findings for type I diabetics, type II diabetics with basal hyperglycemia may also respond to exercise with an increase in plasma glucose concentration caused by an exaggerated increase in counterregulatory hormones. This has been found in response to brief, intense exercise (fig. 1, 2). The first 1–2 h after such exercise insulin sensitivity is decreased, but after 24 h it is increased [13]. The increase is not accompanied by any decrease in basal glucose or insulin levels probably because of increased food intake after exercise.

With repeated exercise (training) healthy subjects develop a reduced pancreatic beta cell response to stimulation. A similar adaptation would immediately appear less expedient in type II diabetics who have reduced beta cell function in advance. We tested the hypothesis that type II diabet-

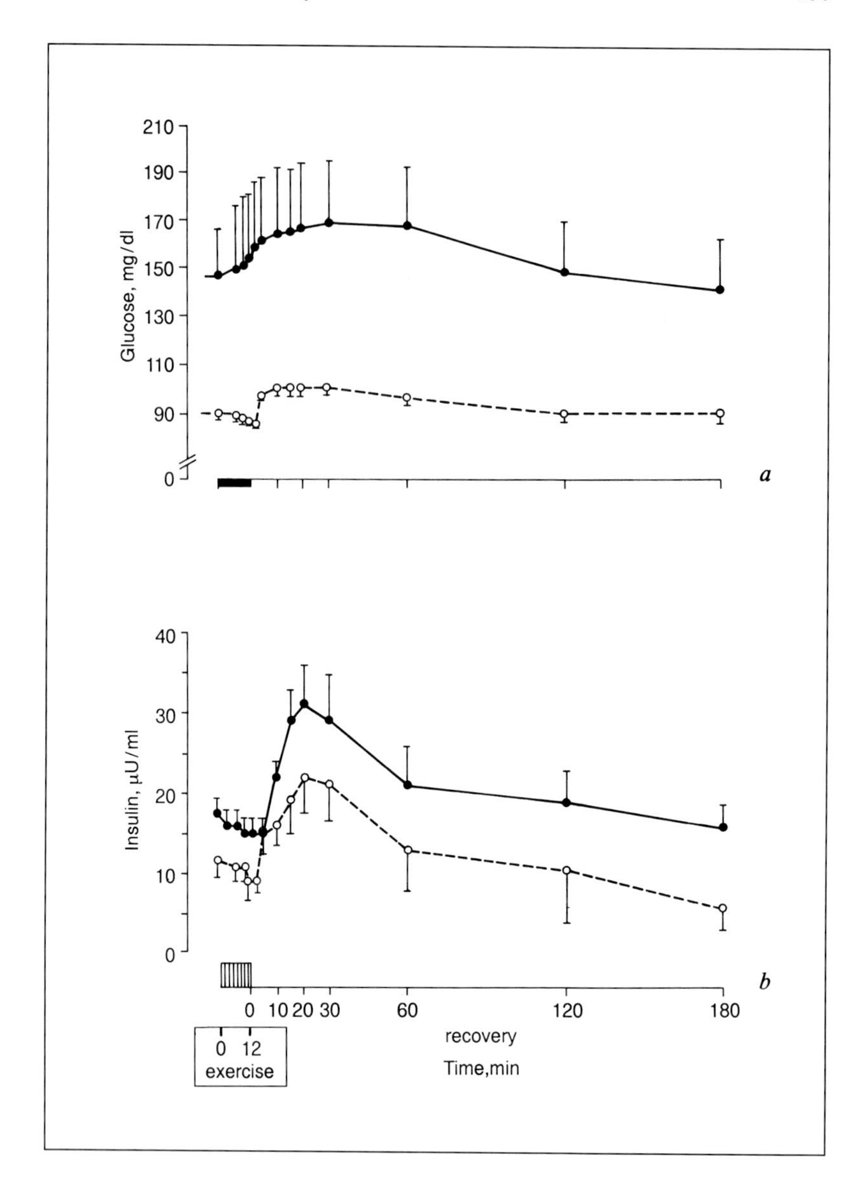

Fig. 1. Glucose *(a)* and insulin *(b)* concentrations in plasma in response to graded bicycle exercise (7 min 60%, 3 min 100%, 2 min 110% $\dot{V}O_2$max) in type 2 diabetics (●; n = 7) and age-matched controls (○; n = 7). [From ref. 13.]

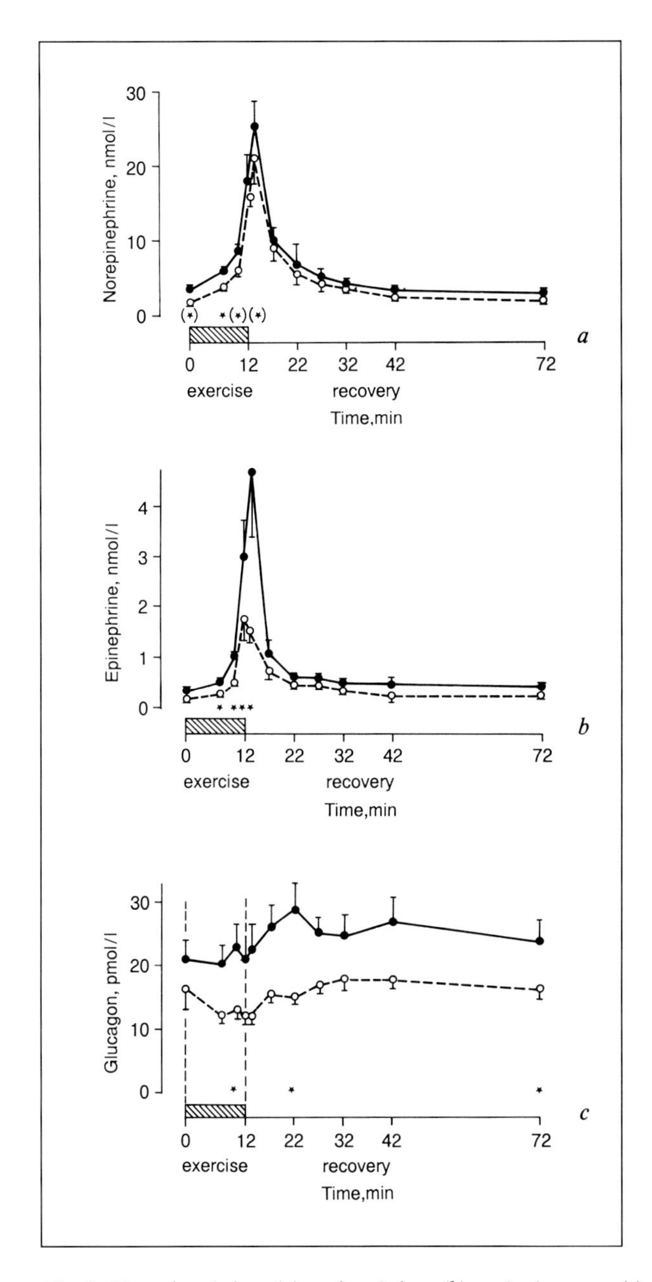

Fig. 2. Norepinephrine *(a),* epinephrine *(b)* and glucagon *(c)* concentrations in plasma in response to graded exercise in type II diabetics (●; n = 7) and controls (○; n = 7). See legend to figure 1. [From ref. 13.]

ics might respond to training in the opposite direction to healthy subjects: glucose concentration, overloaded beta cells might regain secretory capacity. During a 12-week endurance training program, 8 type II diabetics had a marked ($p < 0.05$) increase in $\dot{V}O_{2max}$ [2.5 ± 0.2 (SE) – 3.0 ± 0.3 liters/min] and decrease in heart rate at 100 W (117 ± 6 – 109 ± 5 beats/min). During euglycemic clamping at 100 μU/ml insulin, the rate of glucose infusion was 7.3 ± 1.6 mg/kg/min before and 8.2 ± 2.0 mg/kg/min ($p < 0.1$) after training. Pancreatic beta cell function evaluated by sitmulation with glucose and arginine (fig. 3), with a mixed meal (fig. 4) and with 1 mg glucagon intravenously (C peptide increase in 6 min from 1.5 ± 0.3 to 2.9 ± 0.6 pmol/ml before and from 1.0 ± 0.2 to 1.9 ± 0.4 pmol/ml after; $p > 0.05$), was unaltered by training. That beta cell function was not reduced as seen in young healthy subjects was probably not due to relief of beta cells by lowered glucose levels. This is so as glycosylated hemoglobin A_{1c} levels were unchanged (8.4 ± 0.8% before vs. 8.2 ± 0.7% after). Again, the lack of improvement in glucose homeostasis during training is probably due to increased food intake as indicated by unchanged body weight. It seems that type II diabetics only achieve improvements in glycemic control during training if they have restricted calorie intake and a considerable weight loss [14–16].

In order to clarify whether the increased effect of insulin after training reflects an adaptation in muscle tissue, we studied type II diabetics who performed one-legged ergometer bicycling at 70% $\dot{V}O_{2max}$ 30 min a day 6 days a week for 10 weeks. Glucose uptake was determined in both legs by arterial and femoral venous catheterization and blood flow measurement. Preliminary data are presented in figure 5. It is seen that the effect of insulin on glucose uptake was higher in the trained than in the untrained leg, indicating an adaptation in muscle. Muscle hypertrophy only partly explained this finding. Interestingly, however, the increased insulin action after training did not reflect an increased glucose extraction. In contrast to findings in young healthy subjects, glucose extraction was lower in the trained than in the untrained leg (fig. 5). This may indicate that in type II diabetics, compared to healthy subjects, training-induced adaptations favoring glucose extraction (e.g. increase in capillarization and glucose transporters) are less readily developed relative to increases in mass of contractile proteins. metabolic rate and blood flow. Thus, even though training may enhance the effect of insulin in type II diabetics, the training response seems to be abnormal, a fact which confirms that these patients have a defective muscle glucose metabolism.

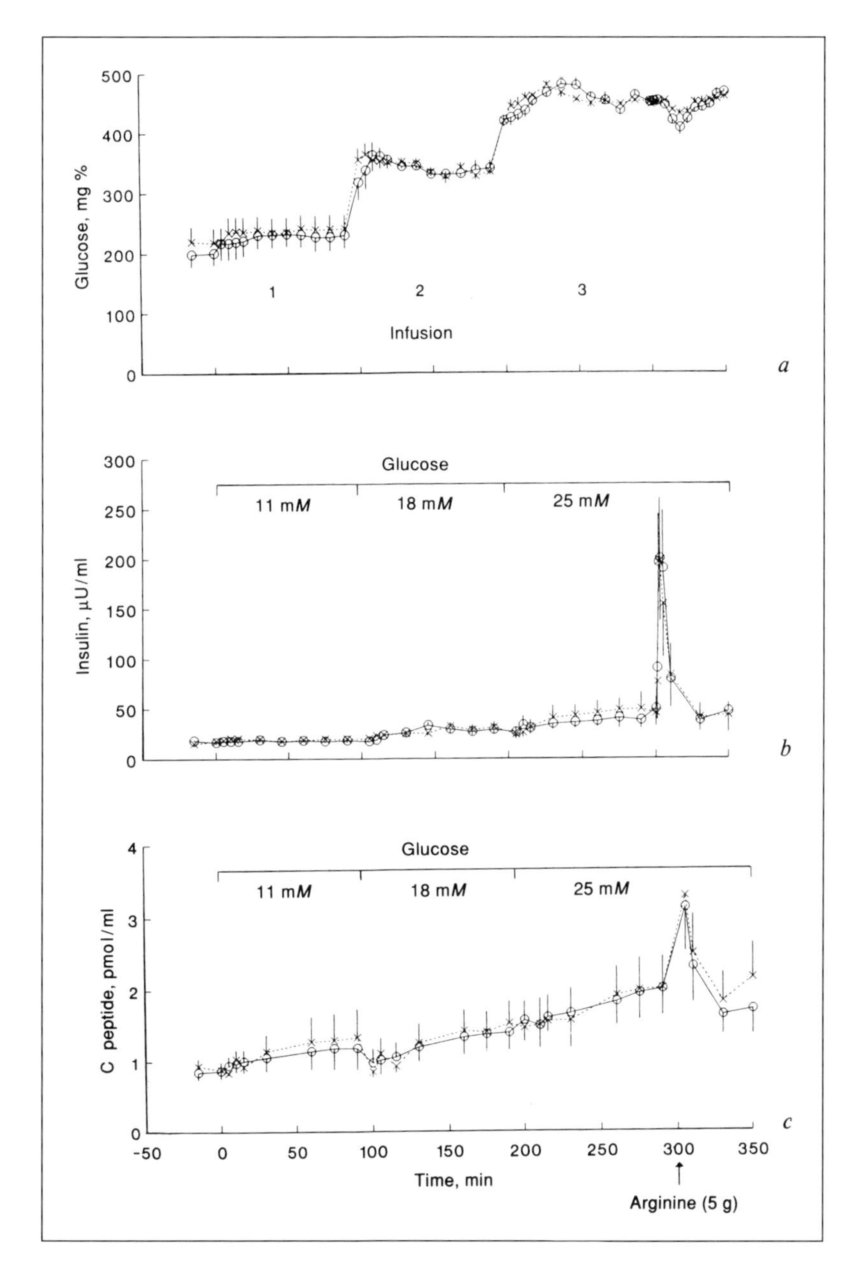

Fig. 3. Glucose *(a),* insulin *(b)* and C peptide *(c)* concentrations in plasma (mean ± SE) during hyperglycemic clamp and arginine bolus in type II diabetics before (○) and after (×) training.

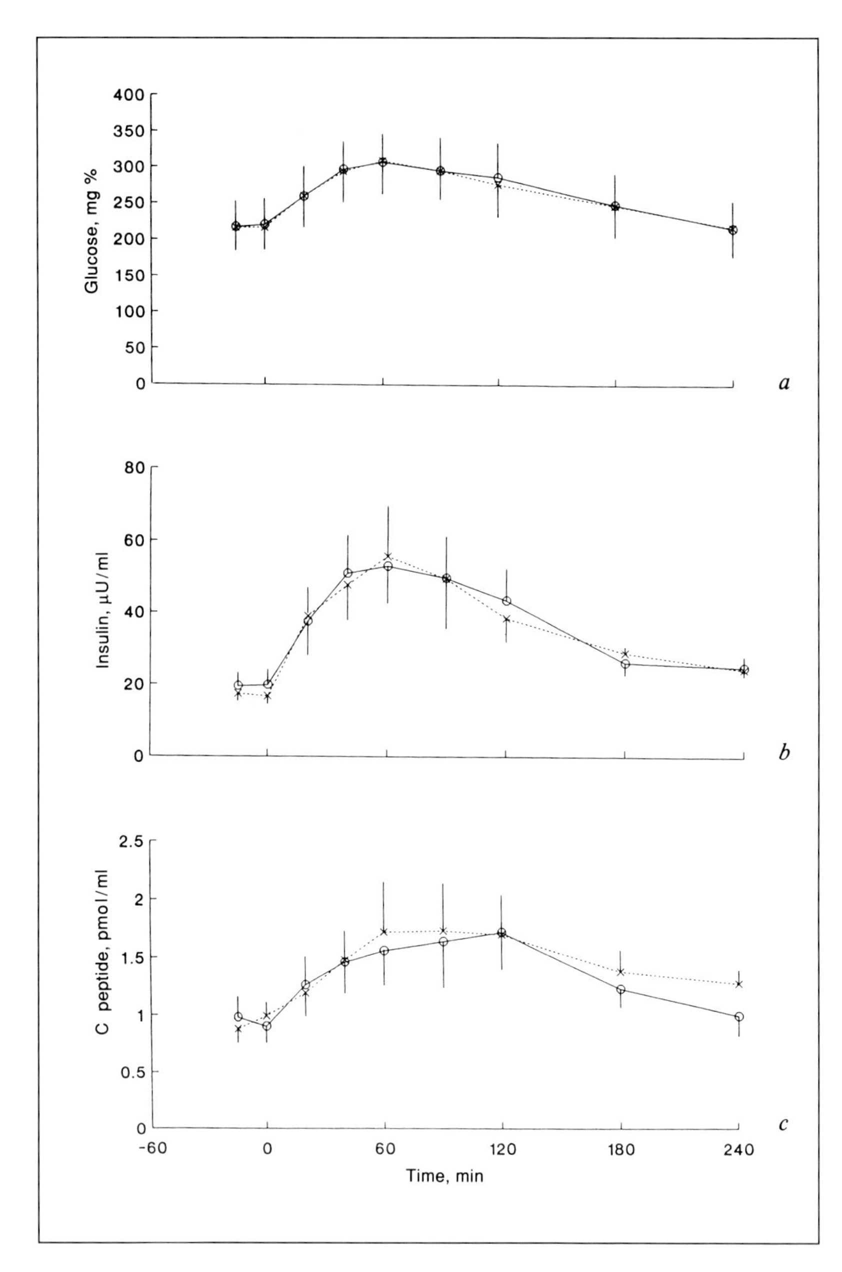

Fig. 4. Glucose *(a)*, insulin *(b)* and C peptide *(c)* concentrations in plasma (mean ± SE) in response to a mixed meal (1,912 kJ) in type II diabetics (n = 6) before (○) and after (×) training.

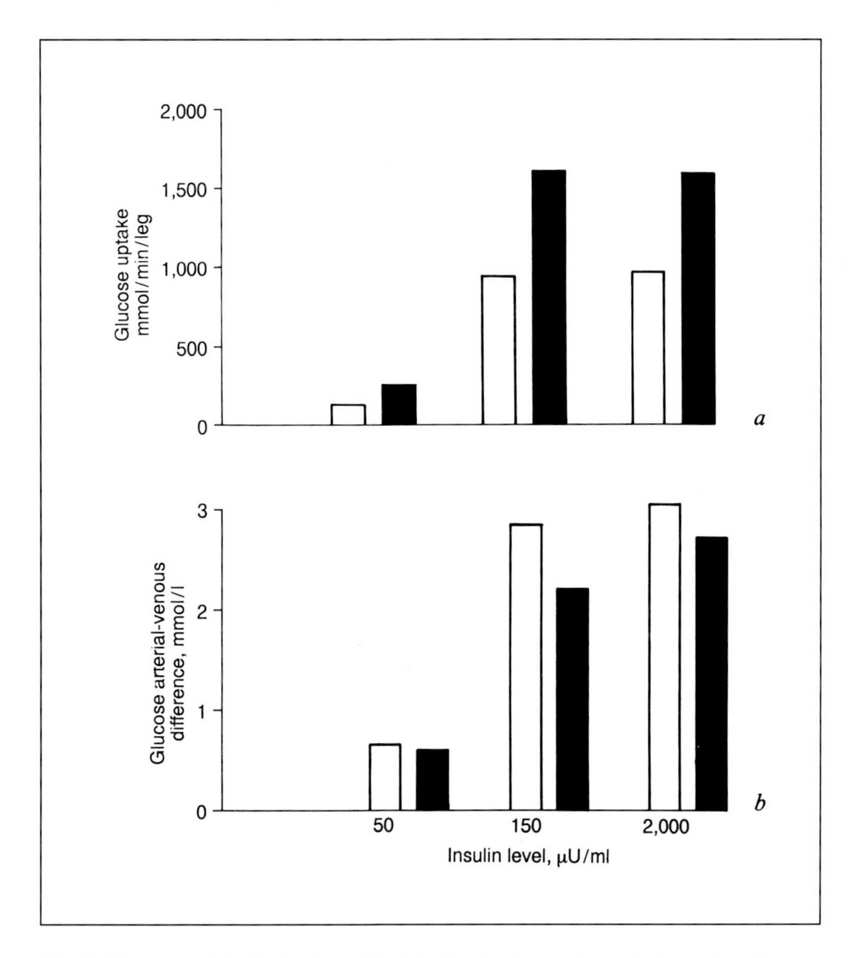

Fig. 5. Glucose uptake in the legs of 2 diabetics during euglycemic, hyperinsulinemic clamp after one-legged training. □ = Untrained leg; ■ = trained leg. *a* Glucose uptake. *b* Glucose arterial-venous difference.

References

1 Galbo H: Exercise and diabetes. Scand J Sports Sci 1988;10:89–95.

2 Koivisto VA, Yki-Järvinen H, DeFronzo RA: Physical training and insulin sensitivity. Diabetes Metab Rev 1986;1:445–481.

3 Rönnemaa T, Koivisto VA: Combined effect of exercise and ambient temperature on insulin absorption and postprandial glycemia in type 1 patients. Diabetes Care 1988;11:769–773.

4 Schiffrin A, Parikh S: Accommodating planned exercise in type 1 diabetic patients on intensive treatment. Diabetes Care 1985;8:337–342.
5 Sonnenberg GE, Kemmer FW, Berger M: Exercise in type 1 (insulin-dependent) diabetic patients treated with continuous subcutaneous insulin infusion. Diabetologia 1990;33:696–703.
6 Shilo S, Sotsky M, Shamoon H: Islet hormonal regulation of glucose turnover during exercise in type 1 diabetes. J Clin Endocrinol Metab 1990;70:162–172.
7 Wasserman DH, Spalding JA, Lacy DB, Colburn CA, Goldstein RE, Cherrington AD: Glucagon is a primary controller of hepatic glycogenolysis and gluconeogenesis during muscular work. Am J Physiol 1989;257:E108–E117.
8 Wasserman DH, Lavina H, Lickley A, Vranic M: Role of beta-adrenergic mechanisms during exercise in poorly controlled diabetes. J Appl Physiol 1985;59:1282–1289.
9 Zander E, Bruns W, Wulfert P, Besch W, Lubs D, Chlup R, Schulz B: Muscular exercise in type 1 diabetics. Exp Clin Endocrinol 1983;82:78–90.
10 Graham C, Lasko-McCarthey P: Exercise options for persons with diabetic complications. Diabetes Educ 1990;16:212–220.
11 Hornsby WG, Boggess KA, Lyons TJ, Barnwell WH, Lazarchick J, Colwell JA: Hemostatic alterations with exercise conditioning in NIDDM. Diabetes Care 1990; 13:87–92.
12 Kemmer FW, Tacken M, Berger M: Mechanism of exercise-induced hypoglycemia during sulfonylurea treatment. Diabetes 1987;36:1178–1182.
13 Kjær M, Hollenbeck CB, Frey-Hewitt B, Galbo H, Haskell W, Reaven GM: Glucoregulation and hormonal responses to maximal exercise in non-insulin dependent diabetes. J Appl Physiol 1990;68:2067–2074.
14 Wing RR, Epstein LH, Paternostro-Bayles M, Kriska A, Nowalk MP, Gooding W: Exercise in a behavioural weight control programme for obese patients with type 2 (non-insulin-dependent) diabetes. Diabetologia 1988;31:902–909.
15 Skarfors ET, Wegener TA, Lithell H, Selinus I: Physical training as treatment for type 2 (non-insulin-dependent) diabetes in elderly men: A feasibility study over 2 years. Diabetologia 1987;30:930–933.
16 Lucas CP, Patton S, Stepke T, Kinhal V, Darga LL, Carroll-Michals L, Spafford TR, Kasim S: Achieving therapeutic goals in insulin-using diabetic patients with non-insulin-dependent diabetes mellitus. Am J Med 1987;83:3–9.

Henrik Galbo, MD, Department of Medical Physiology B, The Panum Institute, Blegdamsvej 3 C, DK–2200 Copenhagen N (Denmark)

Sato Y, Poortmans J, Hashimoto I, Oshida Y (eds): Integration of Medical and Sports Sciences. Med Sport Sci. Basel, Karger, 1992, vol 37, pp 237–242

Is There an Intimate Interplay between Temperature Acclimation and Exogenous Insulin?

With Special Reference to the Participation of Brown Adipose Tissue

H. Yamashita[a], *N. Sato*[b], *M. Yamamoto*[b], *S. Gasa*[c], *J. Nagasawa*[d], *Y. Sato*[d], *Y. Habara*[e], *M. Ishikawa*[f], *M. Segawa*[a], *H. Ohno*[a]

Departments of [a]Hygiene and [b]Biochemistry, National Defense Medical College, Tokorozawa; [c]Biochemistry Laboratory, Cancer Institute, Hokkaido University School of Medicine, Sapporo; [d]Research Center of Health, Physical Fitness and Sports, Nagoya University, Nagoya; [e]Laboratory of Cellular Metabolism, Department of Molecular Physiology, National Institute for Physiological Sciences, Okazaki; [f]Department of Obstetrics and Gynecology, Asahikawa Medical College, Asahikawa, Japan

An association between the incidence of childhood type I (insulin-dependent) diabetes mellitus and the average yearly temperature in different countries has been reported, the incidence being lower in countries with a higher mean temperature [1]. Likewise, animals kept at a higher temperature have a lower incidence of diabetes than those kept at a lower temperature [2]. In addition, glycosylated hemoglobin levels of nondiabetic children and adults are highest at the ends of autumn and lowest at the ends of spring and summer [3]. Our previous study has demonstrated, however, that the immunoreactive insulin level in heat-acclimated (HA) rats is significantly decreased in plasma and not detected in brown adipose tissue (BAT) [4], which is widely recognized as a main site for both cold- and diet-induced thermogenesis [5–7]. BAT mass is remarkably greater in cold-acclimated (CA) rats in terms of weight or weight per unit body weight, whereas it is significantly smaller in HA rats than in warm-adapted (WA) controls [4]. Therefore, the experiments reported here were undertaken to investigate whether an interaction between temperature acclimation and exogenous insulin exists in BAT of rats.

Materials and Methods

A total of 178 male Wistar strain rats (7 weeks old) weighing 150–180 g at the beginning of the experiments were divided into three groups. The first group (n = 91) was reared at an ambient temperature of 34 °C with a relative humidity of 40% for 2 weeks in an artificial climatic room with lights on from 7.00 to 19.00 h daily in individual cages. These rats were referred to as HA rats. They had free access to standard laboratory diet (Oriental MF, Oriental Yeast, Tokyo, Japan) and tap water. The animals were cared for in accordance with the guiding principles in the care and use of animals based upon the Helsinki declaration. The second group (n = 8) was kept at 5 °C (CA rats). The third group, made up of control (WA) rats, was kept at the thermoneutral temperature of 25 °C. Half of the rats from each group were injected subcutaneously (interscapular portion) with 3.62 nmol (equivalent to 0.5 IU) insulin Novo Lente MC (porcine/bovine; Novo Industry, Copenhagen, Denmark) per 125 μl of saline per 100 g body weight at 16.00 h for 13 days and the rest were treated with physiological saline according to the method of Harada and Kato [8]. The rats were fasted for 17–18 h under each climatic condition prior to the experiment but were allowed tap water. On the day before the experiments, neither insulin nor saline was administered because the rats died of extensive hypoglycemia during the starvation period as a result of the hormone injection [9].

After collecting trunk blood by cervical dislocation and exsanguination, the interscapular BAT was removed. In consequence, an apparent effect of insulin treatment was noted only in HA rats, resulting in a remarkable gain in interscapular BAT mass in HA/insulin-treated (HI) rats in terms of weight or weight per unit body weight (table 1). Moreover, since valuable pieces of information are now available on the effects of cold stress on BAT [5, 6], the following experiments were conducted only in HA and WA rats,

Table 1. Changes in body and BAT weights of temperature-acclimated rats

Group	Body weight, g	BAT weight	
		mg	mg/100 g body weight
Warm-adapted			
Saline (40)	216.4 ± 3.3	124.3 ± 8.9	59.2 ± 4.7
Insulin (39)	218.6 ± 2.8	131.8 ± 11.7	61.3 ± 5.6
Heat-exposed			
Saline (37)	189.1 ± 2.1	93.6 ± 7.0	50.2 ± 4.1
Insulin (54)	188.7 ± 1.7	270.9 ± 28.1*	148.7 ± 16.1*
Cold-exposed			
Saline (4)	149.0 ± 8.4	200.8 ± 12.3	134.7 ± 1.8
Insulin (4)	150.0 ± 4.0	199.5 ± 5.3	133.3 ± 4.5

Values are means ± SEM. Figures in parentheses indicate number of animals. * $p < 0.05$: significantly higher than 'saline' value.

except for the mRNA study of several proteins. The contents of protein, DNA and RNA in BAT were determined by routine methods [10, 11]. Total RNA was isolated from BAT using the acidic guanidinium isothiocyanate method [12]. Oligonucleotide probes including uncoupling protein, basic fibroblast growth factor (bFGF), insulin-like growth factor II (IGF-II), hepatocyte growth factor and insulin receptor were synthesized with an Applied Biosystems model 380B DNA synthesizer (Foster City, Calif., USA) according to the manufacturer's instruction, and each mRNA sequence was detected by a polymerase chain reaction method described by Arrigo et al. [13].

In addition, the effect of BAT extract on capillary growth in vitro was examined by co-cultivating the microvessel fragments and myofibroblastic cells from rat epididymal fat pads, as originally developed in our group [14, 15]. The activities of lysosomal enzymes (β-glucuronidase, arylsulfatase A and cathepsin D) were determined by the method of Barrett [16]. Triglyceride content in BAT was assayed by a commercial kit (Triglyceride E-Test Wako, Wako Pure Chemical, Osaka, Japan). To assess cold tolerance, changes in the colonic temperature were measured by exposing the animals to 5 °C for 2 h. The colonic temperature was measured by a thermistor thermometer inserted 5 cm into the rectum.

Statistical tests included Student's t test and variance analysis. Differences were considered significant at $p < 0.05$.

Results and Discussion

As previously stated, a significant effect of insulin treatment was observed only in BAT of HA rats (table 1). The BAT from HI rats had significantly higher levels of protein, DNA and RNA than that from HA/saline-treated (HS) rats, probably growing towards hyperplasia (table 2). Considering the increased level of triglyceride (table 2), it seems likely that

Table 2. Effects of heat exposure on protein, DNA and RNA contents in BAT

Group	Protein content mg/g wet tissue	DNA content μg/g wet tissue	RNA content μg/g wet tissue	Triglyceride content mg/g wet tissue
Warm-adapted				
Saline	22.2 ± 3.2	552.7 ± 61.8	726.8 ± 110.9	480.8 ± 15.5
Insulin	36.8 ± 4.6*	503.2 ± 49.3	953.4 ± 141.3	463.8 ± 18.1
Heat-exposed				
Saline	27.4 ± 3.1	287.4 ± 60.6	503.8 ± 134.8	335.2 ± 40.1
Insulin	43.1 ± 5.7*	589.5 ± 99.4*	992.6 ± 98.8*	478.4 ± 9.3*

Values are means ± SEM obtained from 5 tissues. * $p < 0.05$: significantly higher than 'saline' value.

the growth of BAT in HI rats was mostly due to the anabolic effects of insulin. Indeed, in our previous report on pancreatic exocrine secretion, daily food consumption in HA rats decreased significantly to 68% of that of WA rats, there being no significant difference between food consumption in HS and HI rats [9]. The facts indicate that the marked growth of BAT in HI rats in the present work cannot be attributed to overfeeding.

However, uncoupling protein mRNA was present in BAT of HI rats at a rather depressed level, explaining a corresponding decrease in cold tolerance. Actually the colonic temperature dropped gradually and significantly in HI rats during cold exposure at 5 °C, but not in HS, WA/saline-treated or WA/insulin-treated rats. As expected, mRNA for uncoupling protein was remarkably induced in CA rats, especially in CA/insulin-treated rats. As also expected, the expression of insulin receptor mRNA appeared to be attenuated in BAT of insulin-treated groups (HI, WA/insulin-treated and CA/insulin-treated rats), possibly because of the down-regulation of insulin. The mRNAs for IGF-II and hepatocyte growth factor existed in sufficient quantities in any BAT irrespective of temperature exposure or insulin administration.

On the other hand, we examined the effects of BAT extract on capillary growth in vitro. The addition of BAT extracts from WA/saline-treated, WA/insulin-treated, HS and HI rats, particularly from HI rats, resulted in an increase in length of capillary tubes, the effects being dose-dependent. From the facts that the growth of bovine capillary endothelial cells was not affected by BAT extract from HI rats while the extract increased the number of myofibroblastic cells, the effects of BAT extract on angiogenesis seemed to occur via the proliferation of myofibroblastic cells rather than endothelial cells. The BAT extract from HI rats also stimulated the production of endothelial cell growth factor, which is responsible for in vitro capillary growth [14], by myofibroblastic cells. Moreover, the BAT extract enhanced the synthesis of collagen, which contributes significantly to tubular formation [14], by myofibroblastic cells. No direct effect of anti-bFGF antibody, insulin, IGF-I or IGF-II on the capillary growth was observed. Schweigerer et al. [17] have demonstrated that capillary endothelial cells express bFGF, a mitogen that promotes their own growth. In the present study, bFGF mRNA transcripts could be detected only in BAT from HA rats, which is much greater in HS rats than in HI rats. It appears, thus, that heat temperature induced the expression of bFGF mRNA in BAT of rats irrespective of insulin treatment, but there was no direct evidence linking FGF to the growth of BAT.

Surprisingly, the activities of the three lysosomal enzymes in BAT of HI rats showed remarkable decreases, although the tissue grew to a great mass. This finding was in sharp contrast to that observed in CA rats [18]. The precise physiological meaning of the decreases in BAT activities of lysosomal enzymes remains to be explained. However, in view of the facts that BAT in streptozotocin-diabetic rats is atrophied [19] and that at thermoneutrality (28–32 °C) insulin-induced hypoglycemia brings about an increase in blood supply to BAT [20] – in addition to our findings – one would deduce that some linkage is present between BAT, temperature and insulin. Work is, therefore, presently being carried out to investigate the mechanism of the growth of BAT in HI rats.

References

1 Diabetes Epidemiology Research International Group: Geographic patterns of childhood insulin-dependent diabetes mellitus. Diabetes 1988;37:1113–1119.
2 Williams AJK, Krug J, Lampeter EF, Mansfield K, Beales PE, Signore A, Gale EAM, Pozzilli P: Raised temperature reduces the incidence of diabetes in the NOD mouse. Diabetologia 1990;33:635–637.
3 MacDonald MJ, Liston L, Carlson I: Seasonality in glycosylated hemoglobin in normal subjects: Does seasonal incidence in insulin-dependent diabetes suggest specific etiology? Diabetes 1987;36:265–268.
4 Habara Y, Kuroshima A: Changes in glucagon and insulin contents of brown adipose tissue after temperature acclimation in rats. Jpn J Physiol 1983;33:661–665.
5 Cannon B, Nedergaard J: The biochemistry of an inefficient tissue: Brown adipose tissue. Essays Biochem 1985;20:110–164.
6 Himms-Hagen J: Brown adipose tissue and cold acclimation; in Trayhurn P, Nicholls DG (eds): Brown Adipose Tissue. London, Arnold, 1986, pp 214–268.
7 Lardy H, Shrago E: Biochemical aspects of obesity. Annu Rev Biochem 1990;59: 689–710.
8 Harada E, Kato S: Influence of adrenaline, glucagon, hydrocortisone, thyroxine, or insulin administration on pancreatic exocrine secretion in rats. Jpn J Vet Sci 1982; 44:589–596.
9 Habara Y: Augmentation of secretagogue-induced amylase secretion in pancreatic acini of heat-exposed rats. J Physiol 1989;413:91–105.
10 Lowry OH, Rosebrough NJ, Farr AL, Randall RJ: Protein measurement with the Folin phenol reagent. J Biol Chem 1951;193:265–275.
11 Schneider WC: Phosphorus compounds in animal tissues. I. Extraction and estimation of desoxypentose nucleic acid and of pentose nucleic acid. J Biol Chem 1945; 161:293–303.
12 Chomczynski P, Sacchi N: Single-step method of RNA isolation by acid guanidinium thiocyanate-phenol-chloroform extraction. Anal Biochem 1987;162:156–159.

13 Arrigo SJ, Weitsman S, Rosenblatt JD, Chen IY: Analysis of rev gene function on human immunodeficiency virus type 1 replication in lymphoid cells by using a quantitative polymerase chain reaction method. J Virol 1989;63:4875–4881.

14 Sato N, Sawasaki Y, Sendo A, Fuse Y, Hirano Y, Goto T: Development of capillary networks from rat microvascular fragments in vitro: The role of myofibroblastic cells. Microvasc Res 1987;33:194–210.

15 Sato N, Fukuda K, Nariuchi H, Sagara N: Tumor necrosis factor inhibiting angiogenesis in vitro. J Natl Cancer Inst 1987;79:1383–1391.

16 Barrett AJ: Lysosomal enzymes; in Dingle JT (ed): Lysosomes: A Laboratory Handbook. Amsterdam, North-Holland, 1977, pp 46–135.

17 Schweigerer L, Neufeld G, Friedman J, Abraham JA, Fiddes JC: Capillary endothelial cells express basic fibroblast growth factor, a mitogen that promotes their own growth. Nature 1987;325:257–259.

18 Ohno H, Yahata T, Kuroshima A, Gasa S, Makita A, Kondo T, Fujiwara Y, Yamashita K, Yamamura K: Effects of swimming training on tolerance to cold in rats – with special reference to lysosomal enzymes; In Ueda G, Kusama S, Voelkel NF (eds): High-Altitude Medical Science. Matsumoto, Shinshu University, 1988, pp 431–435.

19 Seydoux J, Chinet A, Schneider-Picard G, Bas S, Imesch E, Assimacopoulos-Jeannet F, Giacobino JP, Girardier L: Brown adipose tissue metabolism in streptozotocin-diabetic rats. Endocrinology 1983;113:604–610.

20 Benzi RH, Girardier L: The response of adipose tissue blood flow to insulin-induced hypoglycemia in conscious dogs and rats. Pflügers Arch 1986;406:37–44.

Prof. H. Ohno, Department of Hygiene, National Defense Medical College,
3-2, Namiki, Tokorozawa 359 (Japan)

Regulation of Substrate Utilization in Muscle during Exercise

Sato Y, Poortmans J, Hashimoto I, Oshida Y (eds): Integration of Medical and Sports Sciences. Med Sport Sci. Basel, Karger, 1992, vol 37, pp 243–251

Glucose Metabolism in Exercising Man

John Wahren, Abram Katz

Department of Clinical Physiology, Karolinska Hospital, Stockholm, Sweden

Glucose is one of the body's key nutrients and it plays a particularly important role in energy metabolism under conditions of exercise. The two tissues primarily responsible for glucose homeostasis during exercise in man are the liver and muscle, the liver releasing and the working muscle utilizing glucose. Here, we will briefly review selected aspects of hepatic glucose production and muscle glucose utilization during exercise in man.

Hepatic Glucose Production during Exercise

The intensity and duration of exercise are the major determinants of the magnitude of the increase in hepatic glucose production as well as the relative contributions of glycogenolysis and gluconeogenesis, respectively. In response to mild (65 W), moderately heavy (130 W) or strenuous exercise (200 W) lasting 40–60 min, hepatic glucose production increases gradually to values 2-, 3- or 5-fold, respectively, above basal [1]. Examination of the rate of hepatic glucose precursor uptake during exercise indicates that, while peripheral release of lactate increases during exercise, splanchnic blood flow decreases, resulting in a largely unchanged hepatic precursor uptake at mild work intensities and during short-term exercise. The estimated contribution made by gluconeogenesis to total glucose output remains at the basal level (25–30%) with light exercise but may fall to 6–15% with moderate to heavy exercise. Consequently, the major source of increased glucose production during short-term exercise, especially at heavy work loads, is glycogen breakdown.

Table 1. Glucose production and uptake of glucose precursors across the splanchnic vascular bed at rest and during prolonged exercise (30% VO_2max) in healthy subjects

	Rest	Exercise			
		40 min	90 min	180 min	240 min
Glucose production	0.82	1.86	1.85	1.92	1.46
Uptake of					
Lactate	0.14	0.17	0.10	0.20	0.33
Amino acids	0.05	0.09	0.13	0.12	0.12
Glycerol	0.02	0.10	0.10	0.17	0.21
Pyruvate	0.01	0.02	0.01	0.01	0.03
Total	0.22	0.38	0.34	0.50	0.69
Percent of glucose production	27	21	18	26	47

Data from Ahlborg et al. [2] are presented as glucose equivalents in millimoles per minute.

An increasing reliance on hepatic gluconeogenesis is observed as exercise extends beyond the first hour. During prolonged exercise at 30% of maximal oxgen uptake (VO_2max), glucose output doubles in the first 40 min and thereafter remains constant for the ensuing 3 h [2]. However, glucose production fails to keep pace with the rate of glucose utilization and a modest decline in blood glucose concentration is observed. The relative contribution from gluconeogenesis to overall hepatic glucose production (as evaluated from splanchnic precursor uptake) increases from 27% in the basal state to 47% during prolonged exercise, representing a 3-fold rise in the absolute rate of gluconeogenesis (table 1). The rise in hepatic uptake of glucose precursors is a result of both augmented fractional extraction and increased arterial concentrations. In the case of alanine, the major amino acid precursor, fractional extraction by the splanchnic bed increases from basal levels of 35–45% to almost 90% during prolonged exercise. The overall importance of gluconeogenesis in prolonged exercise is underscored by the estimation that 50–60 g of liver glycogen is mobilized in 4 h of exercise, representing a depletion of approximately 75% of the total liver glycogen stores [2, 3].

Regulation of Hepatic Glucose Output

Several hormonal changes which accompany the onset of exercise serve to maintain an adequate supply of fuels for working muscle. The plasma insulin concentration decreases during exercise regardless of intensity. Glucagon levels do not change markedly in response to mild or moderate short-term exercise, but increase when exercise is prolonged and accompanied by mild hypoglycemia. In addition, circulating concentrations of both noradrenaline and adrenaline as well as growth hormone rise during exercise. These hormonal changes together with the possible stimulation of hepatic glucose production directly via neural activation of the hepatocyte should be considered the primary regulatory factors responsible for control of hepatic glucose production during exercise.

Studies in intact humans – involving hyperinsulinemia induced by intravenous glucose infusion – have indicated that a normal exercise-induced rise in hepatic glucose production can occur even though insulin does not fall below the basal level [4]. Additional studies have examined the role of glucagon in the stimulation of glucose production during exercise. Hypoglucagonemia was induced by infusion of somatostatin for 10 or 60 min prior to exercise. In either circumstance the exercise-induced rise in glucose production was similar to that of controls [5]. Moreover, with somatostatin infusion and insulin replacement during exercise, the rise in glucose production during exercise was also found to be normal. These studies thus indicate that the glucagon levels normally observed during exercise in humans are not essential for the rise in hepatic glucose production that accompanies short-term exercise. Whether the contribution of gluconeogenesis to hepatic glucose production, particularly during prolonged exercise, is dependent on hypoinsulinemia and/or the presence of basal or elevated levels of glucagon remains to be determined.

Studies designed to examine the role of circulating noradrenaline and adrenaline have indicated that neither β-adrenergic nor α-adrenergic blockade significantly interferes with the exercise-induced rise in glucose production in humans [6]. However, it should be noted that these studies involved exercise after an overnight fast and did not result in hypoglycemia. Under other conditions with an exaggerated catecholamine response, such as during short-term starvation [7] or prolonged exercise with accompanying hypoglycemia [2, 8], it is possible that elevated catecholamine levels play an important role in the stimulation of hepatic glucose production.

Although the possibility that hormones other than insulin, glucagon and catecholamines are involved in the regulation of hepatic glucose production during exercise should not be excluded, it is unlikely that they play a major role in this respect. Changes in circulating hormone levels do not always fully explain the exercise-induced rise in glucose production. It has been suggested that the rise in hepatic glucose production may in part be the result of direct stimulation of the hepatocytes via neural activation. Support for this view has been obtained from animal studies in which electrical stimulation of the splanchnic nerves was found to activate glycogen phosphorylase [9, 10], resulting in hepatic glycogen depletion and hyperglycemia [11]. There is evidence to suggest that human liver cells have an abundant sympathetic innervation and that electrical stimulation of the splanchnic nerves results in augmented glucose production [12], but the possible importance of this phenomenon for the rise in hepatic glucose production during exercise remains to be determined.

Glucose Metabolism in Muscle during Exercise

The first step in muscle glucose utilization is the transport of glucose to the muscle (via the circulation) as well as the diffusion across the capillary wall. The second step is the transport across the cell membrane. The last step is the phosphorylation of glucose to glucose-6-phosphate (G6P), which is catalyzed by hexokinase. G6P can then be further metabolized (i.e. glycolysis, glycogenesis, pentose shunt). Under normal conditions (when glycogen is not severely depleted and the exercise intensity is low to moderate), glucose is glycolyzed and subsequently oxidized in the tricarboxylic acid (TCA) cycle. Below, we will briefly summarize data on muscle glucose metabolism during short- and long-term dynamic exercise in man, and discuss some potential regulatory mechanisms as well as the significance of carbohydrate availability for muscle bioenergetics.

Short-Term Exercise

Glucose uptake by skeletal muscle [leg blood flow × (arterial-femoral glucose concentration)] increases within the first few minutes of exercise [13, 14]. The increase is positively related to exercise intensity and occurs under conditions where the arterial blood glucose concentration is fairly constant [1, 14]. It is, however, not clear whether all of the glucose taken up by the muscle is phosphorylated. Estimates of intracellular glucose in contracting muscle indicate an accumulation at exercise intensities between 60 and 80% of VO_2max [15, 16]. Under these conditions it is clear that neither diffusion across the capillary wall nor transport across the cell membrane are limiting factors for glucose utilization. The accumulation of glucose is probably more dependent on the relative rather than the absolute work intensity. This is illustrated by the finding that increasing the relative work intensity from 50 to ~ 70% of VO_2max (by inspiring 11% O_2), while maintaining the same absolute work load, results in marked accumulation of intracellular glucose at 70% VO_2max but no detectable accumulation at 50% VO_2max [15]. The accumulation of intracellular glucose thus becomes apparent at exercise intensities of ~ 60–80% VO_2max and increases as the exercise intensity is further increased [14–17]. The accumulation of intracellular glucose during heavy short-term exercise will decrease the concentration gradient across the muscle cell membrane and, therefore, it might be expected that transport into the cell as well as net uptake would decrease as the exercise progresses or as the intensity is increased. The experimental evidence, however, indicates that this is not the case. There is no decrease in leg glucose uptake (LGU) even at high exercise intensities (as long as blood flow is not compromised), nor as intense exercise progresses; in fact, LGU increases continuously during maximal dynamic exercise until the point of fatigue [14]. These data suggest that intense exercise is a very potent stimulator of glucose transport into the cell and that this process is not inhibited by large accumulations of intracellular glucose. It cannot be ruled out, however, that LGU would be even higher if the increase in intracellular glucose was attenuated.

Based on available data, a relationship between LGU, leg glucose utilization and relative work load during short-term exercise (5–10) min) has been derived (fig. 1) [18]. This relationship demonstrates that glucose uptake is matched by glucose utilization until ~ 75% VO_2max. Thereafter, LGU continues to increase, whereas utilization decreases progressively, becoming almost insignificant at ~ 100% VO_2max. The dissociation between uptake and utilization is probably due to inhibition of hexokinase. The most potent inhibitors of hexokinase are G6P and glucose-1,6-diphosphate which appear to bind to the same site on the enzyme [19]. During short-term exercise the muscle is heavily dependent on glycogen for ATP production. Glycogenolysis results in formation of G6P, and the accumulation of G6P increases in exponential fashion as a function of work load during short-term exercise [16]. On the other hand, the glucose-1,6-diphosphate content in muscle is relatively low ($<$ 10% of that of G6P), and is not significantly different from the value at rest during short-term incremental exercise [20]. Therefore, the dissociation between LGU and utilization

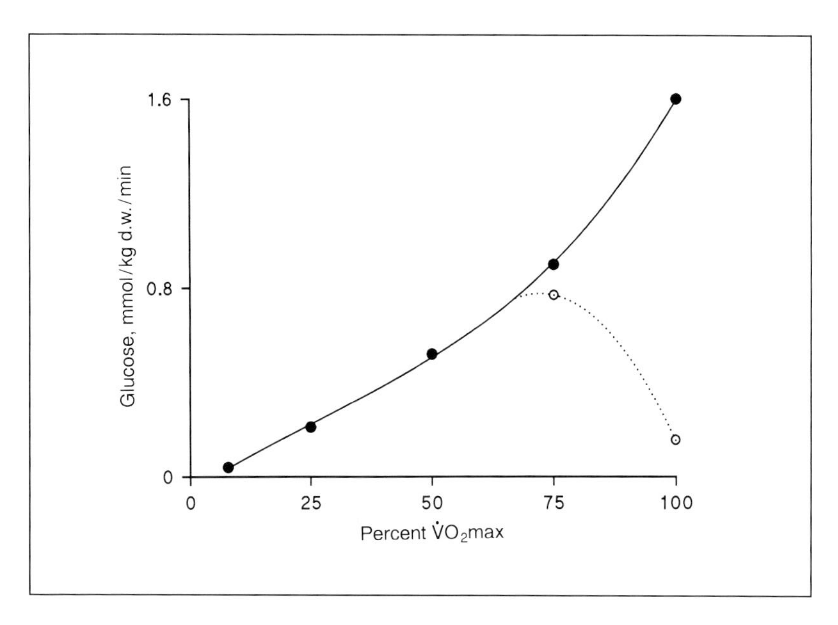

Fig. 1. Leg glucose uptake (——•) and utilization (·····○) during short-term exercise (5–10 min) in man. LGU per kilogram dry weight of muscle is calculated from glucose uptake/(0.0256 × body weight), where 2.56% of body weight is the dry muscle mass of one leg. Glucose utilization = rate of uptake – rate of glucose accumulation. Uptake and utilization are considered to be identical until at least 50% VO_2max, since until this point glucose does not accumulate in the muscle. Values are means from 5–10 subjects. [See ref. 18 for further details.]

during short-term exercise has been attributed to G6P mediated inhibition of hexokinase [14, 15, 21]. It should be noted, however, that other factors, e.g. alterations in binding of hexokinase to the mitochondrial membrane which alters its kinetic characteristics [22], may also play a role in regulation of hexokinase in skeletal muscle.

Prolonged Exercise

Prolonged exercise (>60 min) also results in an increase in muscle glucose uptake which increases as a function of exercise intensity [23]. Depending on the exercise intensity, changes in the LGU rates are variable. For example, at 30–60% VO_2max, LGU increases continuously to a peak at ~90 min, and thereafter falls progressively. The fall is more precipitous at 60 than at 30% VO_2max and is in both cases associated with a fall in the arterial blood glucose concentration [2, 8]. At higher exercise intensities [70–75 VO_2max), LGU increases continuously to a peak at 20–40 min and thereafter remains stable until the point of fatigue (65–75 min) [21, 24]. Under these conditions, the decrease in arterial blood glucose at the point of fatigue is modest (~4.5 m*M*/l), and thus fatigue cannot be attributed to hypoglycemia.

Within the first 5 min of exercise at 75% VO_2max, there is a 7-fold increase in LGU. At the same time there is a large increase in G6P and intracellular glucose [21]. However, as the exercise progresses, muscle glycogenolysis decreases [25]; the levels of G6P and intracellular glucose fall, and LGU continues to increase. By 40 min intracellular glucose is no longer detectable and LGU has peaked. Again, the findings are consistent with the idea that G6P is inhibiting hexokinase, and that during the initial period of exercise neither diffusion across the capillary wall nor transport across the cell membrane limits glucose utilization. During the latter phase of exercise, however, when arterial blood glucose starts to decrease, the limiting step for glucose utilization shifts from phosphorylation to glucose availability. This could mean that either diffusion across the capillary wall or the transport capacity is compromised. The fact that ingestion of exogenous glucose prevents the fall in arterial blood glucose levels, and increases LGU and carbohydrate oxidation [3], suggests that normally the diffusion of glucose across the capillary wall and not the transport capacity limits glucose utilization during the later phase of prolonged exercise.

Significance of Glucose Availability

It is clear that muscle glycogen is the major substrate for ATP production during exercise. During the initial phase of exercise extracellular glucose plays a relatively minor role in the provision of substrate for aerobic ATP production. However, as exercise progresses, and glycogen is depleted, a greater reliance is placed on blood glucose [1, 21]. This is illustrated by the finding that during the last 30 min of a 2-hour exercise bout at ~60% VO_2max, oxidation of plasma glucose can account for ~1/3 of total carbohydrate oxidation [26].

Recent studies suggest that, in addition to providing substrate for the TCA cycle (i.e. acetyl-CoA), carbohydrate availability also plays an important role for stimulation of the TCA cycle via expansion of TCA cycle intermediates [25, 27, 28]. Here, the provision of 3-carbon glycolytic intermediates appears important for increasing flux through anaplerotic reactions. The subsequent increase in TCA cycle intermediates will stimulate oxidation of acetyl-CoA with a lower degree of physiologic stress (as evidenced by an attenuation in the decrease in muscle ATP and increase in inosine monophosphate). Under certain conditions (during the latter part of prolonged submaximal exercise), administration of exogenous glucose can result in higher levels of TCA cycle intermediates and lower levels of inosine monophosphate in human muscle [27]. These findings demonstrate that extracellular glucose is important both as a substrate for and an activator of the TCA cycle.

References

1 Wahren J, Felig P, Ahlborg G, Jorfeldt J: Glucose metabolism during leg exercise in man. J Clin Invest 1971;50:2715–2725.
2 Ahlborg G, Felig P, Hagenfeldt L, Hendler R, Wahren J: Substrate turnover during prolonged exercise in man: Splanchnic and leg metabolism of glucose free fatty acids and amino acids. J Clin Invest 1974;53:1080–1090.
3 Ahlborg G, Felig P: Influence of glucose ingestion on fuel-hormone response during prolonged exercise. J Appl Physiol 1976;41:683–688.
4 Felig P, Wahren J, Hendler R: Influence of physiological hyperglucagonemia on basal and insulin-inhibited splanchnic glucose output in normal man. J Clin Invest 1976;58:761–765.
5 Björkman O, Felig P, Hagenfeldt L, Wahren J: Influence of hypoglucagonemia on splanchnic glucose output during leg exercise in man. Clin Physiol 1981;1:43–57.
6 Simonson DC, Koivisto V, Sherwin RS, Ferrannini E, Hendler R, Juhlin-Dannfeldt A, Defronzo RA: Adrenergic blockade alters glucose kinetics during exercise in insulin-dependent diabetics. J Clin Invest 1984;73:1648–1658.
7 Björkman O, Eriksson LS: Splanchnic glucose metabolism during leg exercise in 60-hour-fasted human subjects. Am J Physiol 1983;245:E443–E448.
8 Ahlborg G, Felig P: Lactate and glucose exchange across the forearm, legs, and splanchnic bed during and after prolonged leg exercise. J Clin Invest 1982;69:45–54.
9 Shimazu T, Amakawa A: Regulation of glycogen metabolism in liver by the autonomic nervous system. IV. Possible mechanism of phosphorylase activation by the splanchnic nerve. Biochim Biophys Acta 1975;385:242–256.
10 Freude KA, Sandler LS, Zieve FJ: Electrical stimulation of the liver cell: Activation of glycogenolysis. Am J Physiol 1981;240:E226–E232.
11 Edwards AV: The glycogenolytic response to stimulation of the splanchnic nerves in adrenalectomized calves, sheep, dogs, cats and pigs. J Physiol 1971;213:741–759.
12 Nobin A, Falc B, Ingemansson S, Jarhult J, Rosengren E: The sympathetic innervation of the liver in man: Possible role in blood glucose regulation. Eur Surg Res 1977;(suppl 1):170.
13 Wahren J: Human forearm muscle metabolism during exercise. IV. Glucose uptake at different work intensities. Scand J Clin Lab Invest 1970;25:129–135.

14 Katz A, Broberg S, Sahlin K, Wahren J: Leg glucose uptake during maximal dynamic exercise in humans. Am J Physiol 1986;251.E65–E70.
15 Katz A, Sahlin K: Effect of hypoxia on glucose metabolism in human skeletal muscle during exercise. Acta Physiol Scand 1989;136:377–382.
16 Sahlin K, Katz A, Henriksson J: Redox state and lactate accumulation in human skeletal muscle during dynamic exercise. Biochem J 1987;245:551–556.
17 Sjogaard G, Saltin B: Extracellular and intracellular water spaces in muscles of man at rest and with dynamic exercise. Am J Physiol 1982;243:R271–R280.
18 Katz A: Regulation of Lactic Acid during Muscle Contraction; thesis, Karolinska Insititute, Stockholm, 1986.
19 Rose IR, Warms JVB: pH dependence of the α-glucose 1,6-diphosphate inhibition of hexokinase II. Arch Biochem Biophys 1975;171:678–681.
20 Katz A, Sahlin K, Henriksson J: Carbohydrate metabolism in human skeletal muscle during exercise is not regulated by G-1,6-P_2. J Appl Physiol 1988;65:487–489.
21 Katz A, Sahlin K, Broberg S: Regulation of glucose utilization in human skeletal muscle during moderate exercise. Am J Physiol 1991;260:E411–E415.
22 Wilson JE: Regulation of mammalian hexokinase activity; in Beitner R (ed): Regulation of Carbohydrate Metabolism. Boca Raton, CRC, 1985, pp 45–85.
23 Björkman O, Wahren J: Glucose homeostasis; in Horton E, Terjung RL (eds): Exercise, Nutrition and Energy Metabolism. New York, Macmillan, 1988, pp 100–115.
24 Broberg S, Sahlin K: Adenine nucleotide degradation in human skeletal muscle during prolonged exercise. J Appl Physiol 1989;67:116–122.
25 Sahlin K, Katz A, Broberg S: Tricarboxylic acid cycle intermediates in human muscle during prolonged exercise. Am J Physiol 1990;259:C834–C841.
26 Coggan AR, Kohrt WM, Spina RJ, Bier DM, Holloszy JO: Endurance training decreases plasma glucose turnover and oxidation during moderate-intensity exercise in men. J Appl Physiol 1990;68:990–996.
27 Spencer MK, Yan Z, Katz A: Carbohydrate supplementation attenuates IMP accumulation in human muscle during prolonged exercise. Am J Physiol 1991;26:C71–C76.
28 Spencer MK, Yan Z, Katz A: Effect of low glycogen on carbohydrate and energy metabolism in human muscle during exercise. Am J Physiol 1992;262:C975–C979.

John Wahren, Department of Clinical Physiology, Karolinska Hospital,
S–104 01 Stockholm (Sweden)

Sato Y, Poortmans J, Hashimoto I, Oshida Y (eds): Integration of Medical and Sports Sciences. Med Sport Sci. Basel, Karger, 1992, vol 37, pp 252–261

Interaction of Fuels in Muscle Metabolism during Exercise

Erik A. Richter

August Krogh Institute, University of Copenhagen, Denmark

Interaction between Glycogen and Glucose Utilization during Muscle Contractions

In an effort to understand the mechanisms that regulate selection of glucose versus muscle glycogen during exercise, we recently performed studies using the isolated perfused rat hindlimb preparation. Rats were preconditioned by different dietary and exercise regimens to have preperfusion muscle glycogen concentrations between 60 and 10 μmol/g wet weight. The hindlimbs were perfused with a standard medium containing 6 m*M* glucose and no insulin at rest and for 15 min of intermittent subtetanic contractions. Glucose uptake increased during muscle contractions in all groups but the increase was larger the lower the muscle glycogen concentration [1]. This is in accordance with findings in man [2]. In fact, a linear negative correlation between initial muscle glycogen concentration and glucose uptake during contractions was obtained (fig. 1). The mechanism behind this phenomenon apparently is multifactorial. In this hindlimb preparation, muscle glycogen breakdown during contractions is linearly related to preexercise muscle glycogen concentrations [1, 3]. The high rate of glycogen breakdown leads to high intramuscular concentrations of glucose-6-phosphate (G6P) [1] which inhibit hexokinase [4]. This in turn results in increased intracellular muscle glucose concentrations [1] which decrease the gradient of glucose from the interstitial space to the cytoplasm, in turn decreasing net glucose uptake. However, this was apparently not the only mechanism by which high preexercise muscle glycogen concentrations decreased muscle glucose uptake. When muscle glucose transport was estimated from muscle uptake of [^{14}C]-3-O-methylglucose during contractions it was found that glucose transport was approximately 25%

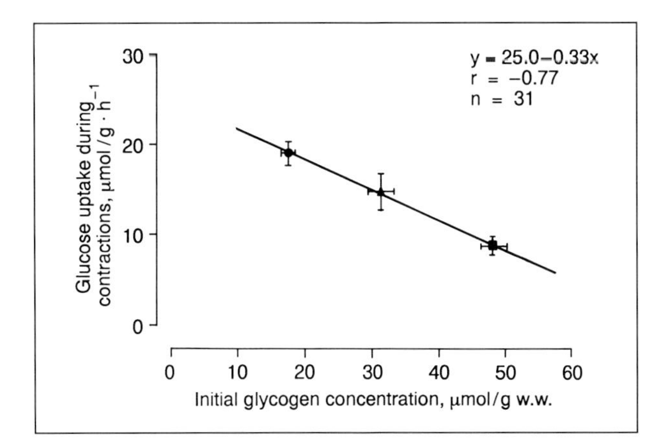

Fig. 1. Correlation between initial muscle glycogen concentration and glucose uptake during muscle contractions in isolated perfused skeletal muscle. w.w. = Wet weight. [Reproduced from ref. 1 with permission.]

lower when glycogen concentrations were high than when they were low [1], suggesting that muscle membrane permeability was affected by the intracellular muscle milieu. This phenomenon could be related to changes in intracellular pH. When glycogen concentrations are high lactate production is higher than when glycogen stores are low [3]. Thus, intracellular pH is probably lower in contracting muscle with high glycogen concentrations than in muscles with low glycogen. Preliminary data in isolated muscle membrane vesicles show that glucose transport is decreased by a fall in pH from the optimal value of pH 7.2 [unpubl. observations]. This is in accordance with findings in erythrocytes [5]. Thus, a larger decrease in pH during contractions in muscles with high glycogen concentrations compared with muscles with low glycogen may in part be responsible for the lower rate of glucose transport observed in the former group.

Effect of Increased Plasma Free Fatty Acid Concentration on Muscle Metabolism during Exercise

The interaction between carbohydrate and lipid metabolism has attracted considerable interest in the 30 years since Randle et al. [6, 7] proposed the glucose-fatty acid cycle. Its operation during exercise in

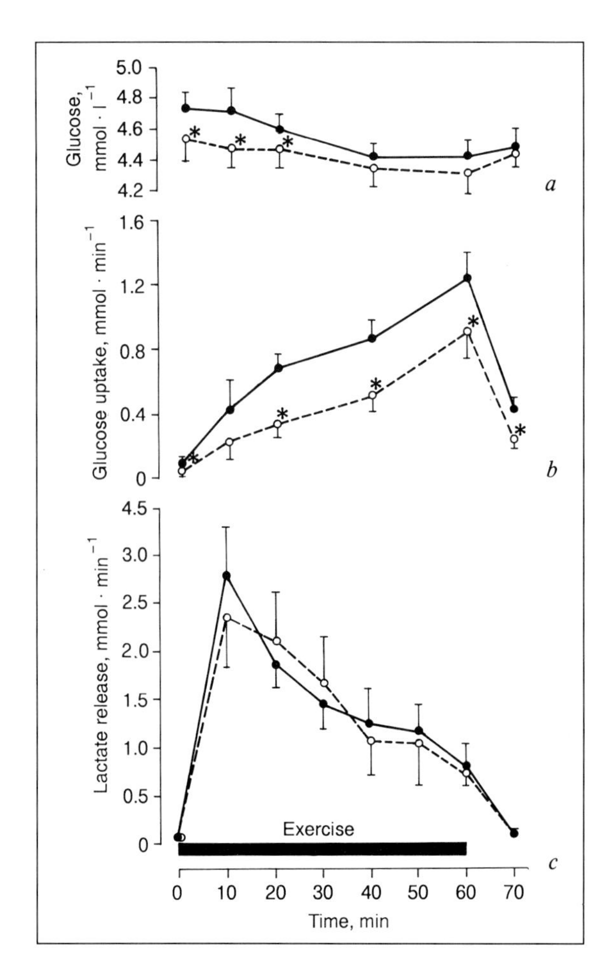

Fig. 2. Arterial concentration of glucose *(a)* glucose uptake *(b)* and lactate release *(c)* before, during and after 1 h of one-legged knee extensions. Subjects exercised with one leg during control conditions (●) and with the other leg during infusion of Intralipid (○). * $p < 0.05$ compared with values during control conditions. [Reproduced from ref. 16 with permission.]

humans is supported by a study [8] in which muscle glycogen breakdown during running was decreased when plasma free fatty acids (FFA) were increased by a fatty meal and heparin injection. Furthermore, muscle glycogen breakdown was increased when FFA were decreased by administration of nicotinic acid [9]. In contrast, there was no effect of increased FFA

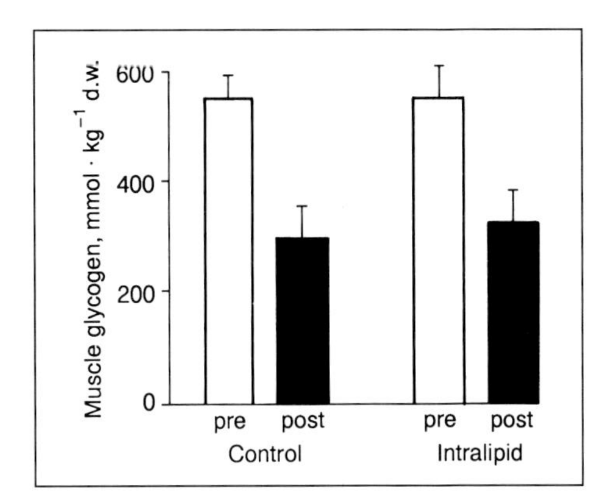

Fig. 3. Concentrations of glycogen in vastus lateralis muscle before and after 1 h of one-legged knee extensions during control conditions and during infusion of Intralipid. d.w. = Dry weight. [Reproduced from ref. 16 with permission.]

on whole-body carbohydrate oxidation during prolonged, moderate-intensity exercise [10]. In rats slower muscle glycogen utilization and higher blood glucose concentrations were found during exercise when plasma FFA were elevated than when they were low [11, 12], whereas in perfused rat muscle controversy exists as to whether FFA decrease muscle glucose uptake and glycogen utilization [13–15].

In an attempt to clarify the influence of FFA on muscle carbohydrate metabolism during exercise, 11 healthy, young males performed one-legged knee extensions for 1 h, after which an infusion of Intralipid and heparin was begun. Thirty minutes later the exercise bout was repeated with the other leg.

Infusion of Intralipid doubled the arterial concentration of FFA from an average of 0.54 ± 0.08 m*M* during control exercise to 1.12 ± 0.09 m*M* (mean ± SE, n = 11). Glucose uptake was significantly lower during infusion of Intralipid both at rest, during exercise and after 10 min of recovery (fig. 2). On the other hand, glycogen breakdown was identical in the two legs (fig. 3) and so was release of lactate (fig. 2), pyruvate and citrate [16]. At the end of the 1-hour knee extension, muscle G6P and glucose were identical whether FFA were raised or not [16]. This makes it difficult to attribute the decrease in glucose uptake to the operation of the glucose-fatty acid cycle, since this would require an increase in both G6P and glucose in muscle and such an increase was not found. Alternatively, if an increase in FFA does not increase G6P and glucose in muscle, then it is tempting to speculate that the decrease in glucose uptake during Intralipid infusion was due to a direct effect of increased FFA on muscle glucose transport. Since plasma concentrations of insulin and catecholamines were similar with and without Intralipid infusion [16], such an effect on transport could not be ascribed to FFA-induced changes in these hormones. In the rat heart, FFA and ketone bodies decrease the rate of efflux of arabinose, suggesting a direct effect on the transport mechanism [7]. However, studies in membrane preparations from skeletal muscle are lacking, and until such experiments have been performed no definite conclusions can be made.

Interaction between Lipid and Ammonia and Amino Acid Metabolism during Exercise

In resting muscle, increased concentrations of FFA have been shown to decrease amino acid release, suggesting that intramuscular proteolysis is inhibited [17]. There are several reports that indicate that net protein catabolism is increased during exercise [18–22]. Bearing in mind the effect of FFA on muscle amino acid release at rest [17], it could be that elevated concentrations of FFA during exercise might decrease net protein degradation. Thus, we also studied the influence of FFA on ammonia and amino acid metabolism in the same subjects in which the effect of FFA on carbohydrate metabolism was investigated [23]. Over the hour, total ammonia release was 4.36 ± 0.64 mmol in the control experiments while it was 2.44 ± 0.46 mmol during Intralipid infusion ($p < 0.05$; fig. 4).

There was a significant net release of all amino acids except for glutamate, which displayed a net uptake. The net exchange for the quantitatively most important amino acids is shown in figure 4. It can be seen that, in contrast to the inhibitory effect on ammonia release, infusion of FFA had no significant effect on release of amino acids.

Frequently, muscle ammonia formation has been attributed to the fast-twitch muscle fibers due to the activity of the purine nucleotide cycle, specifically the AMP deaminase reaction [24, 25]. This appears to hold true for short-term, high-intensity exercise, but it is unlikely that this mechanism is dominant in the present study because exercise of this nature has been shown to recruit both type I and type IIa fibers, with the former predominating [26]. The known positive modulators of AMP deaminase

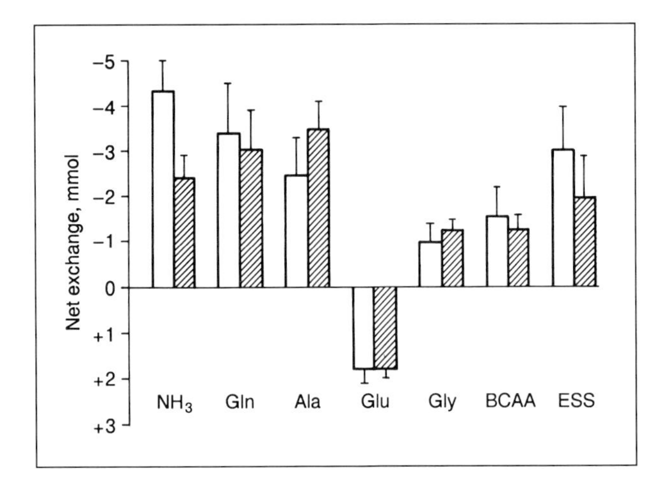

Fig. 4. Net exchange of ammonia and amino acids by the thigh during 1 h of one-legged knee extensions. The histogram summarizes the net total uptake (positive values) or release (negative values) of ammonia (NH_3), glutamine (Gln), alanine (Ala), glutamate (Glu), glycine (Gly), branched-chained amino acids (BBAA) and essential amino acids (ESS). □ = Mean for the control condition; ▨ = mean for the Intralipid infusion. [Reproduced from ref. 23 with permission.]

(ADP, AMP, H^+) have very little alteration in such exercise [27], and while slow-twitch fibers can accumulate IMP there is little, if any, change in either the total adenine nucleotides or IMP for at least 45 min at a moderate exercise intensity [27].

The other potential source of ammonia is amino acid catabolism but the magnitude of the catabolism during exercise and the role that it plays in ammonia production is uncertain. From the above-mentioned considerations of activity of the purine nucleotide cycle, it follows that protein catabolism probably is the major source of ammonia in the present study. However, if this is the case it is strange that Intralipid would decrease ammonia production but have no effect on net amino acid release from the legs during exercise.

The conclusion, therefore, seems to be that a doubling in FFA concentrations in plasma decreases ammonia production but does not significantly affect amino acid exchange in an exercising limb. Thus, it is unclear whether intramuscular proteolysis was influenced by high concentrations of FFA.

Effect of Carbohydrate Availability on Uptake and Oxidation of Free Fatty Acids in Perfused Rat Muscle

During prolonged exercise of a moderately high intensity, exhaustion is often found when muscle glycogen stores are depleted even though fat is readily available [28, 29]. Thus, it appears that when carbohydrate availability is limited fat combustion cannot take place sufficiently fast to deliver the necessary energy for contraction. Recent studies in man have suggested that glycogen depletion may lead to exhaustion because of inability to maintain the Krebs cycle intermediates [29]. However, whether glycogen depletion in fact leads to decreased fatty acid oxidation has not been directly shown.

To further study the role of carbohydrate for uptake and oxidation of FFA, the latter was studied in isolated, perfused rat hindlimbs. The rats were glycogen-depleted by a combination of swimming and lard feeding the day before the perfusion. The perfusate contained glucose at 0, 6 or 20 m*M*, respectively, and a physiological insulin concentration of 100 μU/ml cell free perfusate. Perfusate FFA concentration was either 600 or 2,000 μmol/l, and all of the FFA was palmitate. [^{14}C]-labeled palmitate was added to the perfusate to allow accurate quantification of uptake and oxidation of the palmitate. Uptake and oxidation of palmitate were measured after 15, 22, and 30 min of contractions at 20 trains/min, at which point steady state for these variables was found.

Muscle glycogen analysis showed that during the final 15 min of contractions no net glycogen breakdown took place. Therefore, in those experiments in which no glucose was added to the perfusate, the only carbohydrate fuel available was lactate, which was present in concentrations of around 0.5–1.0 m*M*.

When the palmitate concentration was 600 μmol/l, palmitate uptake was approximately 30% lower at 0 glucose than when glucose was at 20 m*M*. However, because the percentage of the palmitate taken up that was oxidized was higher at 0 glucose than at 20 m*M* glucose, total palmitate oxidation was not significantly different in the two groups. At a palmitate concentration of 2,000 μmol/l the same pattern in uptake was found as when palmitate was at 600 μmol/l, but the difference in percentage oxidation was not large enough to offset the difference in uptake between 0 and 20 m*M* glucose. Thus, at a palmitate concentration of 2,000 μmol/l, palmitate oxidation was significantly decreased by approximately 20% when the perfusate contained no glucose compared to when it contained 20 m*M* glucose. These data indicate that during conditions of abundance of FFA oxidation of FFA is dependent upon carbohydrate availability. In contrast, when FFA availability is less, its oxidation is not impeded by lack of carbohydrate.

In conclusion, the present studies have revealed some new aspects of carbohydrate/lipid/ammonia interaction in metabolism. Whereas we have been unable to confirm the operation of the classic glucose-fatty acid cycle in humans during exercise, an inhibitory effect of increased concentrations of FFA on leg glucose uptake was found. The mechanism behind this finding, however, remains obscure. Moreover, increased FFA concentrations decreased the release of ammonia during exercise, but did not significantly affect leg amino acid exchange. On the other hand, in perfused rat skeletal muscle, we have provided direct evidence that lack of carbohydrate during muscle contractions is not compensated for by an increase in FFA oxidation. This has functional consequences because a more rapid decline in force production was found in the absence than in the presence of carbohydrate fuels. Furthermore, when FFA concentrations are high lack of carbohydrate cannot be properly compensated for by other anaplerotic reactions and in fact causes a decrease in FFA oxidation.

Acknowledgements

The present studies were supported by a grant from the Danish Medical Research Council, grant No. 12-9535, from the Danish Natural Sciences Research Council, grant No. 11-7766, from the Danish Diabetes Association and the Novo Research Foundation. Betina Bolmgreen provided excellent technical assistance.

References

1 Hespel P, Richter EA: Glucose uptake and transport in contracting, perfused rat muscle with different pre-contraction glycogen concentrations. J Physiol 1990;427: 347–359.
2 Gollnick PD, Pernow B, Essen B, Jansson E, Saltin B: Availability of glycogen and plasma FFA for substrate utilization in leg muscle of man during exercise. Clin Physiol 1981;1:27–42.
3 Richter EA, Galbo H: High glycogen levels enhance glycogen breakdown in isolated contracting skeletal muscle. J Appl Physiol 1986;61:827–831.
4 Newsholme EA, Leech AR: Biochemistry for the Medical Sciences. New York, Wiley, 1983, pp 178, 184, 339–346.

5 Brahm S: Kinetics of glucose transport in human erythrocytes. J Physiol 1983;339: 339–354.
6 Randle PJ, Garland RB, Hales CN, Newsholme EA: The glucose-fatty acid cycle: Its role in insulin sensitivity and the metabolic disturbances of diabetes mellitus. Lancet 1963;i:1785–1789.
7 Randle PJ, Newsholme EA, Garland PB: Regulation of glucose uptake by muscle. Biochem J 1964;93:652–665.
8 Costill DL, Coyle E, Dalsky G, Evans W, Fink W, Hoopes D: Effects of elevated plasma FFA and insulin on muscle glycogen usage. J Appl Physiol 1977;43:695–699.
9 Bergström J, Hultman E, Jorfeldt L, Pernow B, Wahren J: Effect of nicotinic acid on physical working capacity and on metabolism of muscle glycogen in man. J Appl Physiol 1969;26:170–176.
10 Ravussin E, Bogardus C, Scheidegger K, Lagrange B, Horton ED, Horton ES: Effect of elevated FFA on carbohydrate and lipid oxidation during prolonged exercise in humans. J Appl Physiol 1986;60:893–900.
11 Hickson RC, Rennie MJ, Conlee RK, Winder WW, Holloszy JO: Effects of increased plasma fatty acids on glycogen utilization and endurance. J Appl Physiol 1977;43:829–833.
12 Rennie MJ, Winder WW, Holloszy JO: A sparing effect of increased plasma fatty acids on muscle and liver glycogen content in the exercising rat. Biochem J 1976; 156:647–655.
13 Reimer F, Löffler G, Henning G, Wieland O: The influence of insulin on glucose and fatty acid metabolism in the isolated perfused rat hindquarter. Hoppe-Seylers Z Physiol Chem 1975;356:1055–1066.
14 Rennie MJ, Holloszy JO: Inhibition of glucose uptake and glycogenolysis by availability of oleate in well-oxygenated perfused skeletal muscle. Biochem J 1977;168: 161–170.
15 Richter EA, Ruderman NB, Gavras H, Belur ER, Galbo H: Muscle glycogenolysis during exercise: Dual control by epinephrine and contractions. Am J Physiol 1982; 242:E25–E32.
16 Hargreaves M, Kiens B, Richter EA: Effect of increased plasma free fatty acid concentrations on muscle metabolism in exercising men. J Appl Physiol 1991;70:194–201.
17 Wicklmayer M, Rett K, Schwiegelshohn B, Wolfram G, Hailer S, Dietze G: Inhibition of muscular amino acid release by lipid infusion in man. Eur J Clin Invest 1987; 17:301–305.
18 Ahlborg G, Felig P, Hagenfeldt L, Hendler R, Wahren J: Substrate turnover during prolonged exercise in man: Splanchnic and leg metabolism of glucose, free fatty acids and amino acids. J Clin Invest 1974;53:1080–1090.
19 Dohm GL, Beecher GR, Warren RQ, Williams RT: Influence of exercise on free amino acid concentrations in rat tissues. J Appl Physiol 1981;50:41–44.
20 Eriksson LS, Broberg S, Bjorkman O, Wahren J: Ammonia metabolism during exercise in man. Clin Physiol 1985;5:325–336.
21 Felig P, Wahren J: Amino acid metabolism in exercising man. J Clin Invest 1971;50: 2703–2714.

ance between the supply and removal via 3 routes: cellular metabolism, the bloodstream and the microdialysis probe itself. Evidently, these factors will differ in importance in different tissues with inherently different metabolic activity as well as for different substances. However, since the probe continuously removes substances from the extracellular space, the concentration of a substance may under several circumstances be unnaturally low close to the probe.

Example of Results

A practical illustration of these considerations are the following results obtained when the influence of dialysis flow rate on the concentration of glucose in the dialysate were compared in dialysates collected from muscle and adipose tissue (n = 8). Thus, when the apparent microdialysis recovery (at 1.0 μl/min) with respect to the extracellular glucose concentration was calculated as the glucose concentration of the dialysate collected at the probe outlet divided by the blood glucose concentration (since at steady state the latter was shown to be equal to the extracellular glucose concentration; fig. 1), we found significantly higher values in adipose tissue than in muscle (37.6 ± 5.9 vs. 17.1 ± 1.8%, mean ± SEM; $p < 0.01$). When the perfusion flow rate was lowered to 0.5 μl/min, however, the recovery increased, but the increase was much greater in skeletal muscle than in adipose tissue. This nearly abolished the difference in apparent in vivo recoveries between the two organs (adipose tissue: 51.3 ± 4.1%; muscle: 50.0 ± 3.1%; n.s.). This finding illustrates the need for caution in interpreting organ-specifc microdialysis results. An approach to experimentally overcome the problem in determining the undisturbed extracellular concentration of a compound was devised by Lönnroth et al. [4]. We have applied this technique to determine muscle extracellular glucose concentration; the results are shown in figure 1.

Microdialysis as a Technique for Monitoring Skeletal Muscle Blood Flow

We have in this laboratory recently developed a microdialysis method for the continuous monitoring of local skeletal muscle blood flow [10]. In summary, changes in blood flow are monitored by the inclusion of a low concentration (5 m*M*) of ethanol in the perfusion medium. Loss of ethanol through the probe into the tissue is related to blood flow since ethanol is metabolically inert in muscle and readily dissolves in the total water space. Ethanol is measured in the perfusion medium and in the collected dialy-

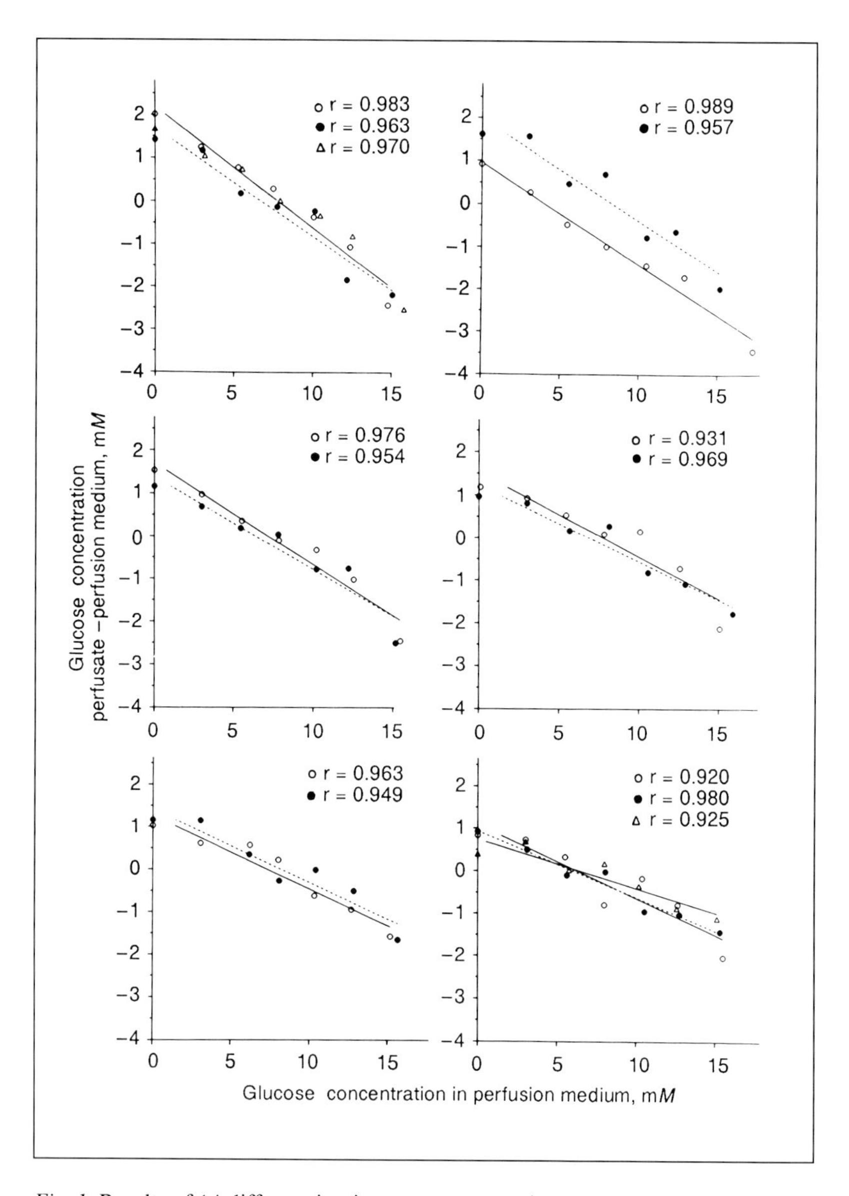

Fig. 1. Results of 14 different in vivo recovery experiments. Probes were placed in the gastrocnemius muscle of the rat and perfused at 1 µl/min with media with different glucose concentrations (x-axis). The y-axis shows the net change in the glucose concentration of the perfusion medium after dialysis has occurred (glucose concentration in the collected dialysate minus the glucose concentration in the infused perfusion medium). The in vivo relative recovery (i.e. the fraction of an absolute change in the diffusion

sates through standard enzymatic fluorometric procedures [14]. The results are expressed as the outflow/inflow ethanol concentration ratio [i.e. $(\text{ethanol})_{\text{collected dialysate}}/(\text{ethanol})_{\text{infused perfusion medium}}$], which is inversely related to blood flow (fig. 2).

Blood flow monitoring appears to be crucial in conjunction with microdialysis experiments. As discussed above, a change in blood flow will alter the balance for a specific compound between delivery to and removal from the extracellular space immediately surrounding the microdialysis membrane. It is, therefore, conceivable that changes in blood flow may be responsible for many of the metabolic changes documented using the microdialysis technique. This is particularly the case when vaso-active pharmacological agents have been included in the perfusion medium. As indicated in figures 2 and 3, changes in dialysate glucose concentration can be due solely to changes in blood flow. For example, dialysate glucose concentrations were increased to 169 ± 33 and 231 ± 11% of baseline (fig. 3) when the ethanol outflow/inflow concentration ratio decreased to 63.0 ± 2.9 and 46.9 ± 8.2% of baseline (fig. 2) during increases of blood flow by external heating and local 2-chloro-adenosine administration, respectively. During reductions in blood flow via leg constriction or local vasopressin administration, dialysate glucose concentration decreased to 44.8 ± 16.4 and 57.3 ± 6.7% of baseline (fig. 3) while outflow/inflow ethanol concentration ratio increased to 128 ± 4 and 132 ± 7% of baseline, respectively (fig. 2).

Microdialysis in Exercising Muscle of Man

In the basal state the outflow/inflow ethanol concentration ratio averaged 49.5 ± 5.5%. This inverse measure of blood flow decreased to 36.6 ± 3.7% (to 74% of basal) and 30.6 ± 3.9% (to 62% of basal), respectively, with intermittent isometric leg contraction of 20 and 40% of the maximal voluntary contraction force ($p < 0.001$; fig. 4). The lactate concentration in the collected muscle dialysates increased to 221 ± 52 and 322 ± 92% of

---

gradient across the dialysis membrane that is recovered as a net change in the glucose concentration of the perfusate collected at the probe outlet) is calculated as the slope of the line ($\Delta y/\Delta x$). The x value at y = 0 (i.e. no net flux of glucose) is taken to represent the tissue extracellular glucose concentration. This value, 6.9 ± 1.1 m*M* (mean ± SD), was not significantly different from the plasma glucose concentration of 7.3 ± 0.6 m*M* (n = 5). [From ref. 8.]

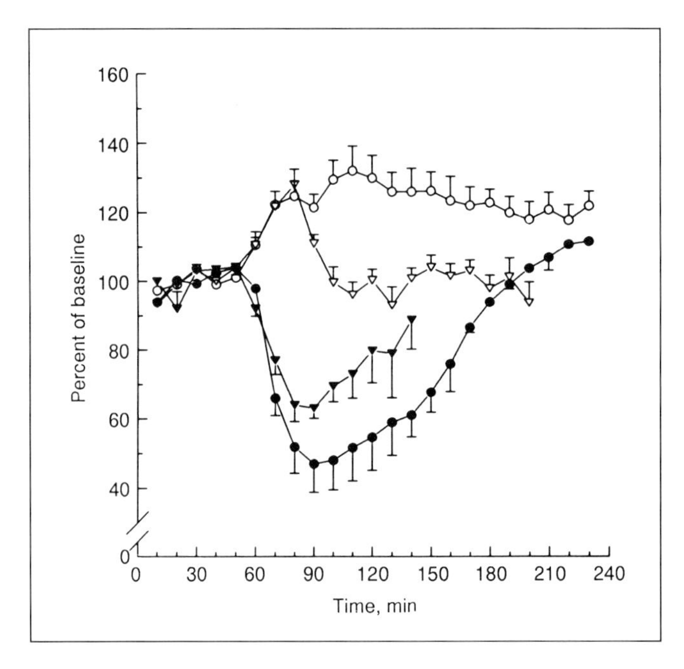

Fig. 2. Ethanol values (outflow/inflow concentration ratio), displayed as percent of average baseline value, obtained from dialysates collected from microdialysis probes placed in rat skeletal muscle. Muscle blood flow was increased at 50 min by either external heating (▼) or microdialysis perfusion with 100 μ*M* 2-chloro-adenosine (●) or decreased by leg constriction (▽) or perfusion with vasopressin, 4 IU/ml (○). Samples collected at times > 80 min for mechanical perturbations (▼, ▽) and > 110 min for pharmacological perturbations (●, ○) represent the post-treatment recovery period (means ± SEM, n = 4). [Modified from ref. 10.]

baseline, respectively ($p < 0.001$), as a result of exercise. No increase was detected in arterialized blood (fig. 4), thus verifying an increased intramuscular production of lactate. The results also indicate that the development of high muscle tension does not preclude the use of dialysis. Therefore, this technique promises to be an important tool in the study of both the metabolic and circulatory adaptations to exercise.

Other Relevant Findings

There is a dose-dependent increase in lactate concentration (up to 4-fold) in collected muscle perfusates when insulin is infused intravenously at 5 different rates (4.4–333 $mU \cdot kg^{-1} \cdot min^{-1}$) in the rat [8, 11]. This mea-

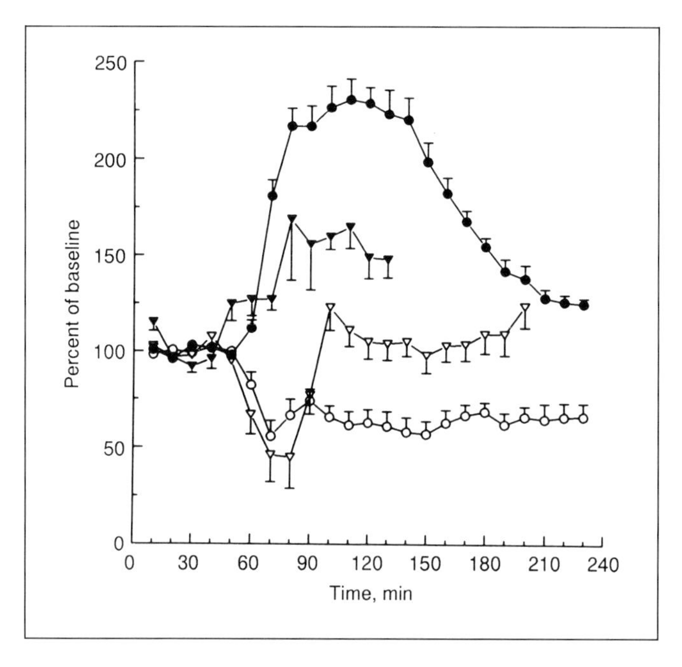

Fig. 3. Glucose concentration, displayed as percent of average baseline value, in dialysates collected from microdialysis probes placed in rat skeletal muscle. Muscle blood flow was increased at 50 min by either external heating (▼) or microdialysis perfusion with 100 μ*M* 2-chloro-adenosine (●) or decreased by leg constriction (▽) or perfusion with vasopressin, 4 IU/ml (○). Samples collected at times > 80 min for mechanical perturbations (▼, ▽) and > 110 min for pharmacological pertubations (●, ○) represent the post-treatment recovery period (means ± SEM, n = 4). [Modified from ref. 10].

sure of muscle insulin sensitivity with respect to glucose metabolism has been found to be significantly increased in muscle, but not in adipose tissue, in trained rats [11]. In another study, Arner et al. [5] studied the effect of a bolus intravenous isoproterenol (isoprenaline) injection in the rat. It was found that increasing doses of isoproterenol caused a sharp transient dose-dependent increase in glycerol in the adipose tissue perfusate. At the same time there was a gradual increase in blood glycerol, but after each isoproterenol dose the glycerol values flattened out at successively higher levels in a staircase manner. These findings seem to indicate that the microdialysis technique may be sensitive enough to pick up tissue-specific differences in the release of a compound although the activator, in

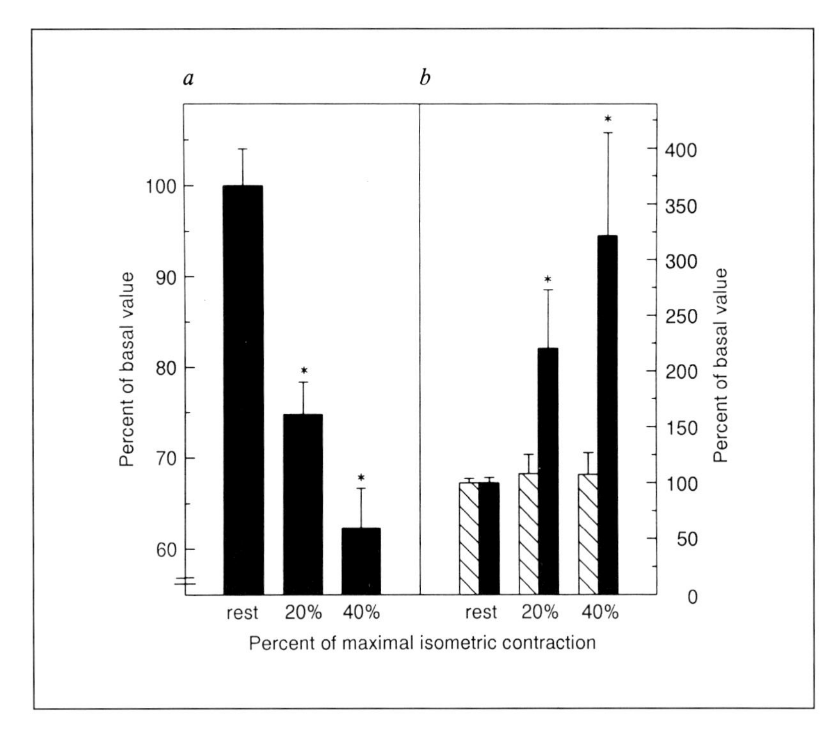

Fig. 4. Ethanol outflow/inflow concentration ratio (*a*) and lactate values (*b*) obtained from dialysates collected from microdialysis probes placed in human quadriceps femoris muscle, vastus lateralis. Probes were perfused at 4.0 µl/min with 5 m*M* ethanol in Krebs-Henseleit buffer, and dialysate samples were obtained at rest, as well as at 20 and 40% of maximal voluntary isometric contraction. Lactate values were also obtained from arterialized (fingertip) blood. Subjects exercised with intermittent contractions (5 s contractions with 10 s recovery) for 30 min at each exercise intensity. ■ = Dialysate, ▧ = blood.

this case insulin or isoproterenol, is administered via the blood. However, in the light of the above results these findings should be verified in conjunction with the monitoring of local blood flow.

Findings by Jansson et al. [15] and supported by Bolinder et al. [16], that the relative kinetics of subcutaneous tissue dialysate glucose concentration closely resemble those of blood glucose, suggest that microdialysis may be used for the monitoring of glycaemic control in man. This would be particularly beneficial in the chronic monitoring of blood glucose in diabetics, potentially leading to a closed-loop system of extracellular space sampling, analysis and insulin infusion. From other experiments based on the microdialysis technique, it has been proposed that lactate production following glucose ingestion occurs in adipose tissue of man [6, 7], which would provide evidence that adipose tissue is a significant source of lactate production in the body. Recently, microdialysis has successfully been used to monitor ethanol following alcohol ingestion in humans with microdialysis probes placed on the skin [17], opening up the possibility of a completely non-invasive technique for monitoring blood levels of various compounds. The above-mentioned advances in microdialysis show that the microdialysis technique is now not only used in brain research, but is also a major research tool to be used in many organs of the body, providing new and exciting breakthroughs in physiology and pharmacology.

Acknowledgements

These studies were supported by the Swedish Medical Research Council [Project B87-14X-07917 (J.H.), 04139 (L.J.)], the Karolinska Institute, the Magnus Bergvall Foundation and the Research Council of the Swedish Sports Federation. We would like to thank Berit Sjöberg and Anne-Chatrine Samuelsson for their technical assistance and Meg Hickner for her help in preparing this manuscript.

References

1 Delgado JMR, DeFeudis FV, Roth RH, Ryugo DK, Mitruka BK: Dialytrode for long-term intracerebral perfusion in awake monkeys. Arch Int Pharmacodyn Ther 1972;198:9–21.
2 Ungerstedt U, Pycock C: Functional correlates of dopamine neurotransmission. Bull Schweiz Akad Med Wiss 1974;1278:1–13.
3 Kendrick KM: Use of microdialysis in neuroendocrinology. Methods Enzymol 1989;168:182–205.
4 Lönnroth P, Jansson P-A, Smith U: A microdialysis method allowing characterization of intercellular water space in humans. Am J Physiol 1987;253:E228–E231.
5 Arner P, Bolinder J, Eliasson A, Lundin A, Ungerstedt U: Microdialysis of adipose tissue and blood for in vivo lipolysis studies. Am J Physiol 1988;255:E737–742.
6 Hagström E, Arner P, Ungerstedt U, Bolinder J: Subcutaneous adipose tissue: A source of lactate production after glucose ingestion in humans. Am J Physiol 1990; 258:E888–E893.
7 Jansson P-A, Smith U, Lönnroth P: Evidence for lactate production by human adipose tissue in vivo. Diabetologia 1990;33:253–256.

8 Henriksson J, Rosdahl H, Fuchi T, Oshida Y, Ungerstedt U: The use of microdialysis for the in vivo study of skeletal muscle glucose metabolism; in Maréchal G, Carraro U (eds): Muscle and Motility. Proc XIXth Eur Conf, Brussels. Andover, Intercept, 1990, vol 2, pp 365–370.
9 Grönlund B, Astrup A, Bie P, Christensen NJ: Noradrenaline release in skeletal muscle and in adipose tissue studied by microdialysis. Clin Sci 1991;80:595–598.
10 Hickner RC, Rosdahl H, Borg I, Ungerstedt U, Jorfeldt L, Henriksson J: Ethanol may be used with the microdialysis technique to monitor blood flow changes in skeletal muscle: Dialysate glucose concentration is blood flow dependent. Acta Physiol Scand 1991;143:355–356.
11 Oshida Y, Ohsawa I, Sato J, Sato Y: Effects of physical training on insulin action in skeletal muscle and adipose tissue of rats by using the microdialysis technique. Bull Phys Fitness Res Inst 1991;77:12–18.
12 Tossman U, Ungerstedt U: Microdialysis in the study of extracellular levels of amino acids in the rat brain. Acta Physiol Scand 1986;128:9–14.
13 Krebs HA, Henseleit K: Untersuchungen über Harnstoffbildung im Tierkörper. Hoppe-Seylers Z Physiol Chem 1932;210:33–66.
14 Bernt E, Gutmann I: Determination with alcohol dehydrogenase and NAD; in Bergmeyer HU (ed): Methods of Enzymatic Analysis. Weinheim, Verlag Chemie, 1974, vol 3, pp 1499–1505.
15 Jansson P-A, Fowelin J, Smith U, Lönnroth P: Characterization by microdialysis of intercellular glucose level in subcutaneous tissue in humans. Am J Physiol 1988;255: E218–E220.
16 Bolinder J, Hagström E, Ungerstedt U, Arner P: Microdialysis of subcutaneous adipose tissue in vivo for continuous glucose monitoring in man. Scand J Clin Invest 1989;49:465–474.
17 Korf J, de Boer J, Van der Kuil JFM, Postema F, Venema K, Flentge F: Methods for the on-line real-time monitoring of lactate, ethanol, glucose or choline using microdialysis and enzyme reactors. Curr Separations 1991;10:108.

Jan Henriksson, MD, Department of Physiology III, Karolinska Institute,
PO Box 5626, S–114 86 Stockholm (Sweden)

Sato Y, Poortmans J, Hashimoto I, Oshida Y (eds): Integration of Medical and Sports Sciences. Med Sport Sci. Basel, Karger, 1992, vol 37, pp 273–280

Effects of Acute and Chronic Exercise on Insulin-Stimulated Glucose Transport Activity in Skeletal Muscle[1]

Eric A. Gulve

Section of Applied Physiology, Department of Medicine, Washington University School of Medicine, St. Louis, Mo., USA

Exercise has several distinct effects on the transport of glucose into skeletal muscle. This review will compare the effects of a single bout of exercise with those induced by endurance exercise-training. Due to severe restrictions on the length of this review, I will concentrate on findings from our recent work.

Pathways for the Transport of Glucose into Mammalian Skeletal Muscle

The transport of glucose into mammalian skeletal muscle appears to be activated by two different pathways. One pathway can be activated by insulin and the other by contractile activity or hypoxia. The concept of separate pathways is strongly supported by numerous studies demonstrating additivity of the maximal effects of insulin and exercise [1–5]. Similarly, the maximal effects of insulin and hypoxia are also additive in skeletal muscle, whereas the maximal effects of hypoxia and contractile activity are not [6]. Thus, muscle contractions and hypoxia activate a common pathway distinct from that stimulated by insulin. The pathway stimulated by contractions and hypoxia appears to be activated by an increase in intracellular calcium concentration [6, 7]. Another recent study [8] using 4 skeletal muscles of different fiber composition demonstrated that the rela-

[1] This research was supported by grants DK18986, AG00425, AG0078 and AG-05404 from the National Institutes of Health.

tive contribution of maximal insulin- and contraction-stimulated uptake to total glucose transport activity differed between muscles, but the effects of contractile activity and insulin were additive in each muscle.

Effects of a Single Bout of Exercise

Acute exercise has multiple effects on glucose transport activity in skeletal muscle. One effect of exercise is the stimulation of glucose transport directly via the contractile-activity-activated pathway (i.e. an enhancement of glucose transport activity measurable in the absence of insulin). A separate action of exercise is to enhance the insulin-stimulated glucose transport pathway. These two effects of exercise have separate time courses. The increase in transport activity measured in the absence of insulin generally reverses within a few hours after cessation of exercise [4, 9]. The increase in insulin action becomes apparent after the increase in glucose transport measured in the absence of insulin has partially or completely worn off [4, 10–12].

Increases in insulin action induced by a single bout of exercise have been demonstrated in a number of different model systems such as the perfused rat hindquarter [13], in vivo during euglycemic clamp studies [14] and in isolated muscles [4, 15]. Acute exercise increases insulin sensitivity and can also enhance insulin responsiveness for glucose transport. An increase in insulin sensitivity is indicated by a lowering of the concentration of insulin required for half-maximal activation of glucose transport. In the absence of complete dose-response curves, enhanced insulin sensitivity is more commonly defined by increased transport activity in the presence of a submaximal insulin concentration without a change in response to a maximally effective insulin concentration, after correction for the effect of exercise on basal transport activity measured in the absence of insulin. An increase in insulin responsiveness is indicated by an increased effect of a maximal insulin stimulus (again taking into account any changes in basal transport activity). When rats are allowed to recover for varying time periods after exercise before measurement of glucose transport activity, the exercise-induced increase in insulin responsiveness persists for much shorter time periods than does the increased insulin sensitivity [12].

As noted above, an increase in insulin action in the epitrochlearis muscle is not evident in the first hour after exercise, but can be demonstrated when muscles are removed 3 h after exercise [4]. In a more recent study, we removed epitrochlearis muscles immediately after exercise and incubated them in vitro for 3 h before measurement of glucose transport activity [16]. Insulin sensitivity was greatly enhanced relative to that measured in muscles from sedentary rats incubated under similar conditions. This finding suggests that, whatever the factor that enhances insulin action, once activated it can then proceed and be maintained in the absence of humoral factors or other factors present in vivo.

Factors Regulating Reversal of Enhanced Insulin Sensitivity after Exercise

When rats were fasted or fed a carbohydrate-free diet, insulin sensitivity remained markedly enhanced for at least 2 days after a bout of exercise [12]. In contrast, when rats were fed rat chow or 100% carbohydrate, the increase in insulin sensitivity reversed to values measured in sedentary rats some time between 3 and 18 h after exercise [12]. The persistent increase in insulin sensitivity was unrelated to caloric intake. In muscles from rats deprived of carbohydrate after exercise, where insulin action remained augmented, glycogen levels remained below or at the levels seen in fasted rats. Complete reversal of the enhanced insulin sensitivity was associated with replenishment of muscle glycogen concentrations above levels measured in sedentary fed rats, i.e. glycogen supercompensation.

These findings suggest that carbohydrate availability regulates the reversal of enhanced insulin sensitivity after exercise. Bergstrom and Hultman [17] demonstrated 25 years ago that a single bout of strenuous exercise results in glycogen supercompensation in the exercised skeletal muscle. The increase in insulin sensitivity induced by exercise, coupled with an adequate source of carbohydrate, is probably an important factor in the regulation of muscle glycogen supercompensation [cf. also ref. 1].

In the study cited above [12], insulin responsiveness was significantly increased (relative to that measured in muscles from sedentary rats) only in the group of rats fed fat for 3 h after exercise. Under these conditions no glycogen was resynthesized. Insulin responsiveness returned to baseline even under conditions where glycogen concentrations remained well below those measured in fasted sedentary rats. This finding suggests that reversal of the transient augmentation of insulin responsiveness in rat muscle is either independent of glycogen resynthesis or that the threshold for reversal is much less than that for reversal of the enhanced insulin sensitivity.

Studies of insulin action in rats fed different diets after exercise cannot discriminate between a direct effect of carbohydrate uptake into muscle or indirect effects of feeding carbohydrate. We examined this question more directly by removing muscles immediately after exercise and incubating them for 3 h with different substrates in the presence of a low concentration of insulin [16]. When muscles were incubated in the absence of carbohydrate after exercise, insulin action remained greatly enhanced relative to that measured in muscles from sedentary rats. Incubation with glucose caused a 50% reversal of the enhanced insulin sensitivity. A reduction of the exercise-enhanced insulin sensitivity was only seen under conditions where muscle glycogen concentrations increased above that measured in sedentary rats. This finding suggested that glucose in combination with low concentrations of insulin can directly regulate the reversal of enhanced insulin sensitivity.

The associations between glycogen repletion and reversal of enhanced insulin action are only correlative, and do not indicate a direct effect of glycogen replenishment. The signal for return of insulin action back toward normal could be generated either during the transport of glucose across the cell membrane or alternatively at some later step in glucose metabolism. We incubated muscles after exercise with 2-deoxyglucose, which is transported across the muscle cell membrane and phosphorylated by hexokinase but is not further metabolized. Transport into muscles of a large amount of 2-deoxyglucose, comparable to the quantity of glucose that caused a reversal of increased insulin action after exercise, did not reverse the enhanced insulin sensitivity [16]. This finding suggests that some step beyond the transport of glucose across the cell membrane is the key event controlling reversal of the enhanced insulin sensitivity.

Exercise Training

Skeletal muscles contain 2 different isoforms of the glucose transporter, referred to as GLUT4 and GLUT1. The location (i.e. muscle vs. non-muscle cells) and function of the GLUT1 isoform are currently under debate. Expression of the GLUT4 isoform of the transporter is apparently limited to tissues in which glucose transport activity is insulin-sensitive [18, 19]. In skeletal muscle the GLUT4 protein comprises the vast majority of glucose transporters [18, 19]. The GLUT4 protein is translocated from an intracellular location to the plasma membrane in skeletal muscle in response to both insulin and contractile activity [19–21]. GLUT4 concentration is closely correlated with maximal glucose transport activity in different skeletal muscles [8, 22], suggesting that GLUT4 protein concentration is an important determinant of a muscle's capacity for glucose transport.

Previous studies have shown that exercise-training increases insulin-stimulated glucose transport into skeletal muscle [15, 23–26]. In this regard the primary effect of exercise-training is an increase in insulin responsiveness. More recently, several groups have demonstrated that exercise-training increases skeletal muscle GLUT4 protein concentration in normal animals [26, 27] and also in insulin-resistant fatty Zucker rats [28]. In recent studies we have compared the relative training-induced changes in glucose transport activity measured in isolated muscles with those of the glucose transporter protein measured in the same muscles.

Wheel Running

In one study [29], rats were trained by voluntary wheel running over a 5-week period. The training program resulted in hypertrophy of the soleus muscle, a unique feature of this mode of training [30]. In the epitrochlearis muscle, which does not hypertrophy in response to wheel running, levels of hexokinase and the oxidative enzyme marker citrate synthase increased by 42 and 33%, respectively, in response to training [29]. Basal glucose transport activity in the epitrochlearis, measured in the absence of insulin, was not altered by wheel running. Training increased maximal insulin-stimulated transport (39% greater than muscles from sedentary rats) and the response to maximal insulin plus contractile stimuli (44% greater). Wheel running also increased GLUT4 protein concentration by 51%. Therefore, conditions that increased concentrations of enzymes involved in carbohydrate metabolism also increased maximal insulin-stimulated glucose transport, total glucose transport activity and GLUT4 protein concentration. These increases occurred in the absence of any changes in epitrochlearis fiber type composition [29].

In the hypertrophied soleus the concentration (expressed per gram protein) of marker enzymes did not increase [29]. Training did not alter transport activity (expressed per milliliter intracellular water) under any of the conditions measured nor did it alter GLUT4 protein concentration (expressed per unit total protein). These indices are all increased in statistically significant fashion when expressed per muscle due to the exercise-induced hypertrophy of the soleus.

Treadmill Training

Wheel running is unique in that it stimulates hypertrophy of the soleus muscle [30]. Another study examined whether glucose transport capacity and GLUT4 concentration could be enhanced after endurance training that does not induce soleus hypertrophy [manuscript in review]. Rats were subjected to treadmill training for 3 weeks (90 min/day). Maximal transport capacity, assessed by the combination of maximal insulin and contractile stimuli, was increased 44% by training. In concert with these changes in transport capacity, training increased GLUT4 protein concentration by 47%. In contrast to the effects of training, an acute bout of treadmill exercise in untrained rats did not result in statistically significant changes in GLUT4 immunoreactivity. To the best of our knowledge, this is the first study to demonstrate that exercise training can increase GLUT4 protein concentration in slow-twitch muscle.

Utilization of Blood Glucose

Despite the fact that endurance training increases maximal insulin-stimulated glucose transport activity, endurance training decreases reliance on carbohydrates as an energy source during submaximal exercise [31]. A recent study in humans using [^{13}C]-labeled glucose demonstrated that training results in large decreases in the rate of blood glucose turnover and oxidation [32]. Thus, at the same absolute exercise intensity, training lowers the utilization of blood glucose. The adaptive increase in glucose transport activity induced by training allows for a greater maximal utilization of carbohydrate, however, enabling a greater maximal work output after training.

References

1 Garetto LP, Richter EA, Goodman MN, Ruderman NB: Enhanced muscle glucose metabolism after exercise in the rat: The two phases. Am J Physiol 1984;246:E471–E475.
2 Nesher R, Karl IE, Kipnis DM: Dissociation of effects of insulin and contraction on glucose transport in rat epitrochlearis muscle. Am J Physiol 1985;249:C226–C232.
3 Zorzano A, Balon TW, Goodman MN, Ruderman NB: Additive effects of prior exercise and insulin on glucose and AIB uptake by muscle. Am J Physiol 1986;251: E21–E26.
4 Wallberg-Henriksson H, Constable SH, Young DA, Holloszy JO: Glucose transport into rat skeletal muscle: Interaction between exercise and insulin. J Appl Physiol 1988;65:909–913.

5 Constable SH, Favier RJ, Cartee GD, Young DA, Holloszy JO: Muscle glucose transport: Interactions of in vitro contractions, insulin, and exercise. J Appl Physiol 1988;64:2329–2332.
6 Cartee GD, Douen AG, Ramlal T, Klip A, Holloszy JO: Stimulation of glucose transport in skeletal muscle by hypoxia. J Appl Physiol 1991;70:1593–1600.
7 Youn JH, Gulve EA, Holloszy JO: Calcium stimulates glucose transport in skeletal muscle by a pathway independent of contraction. Am J Physiol 1991;260:C555–C561.
8 Henriksen EJ, Bourey RE, Rodnick KJ, Koranyi L, Permutt MA, Holloszy JO: Glucose transporter protein content and glucose transport capacity in rat skeletal muscles. Am J Physiol 1990;259:E593–E598.
9 Young DA, Wallberg-Henriksson H, Sleeper MD, Holloszy JO: Reversal of the exercise-induced increase in muscle permeability to glucose. Am J Physiol 1987;253: E331–E335.
10 Richter EA, Garetto LP, Goodman MN, Ruderman NB: Enhanced muscle glucose metabolism after exercise: Modulation by local factors. Am J Physiol 1984;246: E476–E482.
11 Zorzano A, Balon TW, Goodman MN, Ruderman NB: Glycogen depletion and increased insulin sensitivity and responsiveness in muscle after exercise. Am J Physiol 1986;251:E664–E669.
12 Cartee GD, Young DA, Sleeper MD, Zierath J, Wallberg-Henriksen H, Holloszy JO: Prolonged increase in insulin-stimulated glucose transport in muscle after exercise. Am J Physiol 1989;256:E494–E499.
13 Richter EA, Garetto LP, Goodman MN, Ruderman NB: Muscle glucose metabolism following exercise in the rat: Increased sensitivity to insulin. J Clin Invest 1982;69: 785–793.
14 Richter EA, Mikines KJ, Galbo H, Kiens B: Effect of exercise on insulin action in human skeletal muscle. J Appl Physiol 1989;66:876–885.
15 Davis TA, Klahr S, Tegtmeyer ED, Osborne DF, Howard TL, Karl IE: Glucose metabolism in epitrochlearis muscle of acutely exercised and trained rats. Am J Physiol 1986;250:E137–E143.
16 Gulve EA, Cartee GD, Zierath JR, Corpus VM, Holloszy JO: Reversal of enhanced muscle glucose transport after exercise: Roles of insulin and glucose. Am J Physiol 1990;259:E685–E691.
17 Bergstrom J, Hultman E: Muscle glycogen synthesis after exercise: An enhancing factor localized to the muscle cells in man. Nature 1966;210:309–310.
18 James DE, Strube M, Mueckler M: Molecular cloning and characterization of an insulin-regulatable glucose transporter. Nature 1989;338:83–87.
19 Klip A, Paquet MR: Glucose transport and glucose transporters in muscle and their metabolic regulation. Diabetes Care 1990;13:228–243.
20 Hirshman MF, Wallberg-Henriksson H, Wardzala LJ, Horton ED, Horton ES: Acute exercise increases the number of plasma membrane glucose transporters in rat skeletal muscle. FEBS Lett 1988;238:235–239.
21 Douen AG, Ramlal T, Rastogi S, Bilan PJ, Cartee GD, Vranic M, Holloszy JO, Klip A: Exercise induces recruitment of the insulin-responsive glucose transporter: Evidence for distinct intracellular insulin- and exercise-recruitable transporter pools in skeletal muscle. J Biol Chem 1990;265:13427–13430.

22 Kern M, Wells JA, Stephens JM, Elton CW, Friedman JE, Tapscott EB, Pekala PH, Dohm GL: Insulin responsiveness in skeletal muscle is determined by glucose transporter (GLUT4) protein level. Biochem J 1990;270:397–400.
23 Berger M, Kemmer FW, Becker K, Herberg L, Schwenen M, Gjinavcki A, Berchtold P: Effect of physical training on glucose tolerance and on glucose metabolism of skeletal muscle in anaesthetized normal rats. Diabetologia 1979;16:179–184.
24 James DE, Kraegen EW, Chisholm DJ: Effects of exercise training on in vivo insulin action in individual tissues of the rat. J Clin Invest 1985;76:657–666.
25 Mikines KJ, Sonne B, Farrell PA, Tronier B, Galbo H: Effect of training on the dose-resonse relationship for insulin action in men. J Appl Physiol 1989;66:695–703.
26 Ploug T, Stallknecht BM, Pedersen O, Kahn BB, Ohkuwa T, Vinten J, Galbo H: Effect of endurance training on glucose transport capacity and glucose transporter expression in rat skeletal muscle. Am J Physiol 1990;259:E778–E786.
27 Rodnick KJ, Holloszy JO, Mondon CE, James DE: Effects of exercise training on insulin-regulatable glucose-transporter protein levels in rat skeletal muscle. Diabetes 1990:39:1425–1429.
28 Friedman JE, Sherman WM, Reed MJ, Elton CW, Dohm GL: Exercise training increases glucose transporter protein GLUT4 in skeletal muscle of obese Zucker (fa/fa) rats. FEBS Lett 1990:268:13–16.
29 Rodnick KJ, Henriksen EJ, James DE, Holloszy JO: Exercise-training, glucose transporters and glucose transport in rat skeletal muscles. Am J Physiol 1992;262:C9–C14.
30 Rodnick KJ, Reaven GM, Haskell WL, Sims RC, Mondon CE: Variations in running activity and enzymatic adaptations in voluntary running rats. J Appl Physiol 1989;66:1250–1257.
31 Karlsson J, Nordesjo L-O, Saltin B: Muscle glycogen utilization during exercise after training. Acta Physiol Scand 1974;90:210–217.
32 Coggan AR, Kohrt WM, Spina RJ, Bier DM, Holloszy JO: Endurance training decreases plasma glucose turnover and oxidation during moderate-intensity exercise in man. J Appl Physiol 1990;68:990–996.

Eric A. Gulve, PhD, Section of Applied Physiology, Washington University School of Medicine, 4566 Scott Avenue, Campus Box 8113, St. Louis, MO 63110 (USA)

Sato Y, Poortmans J, Hashimoto I, Oshida Y (eds): Integration of Medical and Sports Sciences. Med Sport Sci. Basel, Karger, 1992, vol 37, pp 281–287

The Overtraining Syndrome: Some Biochemical Aspects

Mark Parry-Billings, Eric A. Newsholme

Cellular Nutrition Research Group, Department of Biochemistry, University of Oxford, UK

Overtraining and the Immune System

The overtraining syndrome is a complex clinical condition. It may develop in athletes when training periods are too frequent, too intense and/ or too prolonged, and when training is combined with inadequate nutrition and psychological stress. Such athletes present with a number of symptoms, including central fatigue, depression, frequent viral infections and impaired wound healing [1]. No large-scale, systematic studies of immune function have been performed on overtrained athletes. It is, however, generally agreed that such athletes experience symptoms indicative of immunosuppression. In a study of trained and overtrained athletes, we have shown that the rate of proliferation of peripheral lymphocytes in vitro and the plasma levels of some cytokines were not affected by overtraining [Parry-Billings et al., unpubl. data]. These results suggest that overtraining may not affect the function of immune cells under optimized conditions of in vitro culture, but may cause a state of immunosuppression via a change in the plasma level of a fuel or a non-cytokine regulator of the immune system in vivo.

What plasma factors could be responsible for this immunosuppression? One possibility is glucocorticoids: the plasma concentration of cortisol is elevated in overtrained athletes [2] and glucocorticoids are generally considered to be immunosuppressive. However, this adverse effect of glucocorticoids on the immune system may only be evident at supraphysiological concentrations and a small increase in circulating cortisol levels may in fact *stimulate* the immune system [3].

Another possibility is the amino acid *glutamine*. We propose that frequent, intense and long-duration exercise can decrease the plasma glutamine level and this can result in immunosuppression. The basic biochemical facts underlying this hypothesis are given below.

Nutrition for the Immune System

Glutamine is utilized at a high rate by cells of the immune system and is an important fuel and an important precursor for purine and pyrimidine biosynthesis. It has also been proposed that this high rate of glutamine utilization provides optimal conditions for the precise regulation of the rates of macromolecular biosynthesis. This enables the cells to respond quickly and effectively to an immune challenge [4]. The hypothesis predicts that, if the plasma glutamine level is decreased below the physiological level, the functions of the immune cells may be impaired. A number of experiments have been performed to test this prediction. A decrease in the concentration of glutamine decreases the maximum rate of proliferation of human or rat lymphocytes, decreases the rate of antibody synthesis by lymphocytes and the rate of phagocytosis by macrophages in culture, despite the presence of all other amino acids and fuels [5, 6]. Also, a decrease in the plasma glutamine level in vivo, following administration of the enzyme glutaminase, has been shown to result in immunosuppression [7].

The important point to emerge from this discussion is that the rate of glutamine utilization must be high, even in resting immune cells, to provide optimal metabolic control. Consequently, the plasma glutamine level must be maintained to provide optimal conditions for response to an immune challenge. Therefore, could a decrease in the plasma concentration of glutamine be an explanation for the immunosuppression experienced in the overtraining syndrome?

Effects of Exercise and Overtraining on the Plasma Concentration of Glutamine

The acute response of plasma glutamine to exercise varies according to the duration and intensity. Short-term exercise increases the plasma level of glutamine, whereas endurance exercise decreases the level [8].

Table 1. Effect of overtraining syndrome on plasma amino acid concentrations in man

Subjects	Plasma amino acid concentration, μ*M*			
	glutamine	glutamate	alanine	BCAA
Control (n = 35)	550	125	392	398
Overtrained (n = 40)	503*	161**	379	408

* $p < 0.02$; ** $p < 0.01$; significant difference between control and overtrained means.
BCAA = Branched-chain amino acids.

Few studies have been performed to examine the chronic effects of training on plasma amino acid levels: the plasma glutamine level was decreased following a 6-week training period in rats [9] but unchanged in endurance-trained athletes [10]. There are presently no reports on the effects of the overtraining syndrome on plasma amino acid concentrations. Consequently, we undertook such a study: the plasma concentration of glutamine was *lower* in overtrained compared to that in trained athletes (table 1). Since samples were taken from *resting* subjects some time after performance had been impaired, these results suggest that overtraining has a long-term effect on plasma glutamine levels. Although this change in glutamine concentration is small, it is possible that prolonged exercise in overtrained athletes causes a much greater decrease in plasma glutamine level.

It is hypothesized that the decrease in plasma glutamine concentration in overtraining or following prolonged exercise may contribute to the immunosuppression observed in these conditions. The question arises as to the cause of the decrease in plasma glutamine concentration.

Skeletal Muscle as Part of the Immune System

Skeletal muscle may be quantitatively the most important tissue, in terms of glutamine synthesis and release into the bloodstream. Furthermore, analysis of the process of glutamine release from muscle and glutamine metabolism by the immune system indicates that glutamine release from muscle is the flux-generating step in the pathway of glutamine utilization by cells of the immune system and other tissues [11]. The pathway of glutamine utilization by cells of the immune system is not saturated with this substrate, so that changes in the rate of glutamine release by muscle, via changes in the concentration of glutamine in plasma, will change the rate of glutamine utilization by immune cells. Furthermore, any insult to muscle may lead to an impairment of the rate of glutamine release by muscle. Could this be the explanation for the decrease in plasma glutamine concentration following prolonged exercise or in overtraining?

Overtraining and the Rate of Glutamine Release from Muscle: Effects and Possible Mechanisms

The rate of glutamine release is decereased from incubated soleus and epitrochlearis muscles of the rat following exercise and is decreased from the perfused rat hindlimb following sciatic nerve stimulation [12–14]. Furthermore, the maximum activity of glutamine synthetase in muscle is decreased by training [15]. These findings support the view that the decrease in the plasma glutamine level after prolonged exercise may be caused, at least in part, by an impairment of glutamine metabolism in muscle. However, measurements of arteriovenous differences across skeletal muscle after prolonged exercise and overtraining are essential to provide further evidence.

The mechanism responsible for this decrease in the rate of glutamine from muscle is not known. However, a number of factors have been identified which may be important mediators of the rate of glutamine release and in turn of the plasma concentration of this amino acid. For example, chronic elevation of plasma adrenaline levels following implantation of adrenaline-release tablets in rats causes a decrease in both the rate of glutamine release from muscle and in the plasma glutamine concentration [16].

The rate of glutamine release from incubated muscle of the rat is increased in the presence of physiological concentrations of branched-chain amino acids [Parry-Billings et al., unpubl. data]. This observation is supported by the results of experiments in man. The question arises as to whether the increase in the levels of catecholamines or the decrease in that of the branched-chain amino acids following prolonged exercise is responsible at least in part for the decrease in the rate of glutamine release from muscle.

Development of Treatment Strategies

An extension of this work is the development of treatment strategies to prevent a decline in, or to elevate, the plasma concentration of glutamine. We have shown that supplementation with branched-chain amino acids causes an increase in the plasma concentration of these amino acids and prevents the decrease in plasma glutamine concentration after a marathon race (table 2). This treatment regimen has been employed by others to maintain plasma glutamine levels [17] and enhance immune function [18] after injury. Recently, supplements of free glutamine have been used to increase plasma glutamine levels and to enhance immune function of the gut following injury [19]. These therapeutic interventions have enormous potential in treating hospitalized patients. It is posible that they may also be important in maintaining the functions of the immune system following prolonged exercise and during intense training.

Recently, it has been reported that carbohydrate supplementation prevented the decrease in plasma glutamine concentration following prolonged exercise in athletes [20]. Similarly, we found that the plasma glutamine concentration was decreased following exercise, which would be expected to deplete glycogen levels in muslce [8]. Thus, an adequate amount of carbohydrate in the diet may be one factor influencing the rate of glutamine release from muscle and the development of the overtrained state.

Table 2. Effect of ingestion of branched-chain amino acids (BCAA) on plasma amino acid concentrations before (pre) and after (post) a 42.2-km (marathon) race

Treatment	Time of sample	Plasma amino acid concentration, nmol/ml		
		glutamine	alanine	BCAA
Placebo (n = 24)	pre	592	530	489
	post	495*	388*	404*
BCAA (n = 23)	pre	581	528	478
	post	561	422*	920*

* $p < 0.0005$: significant difference between pre- and post-exercise means.

Conclusion

The plasma glutamine concentration is decreased in overtrained athletes and following prolonged exercise. Given the important role glutamine plays in the nutrition and consequently in the function of the immune system, this change may cause, at least in part, the observed immunosuppression. Nutritional strategies might usefully be developed to prevent this decrease in plasma glutamine levels, possibly by affecting the transport of glutamine out of skeletal muscle.

Acknowledgements

We are indebted to Dr R. Budgett and Dr Y. Koutedakis, British Olympic Medical Centre, Harrow UK, and to Dr E. Blomstrand, Pripps Bryggerier, Bromma, Sweden, for experiments involving human subjects.

References

1 Noakes TD: The Lore of Running. Cape Town, Oxford University Press, 1986.
2 Barron JL, Noakes TD, Levy W, Smith C, Millar RP: Hypothalamic dysfunction in overtrained athletes. J Clin Endocrinol Metab 1985;60:803–806.
3 Jefferies WM: Cortisol and immunity. Med Hypotheses 1991;34:198–208.
4 Newsholme EA, Crabtree B, Ardawi MSM: Glutamine metabolism in lymphocytes: Its biochemical, physiological and clinical importance. Q J Exp Physiol 1985;70: 473–489.
5 Parry-Billings M, Evans J, Calder PC, Newsholme EA: Does glutamine contribute to immunosuppression after major burns? Lancet 1990;336:523–525.
6 Schneider YJ, Lavoix A: Monoclonal antibody production in semi-continuous serum and protein-free culture. J Immunol Method 1990;129:251–268.
7 Brambilla G, Pardodi S, Cavanna M, Caraceni CE, Baldini L: The immunodepressive activity of E. coli *L*-asparaginase in some transplant systems. Cancer Res 1970; 30:2665–2670.
8 Parry-Billings M, Blomstrand E, McAndrew N, Newsholme EA: A communicational link between skeletal muscle, brain and cells of the immune system. Int J Sports Med 1990;11:S122–S128.
9 Beecher GR, Puente FR, Dohm GL: Amino acid uptake and levels: Influence of endurance training. Biochem Med 1979;21:196–201.
10 Einsphar KJ, Tharp G: Influence of endurance training on plasma amino acid concentrations in humans at rest and after intense exercise. Int J Sports Med 1989;10: 233–236.
11 Newsholme EA, Parry-Billings M: Properties of glutamine release from muscle and its importance for the immune system. J Parenter Enter Nutr 1990;14:63S–67S.

12 Parry-Billings M, Blomstrand E, Leighton B, Dimitriadis GD, Newsholme EA: Does endurance exercise impair glutamine metabolism? Can J Sport Sci 1988;13:27P.
13 Goodmann MN, Lowenstein JM: The purine nucleotide cycle: Studies of ammonia production by skeletal muscle in situ and in perfused preparations. J Biol Chem 1977;252:5054–5060.
14 Nie ZT, Lisjo S, Karlson E, Goertz G, Henriksson J: In vitro stimulation of the rat epitrochlearis muscle. Contractile activity per se effects myofibrillar protein degradation and amino acid metablolism. Acta Physiol Scand 1989;135:513–521.
15 Falduto MT, Hickson RC, Young AP: Antagonism by glucocorticoids and exercise of expression of glutamine synthetase in skeletal muscle. FASEB J 1989;3:2623–2628.
16 Parry-Billings M, Leighton B, Dimitriadis GD, Bond J, Newsholme EA: The effects of catecholamines on skeletal muscle glutamine metabolism of the rat. Biochem Soc Trans 1991;19:130S.
17 Johnson DJ, Colpoys M, Smith RJ, Jiang ZM, Kapadia CR, Wilmore DW: Branched chain amino acid uptake and muscle free amino acid concentration predict post-operative muscle nitrogen balance. Ann Surg 1986;204:513–523.
18 Cerra FB, Mazuski JE, Chute E, Nuwer N, Teasley K, Lysne J, Shronts EP, Konstantindes FN: Branched chain metabolic support: A prospective, randomized, double-blind trial in surgical stress. Ann Surg 1984;199:286–291.
19 Souba WW, Klimberg VS, Hautamaki RD, Mendenhall WH, Bova FC, Howard RJ, Bland KI, Copeland EM: Oral glutamine reduces bacterial translocation following abdominal radiation. J Surg Res 1990;48:1–5.
20 Wagenmakers AJM, Beckers EJ, Brouns F, Kuipers H, Soeters PB, Van der Vusses GJ, Saris WHM: Carbohydrate supplementation, glycogen depletion and amino acid metabolism. Am J Physiol 1991;260:E883–E890.

Dr. Mark Parry-Billings, Cellular Nutrition Research Group,
Department of Biochemistry, University of Oxford, South Parks Road,
Oxford OX1 3QU (UK)

Sato Y, Poortmans J, Hashimoto I, Oshida Y (eds): Integration of Medical and Sports Sciences. Med Sport Sci. Basel, Karger, 1992, vol 37, pp 288–298

A Mathematical Model of Regulation of Protein Synthesis by Activation Feedback: Some Reflections on Its Possibilities and Limits in Describing Muscle Mass Adaptations with Exercise

Sylvia Ullmer, Alois Mader

Institute for Circulatory Research and Sports Medicine, Federal German Sports University, Köln, FRG

Introduction

Biological Problem

The cycle of protein breakdown and resynthesis can be divided into two major steps. The synthesis of an RNA copy (mRNA) of the nucleotide sequence in a probably limited region of DNA is catalyzed by the enzyme RNA-polymerase (transcription). Synthesis of proteins (translation) in the cytoplasm depends on the number and activity of ribosomes. Many details of these two processes are well established, while little is known about the feedback mechanism that guarantees synthesis of the specific proteins that have to replace faulty ones at any time in the life of a cell. In vivo, the control of protein synthesis and ribosome biogenesis involves mechanisms that permit rapid changes in its intensity level, finely tuned to functional demands; activation of transcription is protein specific.

Muscle is an extremely versatile tissue capable of acute adaptation to a wide range of work requirements. Various animal models simulating muscle atrophy and hypertrophy in vivo show that altered functional demands are the major stimulus for muscle adaption [2, 4, 5, 8, 11, 25–27, 28]. 'Unlike the growth of muscle that occurs during postnatal development, work-induced growth does not require pituitary growth hormone, thyroid hormones, insulin or androgens' [22]. 'The precise "make up" of the mus-

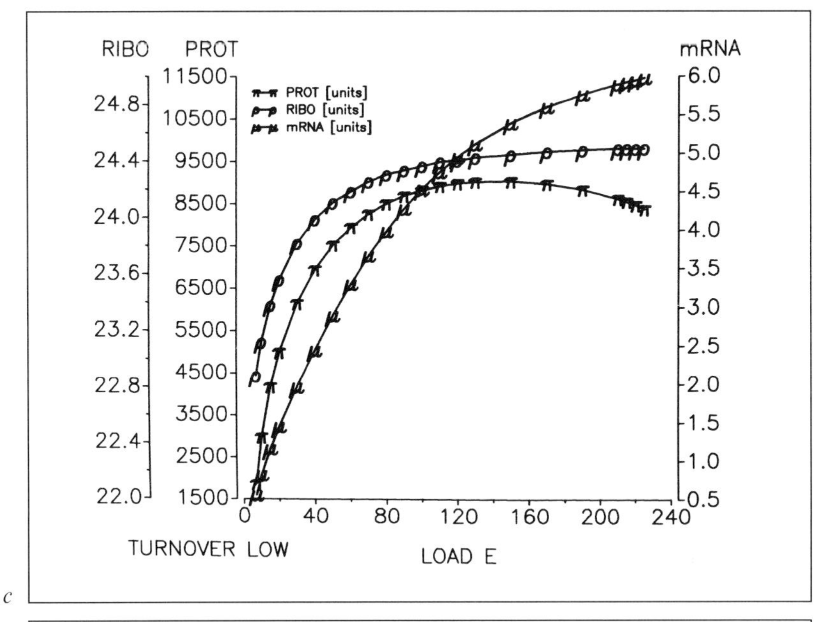

c

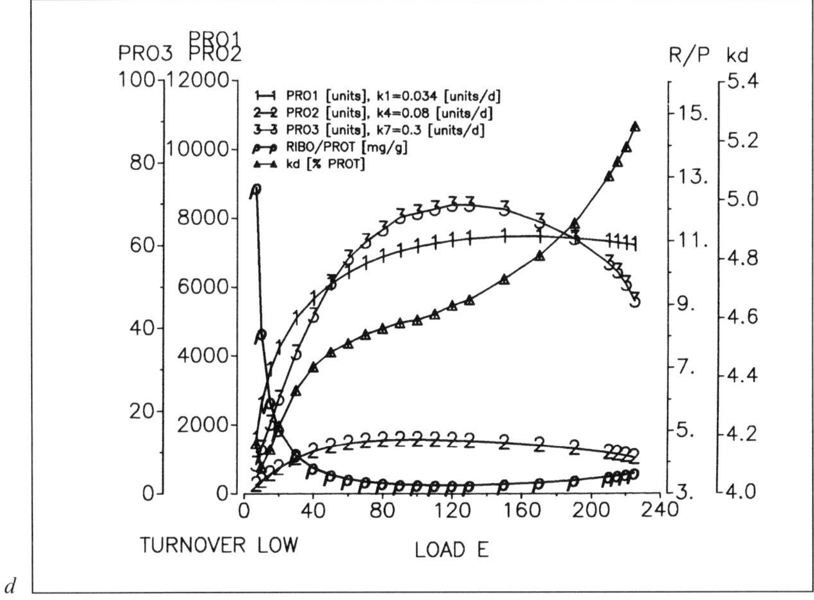

d

Simulated steady state of each of three protein subsystems (1, 2, 3) for high turnover (*b*) and low turnover (*d*) in model units is plotted together with the RNA-protein ratio (ρ) in mg/g and fractional rate of protein breakdown (△) against functional load.

processes that are not modelled here might have an additional effect that is not negligible. The steady state of the tissue varies according to work load. Maximum hypertrophy is reached at an optimum work load that lies below the maximum tolerable work load. The range of chronic overload is characterized by high fractional rates of protein breakdown, while the range of chronic underload is accompanied by a high RNA-protein ratio. Model parameters and physiological constants can be varied to specify a tissue. Difficulties arise from the fact that we have to make assumptions on most of the numbers, which characterize the living system. The number of genes that encode each protein determines the maximum transcription rate. We assumed quite different transcription rates for the three considered protein subsystems. This remains open for discussion. With cross nerve innervation contractile and metabolic characteristics of a differentiated muscle are, to a great degree, reversible [9, 10, 12–14], which might indicate that maximum transcription rates for fast and slow myosins are equal. It is now generally accepted that the expression of specific myofibrillar proteins is under control of innervation, because the neural activity determines the functional activity of the contracting proteins. The frequency of neural activity might influence the parameter that calculates protein attrition due to functional load. Different protein degradation pathways ultimately lead to protein-specific fragments. The effectiveness and speed of various proteolytic pathways under different conditions determines the fraction of protein degradation products that increase the amount of activating factors. We have to make assumptions on physiological constants and model parameters that are plausible and bring out results comparable to living systems. It would be desirable if some of these numbers could be determined by measurements. In conclusion, the presented mathematical model should be able to help to interpret experimental results and perhaps give rise to a working hypothesis for further studies.

Appendix

I. Differential Equations

A.I.1 $dY1/dt = vprot1 - p8 \cdot E2/Y1 - k1 \cdot Y1$,
with $vprot1 = Y3/(Y3 + Y6 + Y9) \cdot p1 \cdot Y10/(1 + p2/Y3^{p3})$ [units/day]

A.I.2 $dY2/dt = p11 \cdot p8 \cdot E2/Y1 + p12 \cdot k1 \cdot Y1 - k2 \cdot Y2$

A.1.3 $dY3/dt = vmRNA1 - k3 \cdot Y3$,
with $vmRNA1 = vm1/(1 + p17/Y2p18)$ [units/day]

A.I.4 $dY4/dt = vprot2 - p9 \cdot E2/Y4 - k4 \cdot Y4$,
with $vprot2 = Y6/(Y3 + Y6 + Y9) \cdot p1 \cdot Y10/(1 + p4/Y6^{p5})$ [units/day]

A.I.5 $dY5/dt = p13 \cdot p9 \cdot E2/Y4 - p14 \cdot k4 \cdot Y4 - k5 \cdot Y5$

A.I.6 $dY6/dt = vmRNA2 - k6 \cdot Y6$,
with $vmRNA2 = vm2/(1 + p19/Y5p20)$ [units/day]

A.I.7 $dY7/dt = vprot3 - p10 \cdot E2/Y7 - k7 \cdot Y7$,
with $vprot3 = Y9/(Y3 + Y6 + Y9) \cdot p1 \cdot Y10/(1 + p6/Y9^{p7})$ [units/day]

A.I.8 $dY8/dt = p15 \cdot p10 \cdot E2/Y7 + p16 \cdot k7 \cdot Y7 - k8 \cdot Y8$

A.I.9 $dY9/dt = vmRNA3 - k9 \cdot Y9$,
with $vmRNA3 = vm3/(1 + p21/Y8p22)$ [units/day]

A.I.10 $dY10/dt = vribo - ER - k10 \cdot Y10$,
with $ER = p27/Y10/(1. + p25/(Y3 + Y6 + Y9)^{p26})$ [units/day]
and $vribo = Vm4/(1. + p23/Y11p24)$ [units/day]

A.I.11 $dY11/dt = ER + k10 \cdot Y10 - k11 \cdot Y11$

II.1. List of Variables

Y1 = protein subsystem 1 [units]
Y2 = specific activator subsystem 1 [units]
Y3 = specific mRNA subsystem 1 [units]
Y4 = protein subsystem 2 [units]
Y5 = specific activator subsystem 2 [units]
Y6 = specific mRNA subsystem 2 [units]
Y7 = protein subsystem 3 [units]
Y8 = specific activator subsystem 3 [units]
Y9 = specific mRNA subsystem 3 [units]
Y10 = ribosomes [units]
Y11 = activator for ribosomes [units]

II.2. List of Constants

k1 = degradation rate of protein subsystem 1 [1/day]
k2 = degradation rate of specific activator subsystem 1 [1/day]
k3 = degradation rate of specific mRNA subsystem 1 [1/day]
k4 = degradation rate of protein subsystem 2 [1/day]
k5 = degradation rate of specific activator subsystem 2 [1/day]
k6 = degradation rate of specifc mRNA subsystem 2 [1/day]
k7 = degradation rate of protein subsystem 3 [1/day]
k8 = degradation rate of specific activator subsystem 3 [1/day]
k9 = degradation rate of specific mRNA subsystem 3 [1/day]
k10 = degradation rate of ribosomes [1/day]
k11 = degradation rate of activator for ribosomes [1/day]
vm1 = maximum transcription rate subsystem 1 [units/day]
vm2 = maximum transcription rate subsystem 2 [units/day]
vm3 = maximum transcription rate subsystem 3 [units/day]
vm4 = maximum transcription rate ribosomes [units/day]

II.3.a List of Parameter Tuning the Nonlinear Formation Rates

p1 = transformation factor; it can be roughly assessed by average protein turnover and ribosome content

p2, p4, p6 = represent specific mRNA at half-maximal translation rate of each of three protein subsystems

p3, p5, p7, p24, p18, p20, p22, p26 = exponents expected to lie between 1.1 and 1.4

p25 = total mRNA with half-maximal degradation of ribosomes because of translational activity

p17, p19, p21 = specific activator with half maximal transcription rate

II.3.b Parameter Adjusting Protein Breakdown and Formation of Specific Activator

p8, p9, p10 (units^2/day) = parameter determining protein attrition due to function

p11, p13, p15 = parameter determining the fraction of protein breakdown that contribute to formation of specific activator

p12, p14, p16 = parameter determining the fraction of normal degraded protein that contributes to formation of specific activator

II.3.c Parameters and Constants 'Low Turnover'

p1 = 21.5, p2 = 1.25 1/units$^{1.4}$, p3 = 1.4, p4 = 0.35 1/units$^{1.4}$, p5 = 1.4, p6 = 0.15 1/units$^{1.4}$, p7 = 1.4, p8 = 5.0 units^2/day, p9 = 1.0 units^2/day, p10 = 0.01 units^2/day, p11 = 2.5, p12 = 0.018, p13 = 2.5, p14 = 0.005, p15 = 2.5, p16 = 0.002, p17 = 35.0 1/units$^{1.1}$, p18 = 1.1, p19 = 15.0 1/units$^{1.1}$, p20 = 1.1, p21 = 3.5 1/units$^{1.1}$, p22 = 1.1, p23 = 0.7 1/units$^{1.4}$, p24 = 1.4, p25 = 2.5 1/units$^{1.4}$, p26 = 1.4

k1 = 0.034 1/day, k2 = 0.45 1/day, k3 = 0.5 1/day, k4 = 0.08 1/day, k5 = 0.45 1/day, k6 = 0.5 1/day, k7 = 0.3 1/day, k8 = 0.45 1/day, k9 = 0.5 1/day, k10 = 0.5 1/day, k11 = 0.9 1/day, vm1 = 2.1 units/day, vm2 = 0.9 units/day, vm3 = 0.22 units/day, vm4 = 23.0 units/day

II.3.d Parameters and Constants 'High Turnover'

p1 = 21.5, p2 = 1.25 1/units$^{1.4}$, p3 = 1.4, p4 = 0.35 1/units$^{1.4}$, p5 = 1.4, p6 = 0.15 1/units$^{1.4}$, p7 = 1.4, p8 = 3.0 units^2/day, p9 = 0.1 units^2/day, p10 = 0.005 units^2/day, p11 = 0.9, p12 = 0.0001, p13 = 0.5, p14 = 0.00001, p15 = 1.5, p16 = 0.0001, p17 = 35.0 1/units$^{1.1}$, p18 = 1.1, p19 = 15.0 1/units$^{1.1}$, p20 = 1.1, p21 = 3.5 1/units$^{1.1}$, p22 = 1.1, p23 = 0.7 1/units$^{1.4}$, p24 = 1.4, p25 = 2.5 1/units$^{1.4}$, p26 = 1.4

k1 = 0.08 1/day, k2 = 0.45 1/day, k3 = 0.5 1/day, k4 = 0.034 1/day, k5 = 0.45 1/day, k6 = 0.5 1/day, k7 = 0.3 1/day, k8 = 0.45 1/day, k9 = 0.5 1/day, k10 = 0.5 1/day, k11 = 0.9 1/day, vm1 = 2.1 units/day, vm2 = 0.9 units/day, vm3 = 0.22 units/day, vm4 = 23 units/day

References

1 Armstrong RB, Phelps RO: Muscle fibre type composition of the rat hindlimb. Am J Anat 1983;171:259–272.
2 Booth FW, Seider MJ: Recovery of skeletal muscle after 3 months of hindlimb immobilization in rats. J Appl Physiol 1979;47:435–439.
3 Booth FW, Watson PA: Control of adaptions in protein levels in response to exercise. Fed Proc 1985;44:2293–2300.
4 Desplanches D, Mayet MH, Sempore B, Flandrois R: Structural and functional responses to prolonged hindlimb suspension in rat muscle. J Appl Physiol 1987;63: 558–563.
5 Desplanches D, Mayet MH, Ilyina-Kakueva EI, Sempore B, Flandrois R: Skeletal muscle adaption in rats flown on Cosmos 1667. J Appl Physiol 1990;68:48–52.
6 Garlick PJ, Maltin CA, Baillie AGS, Delday MI, Grubb DA: Fiber-type composition of nine rat muscles. II. Relationship to protein turnover. Am J Physiol 1989;257: E828–E832.
7 Goldspink DF, Garlick PJ, McNurlan MA: Protein turnover measured in vivo and in vitro in muscles undergoing compensatory growth and subsequent denervation atrophy. Biochem J 1983;210:89–98.
8 Hauschka EO, Roy RR, Edgerton VR: Size and metabolic properties of single muscle fibres in rat soleus after hindlimb suspension. J Appl Physiol 1987;62:2338–2347.
9 Hood DA, Pette D: Chronic long-term electrostimulation creates a unique metabolic enzyme profile in rabbit fast-twitch muscle. FEBS Lett 1989;247:471–474.
10 Kaufmann M, Simoneau J-A, Veerkamp JH, Pette D: Electrostimulation-induced increases in fatty acid-binding protein and myoglobin in rat fast-twitch muscle and comparison with tissue levels in heart. FEBS Lett 1989;245:181–184.
11 Kennedy JM, Eisenberg BR, Reid SK, Zak R: Nascent muscle fibre appearance in overloaded chicken slow-tonic muscle. Am J Anat 1988;181:203–215.
12 Kirschbaum BJ, Heilig A, Härtner K-T, Pette D: Electrostimulation-induced fast-to-slow Transitions of myosin light and heavy chains in rabbit fast-twitch muscle at the mRNA level. FEBS Lett 1989;243:123–126.
13 Kirschbaum BJ, Schneider S, Izumo S, Mahdavi Vijak, Nadal-Ginard B, Pette D: Rapid and reversible changes in myosin heavy chain expression in response to increased neuromuscular activity of rat fast-twitch muscle. FEBS Lett 1990;268: 75–78.
14 Kirschbaum BJ, Simoneau J-A, Bär A, Barton PJR, Buckingham ME, Pette D: Chronic stimulation-induced changes of myosin light and heavy chains at the mRNA and protein levels in rat fast-twitch muscle. Eur J Biochem 1988;179:23–29.
15 Lewis SEM, Kelly FJ, Goldspink DF: Pre- and postnatal growth and protein turnover in smooth muscle, heart and slow- and fast-twitch skeletal muscles of rat. Biochem J 1984;217:517–526.
16 Mader A: A transcription-translation activation feedback circuit as a function of protein degradation, with the quality of protein mass adaption related to the average functional load. J Theor Biol 1988;134:135–157.

17 Mader A: Aktive Belastungsadaptation und Regulation der Proteinsynthese auf zellulärer Ebene. Dtsch Z Sportmed 1990;41:40–58.
18 McMillan DN, Reeds PJ, Lobley GE, Palmer RM: Changes in protein turnover in hypertrophing plantaris muscles of rats: Effect of fenbufen: An inhibitor of prostaglandin synthesis. Protaglandins 1987;34:841–852.
19 Millward DJ, Garlick PJ, Nnamyelugo DO, Waterlow JC: The relative importance of muscle protein synthesis and breakdown in regulation of muscle mass. Biochem J 1976;156:185–188.
20 Millward DJ, Garlick PJ, Stewart RJ, Nnamyelugo DO, Waterlow JC: Skeletal muscle growth and protein turnover. Biochem J 1975;150:235–243.
21 Pette D, Staron RS: Cellular and molecular diversities of mammalian skeletal muscle fibres. Rev Physiol Biochem Pharmacol 1990;116:1–76.
22 Poortmanns J: Protein metabolism: Effects of exercise and training. Med Sport 1981;13:66–76.
23 Rennie MJ, Edwards RHT, Kreywawych S, Davies CTM, Haliday D, Waterlow JC, Millward DJ: Effect of exercise on protein turnover in man. Clin Sci 1981;61:627–639.
24 Roy RR, Bello MA, Bouissou P, Edgerton VR: Size and metabolic properties of fibres in rat fast-twitch muscles after hindlimb suspension. J Appl Physiol 1987;62: 2348–2357.
25 Stump CS, Overton JM, Tipton CM: Influence of single hindlimb support during simulated weightlessness in the rat. J Appl Physiol 1990;68:627–634.
26 Thomason DB, Herrick RE, Surdyka D, Baldwin KM: Time course of soleus muscle myosin expression during hindlimb suspension and recovery. J Appl Physiol 1987; 63:130–137.
27 Thomason DB, Herrick RE, Baldwin KM: Activity influences on soleus muscle myosin during rodent hindlimb suspension. J Appl Physiol 1987;63:138–144.
28 Witzmann FA, Kim DH, Fitts RH: Recovery time cours in contractile function of fast and slow skeletal muscle after hindlimb immobilization. J Appl Physiol 1982; 52:677–682.

Dipl. Oz. S. Ullmer, Deutsche Sporthochschule Köln,
Insitut für Kreislaufforschung und Sportmedizin, Carl-Diem-Weg,
D-W-5000 Köln 41 (FRG)

Sato Y, Poortmans J, Hashimoto I, Oshida Y (eds): Integration of Medical and Sports Sciences. Med Sport Sci. Basel, Karger, 1992, vol 37, pp 299–308

Exercise Induction of 5′-Aminolevulinate Synthase: A Mitochondrial Enzyme in the Heme Biosynthetic Pathway

David A. Essig

College of Kinesiology, Division of Exercise Molecular Biology, University of Illinois at Chicago, Ill., USA

It has long been recognized that in muscle the amount of mitochondria is modulated to respond to changing functional demands. For example, the density of mitochondria in skeletal muscle is increased following exercise-training [1, 2]. A similar adaptation occurs following other types of muscular activity including chronic muscle overload or stretch [3] and repetitive nerve stimulation [4, 5]. However, the mechanisms which quantitatively regulate the change in mitochondrial density are not well understood.

Increases in mitochondrial assembly are reflected by proportional increases in the tissue concentrations of the inner membrane cytochromes and heme-associated proteins such as cytochrome oxidase [2]. Several investigations have indicated that changes in the cytochrome content are due in part to an increase in cytochrome synthesis rate [6, 7]. Since heme is required as a prosthetic group or for assembly of these mitochondrial proteins [8, 9], an increase in heme biosynthesis must occur. Early work by Holloszy and Winder [10] and Abraham and Terjung [11] confirmed this by the observation of a rapid induction of 5′-aminolevulinate synthase (ALAS), an enzyme in the heme biosynthetic pathway. Our laboratory has focused upon identifying the regulatory sites controlling the expression of ALAS both in normal and exercised muscle.

5′-Aminolevulinate Synthase Is a Key Enzyme in Mitochondrial Assembly

The pathway for heme biosynthesis involves 8 enzymes distributed in the mitochondrial matrix and cytosolic compartments. In most tissues including muscle, the activity of ALAS, which catalyzes the first step in the pathway, determines the rate of heme synthesis [12, 13]. In the context of mitochondrial biogenesis, ALAS serves an important role in the provision of heme necessary for import and assembly of the inner-membrane cytochromes and cytochrome oxidase. The most dramatic illustration of this can be appreciated from investigations using yeast mutants which are deficient in ALAS activity. In these mutants, the native cytochrome structures are not formed and selected subunits are lacking [8]. These data suggest that understanding the mechanisms involved in the regulation of ALAS activity will yield insight into the overall control of mitochondrial biogenesis.

Alterations in the level of ALAS activity appear to be determined by proportional changes in the concentration of the enzyme protein [12]. Since the half-life of ALAS is very short (90–120 min [reviewed in ref. 12]), a high rate of protein synthesis is required to sustain normal steady-state levels of the enzyme activity. ALAS is a mitochondrial matrix enzyme and the newly translated polypeptide exists in precursor form with an amino-terminal prepeptide necessary for import into the mitochondria. As a result, many regulatory sites in the synthesis of the enzyme are possible including transcription, mRNA stability, import of the preprotein precursor and translation.

Muscle-Specific Regulation of 5′-Aminolevulinate Synthase Activity

The expression of ALAS in normal muscle has been given little attention. Most work has been performed in tissues such as liver and erythrocyte where the highest levels of the enzyme are expressed. The results of recent investigations suggest that muscle ALAS is subject to tissue-specific and developmental modes of regulation [14].

Muscle Fiber Type

There is an apparent range of ALAS activity in skeletal muscle fiber types and heart [10, 11, 14]. The basis for fiber type regulation of ALAS activity in muscle tissue appears to be a function of the different masses of mitochondria. Evidence for this is based upon a strong correlation (r = +0.99) between ALAS activity and cytochrome oxidase and citrate synthase activities in chicken muscle fiber types [14]. Both citrate synthase [5] and cytochrome oxidase [2] have been shown to be reliable predictors of mitochondrial mass in muscle under steady-state conditions.

The range of muscle type variation in the levels of ALAS activity is highly related to the content of the 'liver'-type ALAS mRNA (r = +0.94 [14]). Using isoform-specific probes and Northern analysis, our results [14] indicated that in adult and developing skeletal and cardiac muscles, only the liver mRNA and not the erythroid mRNA [15] isoform is expressed. This confirmed and extended earlier work which had shown the expression of the liver mRNA isoform in cardiac muscle of rats [16]. These relationships imply that during the ontogeny of the different fiber types modulation of the steady-state levels of ALAS liver mRNA was geared to the signals which also determine the different masses of mitochondria and heme protein content in the adult muscle tissues.

Development

The hypothesis above, based on fiber type analysis, was confirmed by our results from a developmental analysis of ALAS activity in the chicken pectoralis muscle [14]. In the interval between 5 and 84 days after hatching, the decrease in ALAS activity was in parallel with the relative changes in the activities of cytochrome oxidase and citrate synthase and ALAS mRNA content. This suggests that during this time frame the corresponding decrease in ALAS mRNA is matched to the demands of normal phenotypic changes in mitochondrial and cytoplasmic (myoglobin) heme protein synthesis and turnover. It should be noted that the decrease in mitochondrial enzyme activities observed with the pectoralis is consistent with the fast-twitch glycolytic phenotype of this muscle in the adult [17]. Other developmental studies using similar types of muscles [18] have observed a comparable posthatch decrease in mitochondrial enzymes including citrate synthase. It is likely that in heart or the slow tonic muscles which have higher mitochondrial enzyme levels the posthatch pattern of ALAS expression may differ quantitatively. Our data would suggest that the pretranslational regulation of ALAS activity, however, would still exist.

Tissue-Specific Differences

The mechanisms by which a single isoform of the ALAS enzyme is regulated in muscle fiber types differ in comparison to liver. First, higher levels of ALAS activity in liver compared to muscle are not a simple function of a greater abundance of ALAS mRNA since our results showed that, in liver, the content of ALAS mRNA per gram tissue protein was similar to ventricle, a muscle in which maximal enzyme activity was 7.5-fold lower than liver [14]. These data suggest that the higher basal levels of enzyme activity in liver in comparison to muscle are due to processes located at one or more steps after gene transcription. Secondly, high levels of ALAS activity in liver were associated with significantly less oxidative enzyme activity than in muscle [14]. This likely represents a form of mitochondrial differentiation in liver to accommodate the high demands for biosynthesis of microsomal cytochrome P_{450} heme proteins [12]. These observations suggest that ALAS may belong to a class of nuclear encoded proteins whose expression may be modulated across different tissues to create a distinct mitochondrial phenotype in consonance with the tissue energy production and metabolic function.

Induction by Exercise

Exercise Disrupts Coordinate Regulation

As described above, under basal conditions, ALAS activity is expressed in concert with enzymes involved in oxidative metabolism. In response to a variety of exercise stimuli, one of the most consistent findings is that ALAS activity is induced at a much more rapid rate than that of heme-dependent proteins such as cytochrome oxidase and cytochrome *c*. For example, Holloszy and Winder [10] showed that in previously untrained rats ALAS activity was increased by 2-fold in the red portion of the vastus lateralis muscle 17 h following a vigorous bout of exercise. No change in the level of cytochrome *c* content was noted in this time frame and subsequent experiments showed that cytochrome *c* did not increase until after 6 consecutive days of treadmill running. Similar results were observed by us using the chicken wing overload model [19]. ALAS activity was induced after 3 days of overload but no significant increase in cytochrome oxidase was noted until after 7 days. In a recent training study, rats undergoing 28 days of treadmill running showed significant increases in plantaris muscle ALAS activity after 3, 7 and 28 days, but cytochrome oxidase activity did not increase until after 7 days of exercise [G. Town and D. Essig, unpubl. observations].

The results suggest that the exercise stimulus can interrupt the developmental program of ALAS expression which normally would constrain the expression of ALAS with that of other enzymes involved in oxidative metabolism. In addition, there must be relatively low levels of free heme in the resting muscle cell since ALAS activity was increased so soon in response to the increased metabolic demands. Resting levels of free heme may have to be increased early on so as not to limit any pending increase in heme-dependent proteins.

Induction of 5′-Aminolevulinate Synthase Is Transient

With discontinuous exercise models such as the running rat, the induction of ALAS in the postexercise recovery period is rather short-lived. In ventricular muscle of rats exercised to exhaustion by running, ALAS activity peaked at 6 h after exercise and declined to control baseline by 16 h [11]. A similar result was also observed in skeletal muscle in which ALAS activity declined to control value within 48 h after a single vigorous bout of exercise [10]. These results were extended in a recent study using 3 h of daily 10 Hz stimulation of rat tibialis anterior muscle [D. Essig and D. Hood, unpubl. observations]. Compared to control (contralateral) unstimulated values, there was a significant 70% increase in cytochrome oxidase activity which was evident measured either at 0, 18 or 48 h after 7 days of stimulation. In contrast, ALAS activity was the same as control immediately after stimulation, but increased to 2-fold higher than contralateral control 18 h after stimulation. This induction was reduced at 48 h of recovery, but remained 1.6-fold greater than the contralateral, nonstimulated muscle. The simplest interpretation of these data is that cytochrome oxidase had undergone a gradual increase in activity over the 7-day period, while ALAS activity was likely re-induced in each recovery period between successive days of 3 h stimulation. The dissociation between the pattern of changes in these two enzymes is likely due to the markedly faster rate of turnover of ALAS compared to cytochrome oxidase. The data from this and the above studies illustrate that the turnover of the ALAS enzyme may be a factor which limits the extent of induction in the absence of additional repetitive contractile activity.

Regulation of 5′-Aminolevulinate Synthase Induction by Exercise

Thus far, only one study has determined the changes in ALAS mRNA expression which accompany induction by repetitive contractile activity [19]. The regulation of ALAS activity during chronic weight-bearing activity (overload) in chicken skeletal muscle was investigated. Maximal enzyme activity was increased 2.5- and 4.0-fold after 3 and 7 days of over-

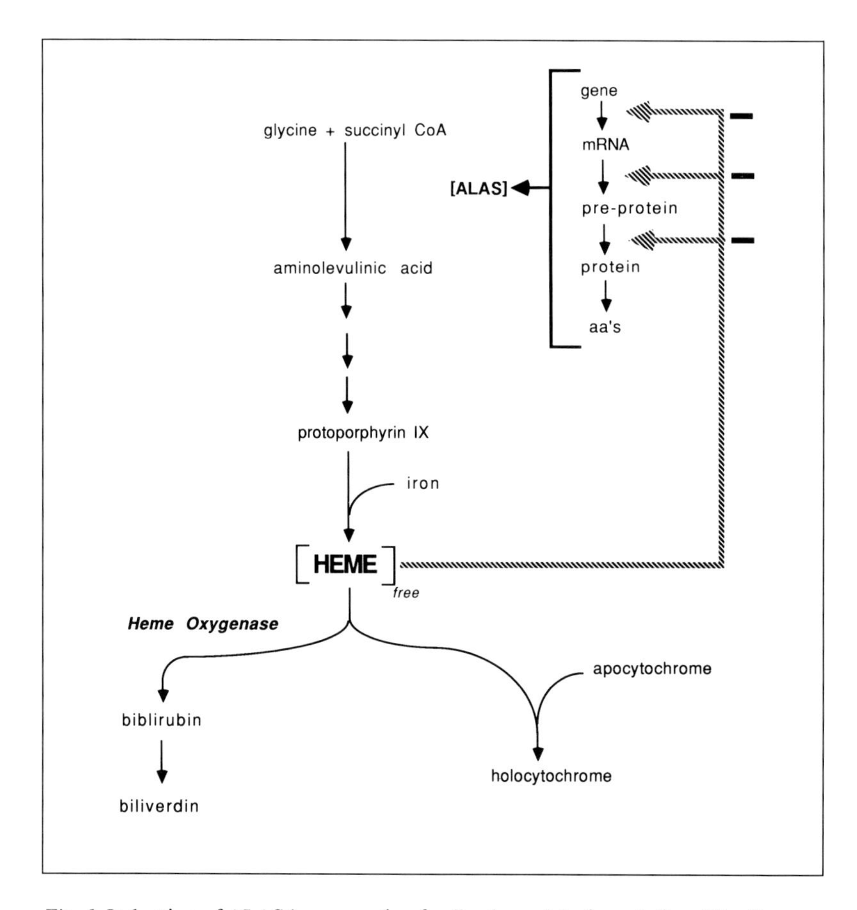

Fig. 1. Induction of ALAS by a negative feedback model of regulation. The diagram is the author's interpretation of the model. The evidence for the model is reviewed extensively by May et al. [12]. The synthesis of free heme under normal resting conditions is postulated to feedback-inhibit or repress the synthesis of ALAS, the first and rate-limiting enzyme in the heme biosynthetic pathway. Heme oxygenase catalyzes the initial reaction in the degradation of heme into bilirubin. The brackets at heme and ALAS denote concentration. aa's = Amino acids.

load. The content of ALAS mRNA (ng/mg total RNA) was not changed after 3 days but increased significantly after 7 days of overload. Normalizing the content of ALAS mRNA relative to the increase in total RNA indicated that ALAS mRNA increased by 1.6- and 2.0-fold at 3 and 7 days, respectively. On this basis, the increase in enzyme activity per gram pro-

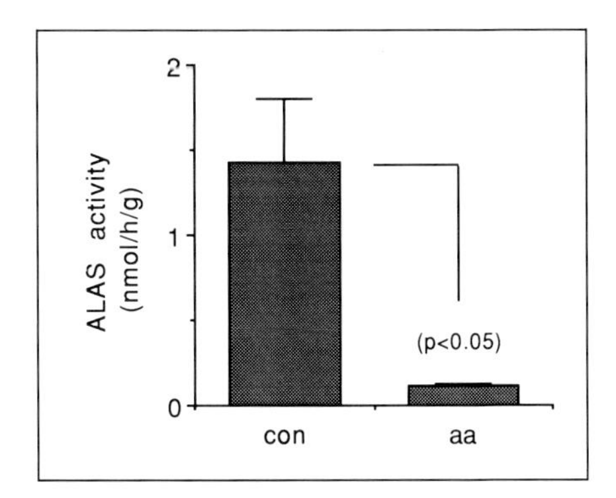

Fig. 2. Injection of heme precursor (aminolevulinate) inhibits plantaris muscle ALAS activity. A bolus intraperitoneal injection (1.7 mg/g) of aminolevulinate (aa group, n = 5) or saline (con group, n = 5) was administered to rats 12 h prior to sacrifice. The plantaris muscles from each rat were then assayed for ALAS activity as described [19].

tein exceeded the increase in mRNA content per gram protein by 60–70%. Thus, induction of ALAS activity was regulated largely by processes at the translational or posttranslational steps in the protein expression pathway. This type of regulation is separate from, but can interact with, the apparent transcriptional control of ongoing developmental and tissue-specific regulation of ALAS expression. Further research will be necessary to clarify the exact site of this posttranscriptional regulation.

Negative Feedback Model of 5′-Aminolevulinate Synthase Induction

The induction of ALAS in liver cells has been explained by a negative feedback model [reviewed in ref. 12] (fig. 1). The main evidence supporting this model is that exogenous heme or its precursors inhibit basal or induced ALAS activity. Furthermore, estimates of normal cellular levels of free heme suggest that ALAS activity is repressed. Induction is thought to occur by de-repression when cellular heme levels are reduced by increased formation of heme-dependent proteins and/or increased degradation of heme by heme oxygenase.

We have confirmed the existence of this pathway in muscle by examining the changes in ALAS activity 12 h after injection of the heme precursor aminolevulinate (1.7 mg/g, i.p.) into rats. The results showed a dramatic 90% decrease in plantaris muscle ALAS activity in comparison to saline-injected animals (fig. 2). These data provide preliminary evidence that a negative feedback pathway exists in skeletal muscle and extend previous work in rat heart [16].

If this general model is applicable to synthesis of mitochondrial cytochromes, one would have predicted cytochrome oxidase or other heme proteins to be induced prior to ALAS. However, as described above (see section on Exercise Disrupts Coordinate Regulation), heme proteins in muscle are induced at the protein level several days after ALAS activity. This contrasts with findings in liver tissue in response to porphyrinogenic drugs in which cytochrome P_{450} proteins and mRNAs are co-induced with ALAS protein and mRNA [20]. In muscle, unlike liver, the free heme pool may not be decreased due to increased mitochondrial cytochrome formatin. Instead, ALAS may be induced by a more direct mechanism to increase the synthesis of heme in preparation for the subsequent demands of cytochrome formation. This hypothesis must be regarded, however, as tentative since it is also possible that levels of free heme could have been lowered by other mechanisms such as an increase in myoglobin synthesis [21] or by an increase in the rate of heme or heme-protein turnover. Another possibility is that a large pool of apocytochromes may exist in the cytosol which rapidly bind the available free heme and thereby induce ALAS. A final and intriguing speculation is that heme oxygenase may be initially induced by exercise such that ALAS is de-repressed as a consequence of the lowered free heme pool. Interestingly, heme oxygenase has been shown to be highly inducible in liver cells and is also classified as a stress protein [22].

In summary, a negative feedback pathway would appear to operate in skeletal muscle cells. The negative feedback model in its present form may not explain induction according to the mechanism observed in liver cells. However, it is too early to rule out the general applicability of the model to muscle and exercise. The alternative mechanisms suggested above provide a basis for testable hypotheses in future experimentation.

Overall Conclusions

ALAS activity is controlled in normal muscle primarily by corresponding changes in the level of the liver isoform mRNA. This type of control would appear to be important in determining the regulation of the enzyme in response to developmental and fiber-type-specific variances in ALAS activity. This mode of regulation is muscle-specific since the higher basal levels of ALAS activity in liver mitochondria are associated with disproportionately less oxidative enzyme activity and less of the liver ALAS isoform mRNA than in muscle. With prolonged daily endurance exercise of adult muscle, a rapid induction-de-induction of ALAS occurs during the recovery period and temporarily disrupts the normal developmental regulation of ALAS activity in relation to other enzymes of the Krebs cycle and oxidative phosphorylation. Evidence was presented suggesting that induction of ALAS is regulated largely at steps after transcription of the mRNA. The possibility that exercise may lower the free heme pool and thereby derepress ALAS activity by a negative feedback loop was considered.

Future Directions

The regulation of ALAS expression in muscle tissues is complex and at least two steps in the pathway are utilized to affect changes in the synthesis of the enzyme in normal and exercised muscle. A serious effort needs to be made to assess which regulatory elements are important in muscle-specific regulation of the gene during development. A second major emphasis will be to determine the exact site of the posttranscriptional regulation utilized to induce the enzyme by endurance exercise.

References

1 Holloszy JO, Booth FW: Biochemical adaptations to endurance exercise in muscle. Annu Rev Physiol 1976;38:273–291.
2 Davies KJA, Packer L, Brooks GA: Biochemical adaptation of mitochondria, muscle and whole-animal respiration to endurance training. Arch Biochem Biophys 1981;209:539–554.
3 Holly RG, Barnett JG, Ashmore CR, Taylor RG, Molé PA: Stretch induced growth in chicken wing muscles: A new model of stretch hypertrophy. Am J Physiol 1980; 238:C62–C71.
4 Eisenberg BR, Salmons S: The reorganization of subcellular structure in muscle undergoing fast-to-slow type transformation. Cell Tissue Res 1981;220:449–471.
5 Williams RS, Salmons S, Newsholme EA, Kaufman RE, Mellor J: Regulation of nuclear and mitochondrial gene expression by contractile activity in skeletal muscle. J Biol Chem 1986;261:376–380.
6 Booth FW, Holloszy JO: Cytochrome *c* turnover in rat skeletal muscles. J Biol Chem 1977;252:416–419.
7 Terjung RL: The turnover of cytochrome *c* in different skeletal muscle fibre types of the rat. Biochem J 1979;178:569–574.

8 Saltzgaber-Muller J, Schatz G: Heme is necessary for the accumulation and assembly of cytochrome *c* oxidase subunits in *Saccharomyces cerevisiae.* J Biol Chem 1978;253:305–310.

9 Nelson D: Biogenesis of mammalian mitochondria. Curr Top Bioenerg 1987;15: 221–272.

10 Holloszy JO, Winder WW: Induction of δ-aminolevulinic acid synthetase in muscle by exercise or thyroxine. Am J Physiol 1979;236:R180–R184.

11 Abraham WM, Terjung RL: Increased δ-aminolevulinic acid synthase activity in rat ventricle after acute exercise. J Appl Physiol 1978;44:507–511.

12 May BK, Borthwick IA, Srivastava G, Pirola BA, Elliott WH: Control of 5′-aminolevulinate synthase in animals. Curr Top Cell Regul 1987;28:233–262.

13 Sedman R, Ingall G, Rios G, Tephly TR: Heme biosynthesis in the heart. Biochem Pharmacol 1982;32:761–766.

14 Essig DA, McNabney LA: Muscle-specific regulation of the heme biosynthetic enzyme 5′-aminolevulinate synthase. Am J Physiol, in press.

15 Riddle R, Yamamoto DM, Engel JD: Expression of δ-aminolevulinate synthase in avian cells: Separate genes encode erythroid-specific and nonspecific isozymes. Proc Natl Acad Sci USA 1989;86:792–796.

16 Srivastava G, Borthwick IA, Maguire DJ, Elfernick VJ, Bawden MJ, Mercer JFB, May BK: Regulation of 5′-aminolevulinate synthase mRNA in different rat tissues. J Biol Chem 1988;263:5202–5209.

17 Barnard EA, Lyles JM, Pizzey JA: Fibre types in chicken skeletal muscles and their changes in muscular dystrophy. J Physiol 1982;331:333–354.

18 Bass A, Lusch G, Pette D: Postnatal differentiation of the enzyme activity pattern of energy supplying metabolism in slow (red) and fast (white) muscles of the chicken. Eur J Biochem 1970;11:289–292.

19 Essig DA, Kennedy JM, McNabney LA: Regulation of 5′-aminolevulinate synthase activity in overloaded skeletal muscle. Am J Physiol 1990;259:C310–C314.

20 Hamilton JW, Bement WJ, Sinclair PR, Sinclair JF, Wetterhahn KE: Expression of 5′-aminolevulinate synthase in cytochrome P-450 mRNAs in chicken embryo hepatocytes in vivo and in culture: Effects of porphyrinogenic drugs and haem. Biochem J 1988;255:267–275.

21 Underwood LE, Williams RS: Pretranslational regulation of myoglobin gene expression. Am J Physiol 1987;252:C450–C453.

22 Shibahara S, Muller RM, Taguchi H: Transcriptional control of rat heme oxygenase by heat shock. J Biol Chem 1987;262:12889–12892.

David A. Essig, PhD, College of Kinesiology, M/C 194, University of Illinois at Chicago, 356 PEB, PO Box 4348, Chicago, IL 60680 (USA)

Sato Y, Poortmans J, Hashimoto I, Oshida Y (eds): Integration of Medical and Sports Sciences. Med Sport Sci. Basel, Karger, 1992, vol 37, pp 309–317

Changes in Myosin Heavy-Chain and Light-Chain Isoforms following Sustained Exercise

Masanobu Wada[a], *Kunio Kikuchi*[a], *Shigeru Katsuta*[b]

[a]Faculty of Integrated Arts and Sciences, Hiroshima University, Hiroshima-shi, Hiroshima-ken; [b]Institute of Health and Sport Sciences, University of Tsukuba, Tsukuba-shi, Ibaraki-ken, Japan

Many studies on the effect of increased or decreased contractile activity have described that, in spite of their high degree of specialization, skeletal muscles have a pronounced capacity for accommodating changes in demand. Chronic low-frequency stimulation of rabbit fast-twitch muscles evokes an orderly sequence of changes which chiefly result in a complete transformation to slow-twitch muscles [1]. Early changes in stimulated muscles are transformations of the energy metabolism and calcium-regulating system. The transition of the subunits of the myosin molecule, which mainly determines the contractile properties of muscle, follows these changes.

Although it has been widely demonstrated that endurance training elicits increases in enzyme activity of the aerobic metabolism and in capillary density [2, 3], only a few studies have shown that a transformation between major fiber types is brought about in a physiological manner such as by training [4]. In view of the findings from electrical stimulation experiments, if it is possible to increase remarkably the contractile activities of fast-twitch fibers, a shift of myofibrillar proteins to the slow type could be evoked by training.

This prompted us to design a first experiment where the major interest was to investigate the possibility of training-induced muscle fiber transformation. This experiment conducted with rats indicated that specific endurance training could constitute a stimulus for a shift of myosin heavy-chain (HC) and light-chain (LC) isoforms. Next, we were interested in examining in more detail changes in myofibrillar proteins of muscles subjected to sustained exercise. The second experiment was undertaken in order to follow the time course of changes in HC and LC isoforms.

Materials and Methods

First Experiment

Animals and Training. Fourteen male Wistar rats were used in the first experiment and randomly divided into a training and control group. Animals were introduced to treadmill running and trained for a period of 16 weeks. Running speed and duration were progressively increased as shown in table 1. After 13 weeks of training, animals were running at 40 m/min for 240 min/day. The soleus and extensor digitorum longus (EDL) muscles were removed about 24 h following the last training.

Histochemistry. Serial cross-sections (10 μm thick) were cut with a cryostat at −20 °C and stained for myofibrillar actomyosin ATPase (mATPase) after preincubation at pH 10.3 [5]. Fiber types were classified into slow-twitch type I and fast-twitch type II fibers.

Table 1. Protocol of the progressive endurance training

Age, weeks	Speed, m/min	Duration, min/day
5	30	20
6	30	30
7	30	60
8	40	60
9	40	75
10	40	90
11	40	100
12	40	120
13	40	120
14	40	120
15	40	180
16	40	210
17	40	240
18	40	240
19	40	240
20	40	240

Values are based on the final run for each week. Sessions were performed 5 times per week.

Electrophoresis. For LC analysis, sample preparation and two-dimensional electrophoresis were performed as described by Hirabayashi [6]. Isoelectric focusing was made in cylindrical agarose gel where pH ranged from 2.5 to 10.5. The second dimension consisted of polyacrylamide gradient (12–20%) gel electrophoresis. After staining with Coomassie blue, the relative distribution of each LC isoform was evaluated by integrating densitometry.

Second Experiment

Animals and Surgery. Adult male Wistar rats weighing 300–350 g were used in the second experiment. Since it has previously been revealed that a removal of synergists is responsible for a transformation of type II to type I fibers in the remaining muscle [7, 8], the ablation technique was used as model for increased contractile activity in this study. The gastrocnemius and soleus muscles were removed from the left hindlimb to product an overload in the remaining plantaris muscle. Muscle from the right limb was used as control. After 2, 4, 6 or 8 weeks of overload, the plantaris muscles were removed from both the control and experimental limbs.

Electrophoresis. Two-dimensional electrophoresis for LC analysis was performed by using the same method described in the first experiment. Myosin HC isoforms were separated by polyacrylamide gradient (5–8%) gel electrophoresis [9]. Proteins were silver-stained and evaluated densitometrically.

Results and Discussion

Most studies performed to date on human and animal muscles provide no evidence of transformation between type I and II fibers. Moreover, Komi and Karlsson [10] have shown that monozygous twins have an essentially identical muscle fiber composition. The high percentage of type I fibers in muscles of endurance athletes and the low percentage of this fiber type in sprinters have been considered to be the result of genetic factors. As judged by the histochemical staining (fig. 1a), training evoked an increase in type I fibers of the soleus muscle in this study. No evidence of muscle fiber degradation was seen and studies concerning skeletal muscle fiber numbers point out that hyperplasia is not an important factor in response to sustained exercise [8, 11]. Therefore, the present observations (fig. 1) imply the replacement of fast HCII by slow HCI in the existing fibers, since mATPase-based fiber types reflect the complement of HC isoforms [12]. The fact that long-term, high-intensity training rather than another type of training causes the shift of the major fiber types is in agreement with the later changes in proteins of thick filaments induced by long-term stimulation of rabbit fast muscle.

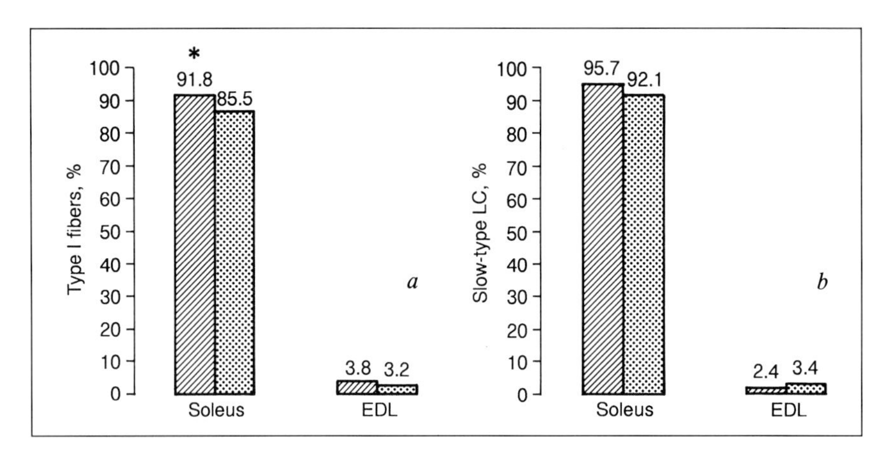

Fig. 1. Effects of endurance training on the relative distribution of type I fibers (*a*) and slow-type myosin LC isoforms (LC1s and LC2s; *b*) in the soleus and EDL muscles. * $p < 0.05$: significant difference between trained (▨) and control (▩) groups.

In spite of the increased percentage of type I fibers, appreciable changes in slow-type LC (LC1s and LC2s) were not observed in the soleus muscles (fig. 1b). It is evident from the findings of the electrical stimulation experiments that the transformation of various systems in muscle fibers does not occur simultaneously in response to increased contractile activity. Changes in HC isoforms seem to precede those of LC in rabbit muscle [13]. The failure of changes in slow-type LC of the soleus muscle suggests that, with regard to transition of fast to slow HC and LC isoforms, a response of rat muscle to sustained exercise could follow a similar time course to that previously reported for rabbits.

In the EDL muscle, training induced a change in fast alkali LC with a decrease in the LC3f/LC1f ratio, as shown in figure 2. Most studies performed so far have not documented such a shift of alkali LC resulting from endurance training, with the exception of the study of Green et al. [4] who observed a similar alteration in the vastus lateralis muscle of rats. However, a recent report [14] has indicated that, in chronically stimulated rat fast-twitch muscle, the major changes consist in an increase of LC1f at the expense of LC3f. It is unknown to what extent such a change in alkali LC affects the contractile property of muscle, since divergent results have been reported with respect to the function of alkali LC isoforms for mATPase [15, 16].

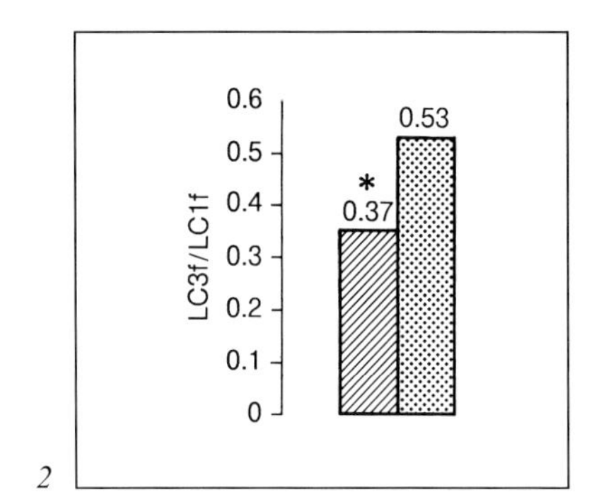

2

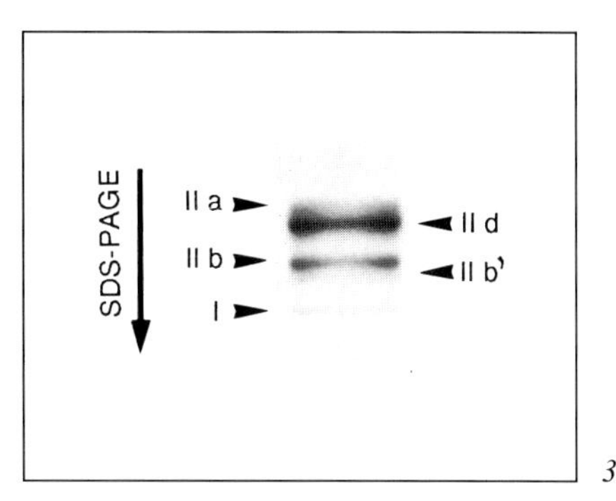

3

Fig. 2. Effects of endurance training on the LC3f/LC1f ratio in the EDL muscle. * $p < 0.05$: significant difference between trained (▨) and control (▩) groups.

Fig. 3. Electrophoretically separated myosin HC isoforms in adult rat plantaris muscle.

Four HC isoforms have been detected to date in fast-twitch muscle of adult rats, i.e. slow HCI and fast HCIIa, IIb and IId [12]. We were able to separate an additional, previously undetected HC isoform by electrophoresis in the present study (fig. 3). This isoform was not detected in the soleus muscles consisting of type I and IIa fibers, and its concentration was higher in the muscle region composed largely of type IIb fibers when compared to the muscle region composed mainly of type IIa fibers (results not shown). Therefore, it is speculated that, with regard to the staining pattern for mATPase activity, muscle fibers containing the new HC isoforms could resemble type IIb fibers. Thus, it was tentatively designated as HCIIb′.

A time course of the change in slow-type LC was similar to that of HCI (fig. 4); they increased after 6 and 8 weeks of overload. Although the earlier replacement of fast by slow isoform in HC than in LC is suggested as described above, there was not such a lag time. The difference between the rates of increases in slow LC and HCI may be small so that the time intervals used were too long to detect the existence of the time lag.

The transformation following the order HCIIb → HCIId → HCIIa appears to be brought about by the influence of chronic contractile activity [17]. The alterations in the fast HC isoform presented in figure 5 support this notion. Interestingly, the changes in HCIIb and HCIId occurred as early as 2 weeks after surgery (fig. 5b, c), but the content of HCIIa did not alter at this time (fig. 5a). The transformation of the HCIIb to the HCIId isoform seems to be evoked more readily than that of HCIId to HCIIa in

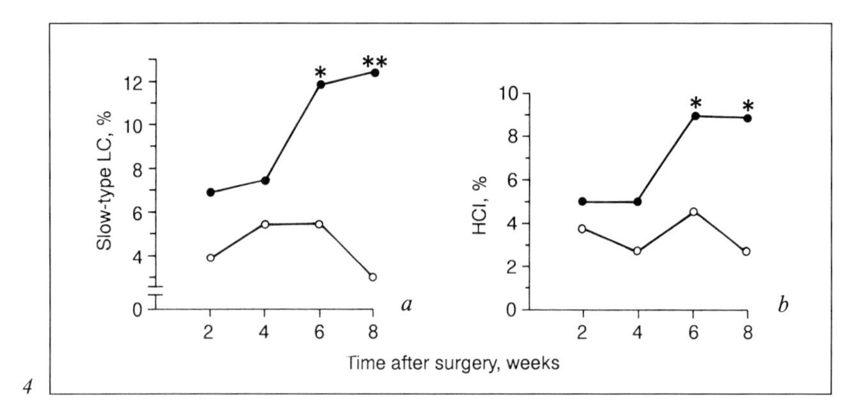

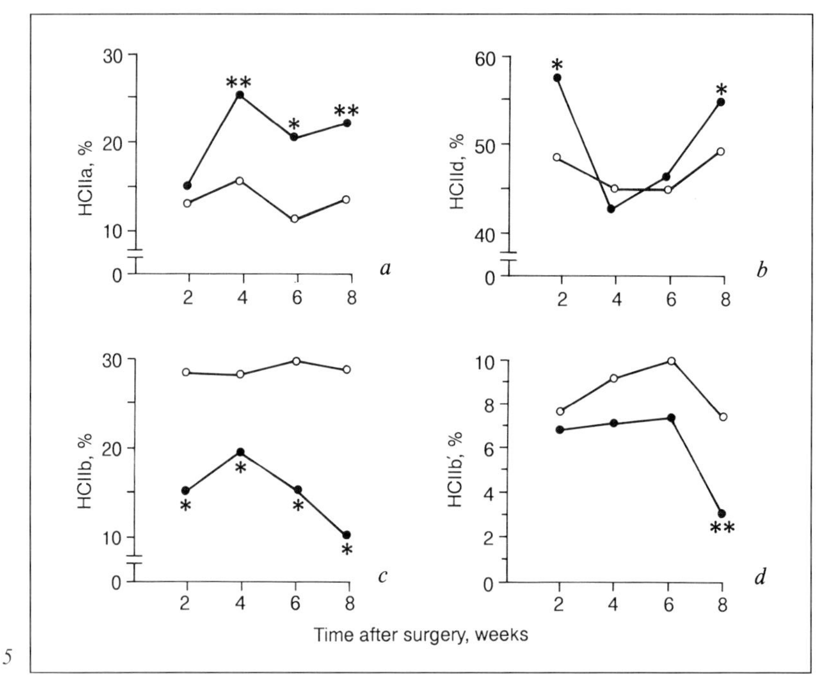

Fig. 4. Changes in the relative distribution of slow-type LC (LC1s and LC2s; *a*) and HCI isoforms (*b*) in the plantaris muscle as a function of time following removal of synergists. ● = Experimental muscle; ○ = control muscle. * $p < 0.05$; ** $p < 0.01$: significant differences as compared to control muscle.

Fig. 5. Changes in the relative distribution of fast-type HCIIa (*a*), HCIId (*b*), HCIIb (*c*) and HCIIb′ (*d*) isoforms in the plantaris muscle as a function of time following removal of synergists. ● = Experimental muscle; ○ = control muscle. * $p < 0.05$; ** $p < 0.01$: significant differences as compared to control muscle.

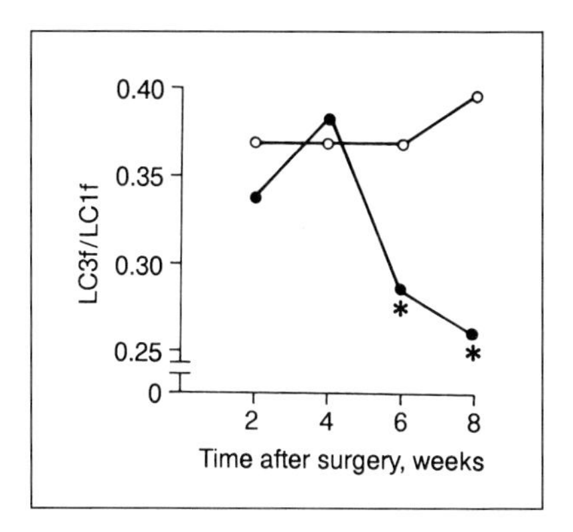

Fig. 6. Changes in the LC3f/LC1f ratio of the plantaris muscle as a function of time following removal of synergists. ● = Experimental muscle; ○ = control muscle. * $p < 0.05$: significant difference as compared to control muscle.

response to this model of increased contractile activity. A change in the HCIIb′ isoform was not evident until 6 weeks following surgery; 8 weeks after surgery, this isoform was significantly decreased (fig. 5d). Unfortunately, the ablation experiment cannot provide insights as to where the HCIIb′ isoform is located in the process of changes in the HC phenotype.

A decrease in the LC3f/LC1f ratio was shown after 6 and 8 weeks of overload (fig. 6) and occurred later than changes in fast HC isoforms with the exception of HCIIb′ (fig. 5). In electrically stimulated rat fast-twitch muscle, the reduction of LC3f is more pronounced at the protein level than at the mRNA level [18]. In addition, the content of LC3f protein in the soleus muscle is less than expected from that of the mRNA [19]. This raises the possibility that, in muscles endowed for sustained contractile activity, LC3f is more rapidly degraded than in muscles which are not frequently recruited. The rate of degradation of LC3f appears to be related to the content of HC isoforms in muscle. In the rat, muscles rich in HCIIa display a higher content of LC1f as compared to muscles rich in HCIIb, suggesting the existence of preferential affinities between LC1f and HCIIa and between LC3f and HCIIb. With regard to a reduction in LC3f induced by increased contractile activity, Kirschbaum et al. [18] have implied that this is evoked by an increase in the free form of LC3f resulting from the replacement of HCIIb by HCIIa. The lag time between changes in LC and HC (fig. 5, 6) supports this implication.

Chronic stimulation of rabbit fast-twitch muscles does not elicit substantial changes in LC3f [1]. Therefore, it is unlikely that the possible mechanism suggested by Kirschbaum et al. [18] is applicable to all mammalian skeletal muscles. It remains unclear whether enhanced contractile activities cause a decrease in LC3f in human skeletal muscle. However, the mechanism suggested for the rat might act for the human if a decrease in LC3f occurs, because human skeletal muscles are characterized by the lower affinity of LC3f for HCIIa than for HCIIb [20].

References

1 Pette D: Activity-induced fast to slow transitions in mammalian muscle. Med Sci Sports Exerc 1984;16:517–528.
2 Andersen P, Henriksson J: Capillary supply of the quadriceps femoris muscle of man: Adaptive response to exercise. J Physiol 1977;270:677–690.
3 Gollnick PD, Armstrong RB, Saubert CW VI, Piehl K, Saltin B: Enzyme activity and fiber composition in skeletal muscle of untrained and trained men. J Appl Physiol 1972;33:312–319.
4 Green HJ, Klug GA, Reichmann H, Seedorf U, Wiehrer W, Pette D: Exercise-induced fibre type transitions with regard to myosin, parvalbumin, and sarcoplasmic reticulum in muscles of the rat. Pflügers Arch 1984;400:432–438.
5 Brooke MH, Kaiser KK: Three 'myosin adenosine triphosphatase' systems: The nature of their pH lability and sulfhydryl dependence. J Histochem Cytochem 1970; 18:670–672.
6 Hirabayashi T: Two-dimensional gel electrophoresis of chicken skeletal muscle proteins with agarose gels in the first dimension. Anal Biochem 1981;117:443–451.
7 Roy RR, Baldwin KM, Martin TP, Chimarusti SP, Edgerton VR: Biochemical and physiological changes in overloaded rat fast- and slow-twitch ankle extensor. J Appl Physiol 1985;59:639–646.
8 Timson BF, Bowlin BK, Dudenhoeffer GA, George JB: Fiber number, area, and composition of mouse soleus muscle following enlargement. J Appl Physiol 1985;58: 619–624.
9 Sugiura T, Murakami N: Separation of myosin heavy chain isoforms in rat skeletal muscles by gradient sodium dodecyl sulfate-polyacrylamide gel electrophoresis. Biomed Res 1990;11:87–91.
10 Komi PV, Karlsson J: Physical performance, skeletal muscle enzyme activities and fiber types in monozygous and dizygous twins of both sexes. Acta Physiol Scand [Suppl] 1979;462.
11 Gollnick PD, Timson BF, Moore RL, Riedy M: Muscular enlargement and number of fibers in skeletal muscles of rats. J Appl Physiol 1981;50:936–943.
12 Termin A, Staron RS, Pette D: Myosin heavy chain isoforms in histochemically defined fiber types of rat muscle. Histochemistry 1989;92:453–457.

13 Brown WE, Salmon S, Whalen RG: The sequential replacement of myosin subunit isoforms during muscle type transformation induced by long term electrical stimulation. J Biol Chem 1983;258:14686–14692.
14 Bär A, Simoneau JA, Pette D: Altered expression of myosin light-chain isoforms in chronically stimulated fast-twitch muscle of the rat. Eur J Biochem 1989;178:591–594.
15 Sivaramakrishnan M, Burke M: The free heavy chain of vertebrate skeletal myosin subfragment 1 shows full enzymatic activity. J Biol Chem 1982;257:1102–1105.
16 Weeds A: Myosin light chains, polymorphism and fibre types in skeletal muscles; in Pette D (ed): Plasticity of Muscle. 1980, De Gruyter, New York, pp 55–68.
17 Termin A, Staron RS, Pette D: Changes in myosin heavy chain isoforms during chronic low-frequency stimulation of rat fast hindlimb muscles. Eur J Biochem 1989;186:749–754.
18 Kirschbaum BJ, Simoneau JA, Bär A, Barton PJR, Buckingham ME, Pette D: Chronic stimulation-induced changes of myosin light chains at the mRNA and protein levels in rat fast-twitch muscle. Eur J Biochem 1989;179:23–29.
19 Barton PJR, Buckingham ME: The myosin alkali light chain proteins and their genes. Biochem J 1985;231:249–261.
20 Wada M, Katsuta S, Doi T, Kuno S: Favourable associations between the myosin heavy-chain and light-chain isoforms in human skeletal muscle. Pflügers Arch 1990; 416:689–693.

Masanobu Wada, MD, Faculty of Integrated Arts and Sciences,
Hiroshima University, 1-1-89 Naka-ku, Hiroshima-shi, Hiroshima-ken 730
(Japan)

Sato Y, Poortmans J, Hashimoto I, Oshida Y (eds): Integration of Medical and Sports Sciences. Med Sport Sci. Basel, Karger, 1992, vol 37, pp 318–323

Non-Exercise-Related Stimulation of Mitochondrial Protein Synthesis in Creatine-Depleted Rats[1]

Y. Ohira[a], *T. Wakatsuki*[a], *N. Inoue*[a], *K. Nakamura*[b], *T. Asakura*[b], *K. Ikeda*[c], *T. Tomiyoshi*[d], *M. Nakajoh*[c]

[a]Department of Physiology and Biomechanics, National Institute of Fitness and Sports, Kanoya City; Departments of [b]Neurosurgery and [c]Radiology, Faculty of Medicine, and [d]University Hospital, Kagoshima University, Kagoshima City, Japan

Introduction

Elevation of mitochondrial enzyme levels is a well-known phenomenon which is induced by intensive endurance training [1]. However, the mechanism responsible for this muscular adaptation to exercise is still unclear, since various factors are influenced, and it is difficult to examine the effect of a specific factor alone in vivo. Non-iron-dependent respiratory enzyme activities in muscles are also increased in rats with severe iron deficiency anemia associated with decreased high-energy phosphates [2]. However, mitochondrial function in such muscles is impaired because synthesis of iron-dependent proteins is inhibited. Our previous study [3] also showed that mitochondrial enzyme activities are similarly increased in response to chronic cold exposure. In the current study, effects of creatine depletion by feeding the creatine analogue β-guanidinopropionic acid (β-GPA) [4-6] on mitochondrial enzyme levels were studied in rats. Body fat

[1]This study was, in part, supported by T. Nanba Memorial Health Care Foundation. We also thank Dr. J.O. Holloszy, Department of Medicine, Washington University School of Medicine, St. Louis, Mo., USA, for his careful reading of the manuscript and criticism.

content, metabolic rate, body temperature, and daily voluntary activity level were also investigated to study the possible mechanism of metabolic muscular responses.

Methods

Newly weaned male Wistar rats were randomly divided into two groups and housed individually. Rats in the control group were fed a powdered diet (CE-2, Nihon CLEA, Tokyo). The experimental group was fed the same diet containing β-GPA (1%). They were pair-fed and the amount of food was gradually increased following growth until the 4th week. After that, each rat was fed 20 g daily. Water was supplied ad libitum. The rats were kept in a room with a 12:12 h light:dark cycle. Temperature and humidity were maintained at approximately 23 °C and 55%.

Daily voluntary activity was measured between weeks 9 and 11. Rats were housed individually in 35 × 24 × 20 cm cages equipped with a running wheel (10 cm wide and with a diameter of 37 cm), thereby permitting the rat to run at will. The numbers of wheel revolutions were registered on a counter and checked daily at 6 a.m. and 6 p.m.

Resting metabolic rate and heart rates were measured as described previously [7, 8]. Both oxygen consumption ($\dot{V}O_2$) and production of carbon dioxide ($\dot{V}CO_2$) as well as electrocardiogram were recorded for approximately 3 h continuously and stable values reached after approximately 30 min were utilized.

Resting rectal temperature was measured by using a thermister (Nihon Kohden, MGA III-219, Tokyo) in unanesthetized rats. Determination of skin temperature (calf, foot, and tail) was performed using a thermography system (Nippon Avionics TVS-2000, Tokyo) after shaving the right hindlimb of rat lightly anesthetized with ether inhalation.

Distribution of phosphorus compounds in the dorsal region of thigh muscles of rats anesthetized by i.p. injection of sodium pentobarbital (5 mg/100 g body weight) was evaluated by ^{31}P-nuclear magnetic resonance (NMR) spectroscopy (Ohtsuka Electronics, BEM 170/200, Tokyo). For biochemical analyses, hindlimb muscles of rats anesthetized with sodium pentobarbital were exposed, keeping the nerve and blood supplies intact. The rat was allowed recovered for approximately 10 min on a heating pad. The muscles were covered with gauze moistened with Krebs-Ringer solution. The soleus and plantaris muscles were freeze-clamped with aluminum tongs cooled in liquid nitrogen. The samples were pulverized and homogenized in $HClO_4$ at –20 °C. After centrifugation at 3,000 *g* for 10 min at 4 °C, the supernatant was neutralized and used for determination of ATP [9], phosphocreatine (PC) [10], and inorganic phosphate [11].

The tibialis anterior was homogenized in 175 m*M* KCl containing 10 m*M* glutathione (reduced form), and 2 m*M* EDTA (pH 7.2) using a Polytron. Activities of citrate synthase [12], β-hydroxyacyl CoA dehydrogenase [13], cytochrome oxidase [14], and lactate dehydrogenase (LDH) [15] were measured spectrophotometrically.

In some rats, body fat content was estimated by underwater weighing. The interscapular brown adipose tissue (BAT) was dissected out and wet weight was measured. The wet weight of the adrenals was also determined.

Results and Discussion

Creatine in skeletal muscle is severely depleted by β-GPA feeding [16]. As previously studied [4–6], high-energy phosphate contents in skeletal muscles were significantly decreased in response to β-GPA feeding. For example, ATP and PC contents in soleus of β-GPA-fed rats were approximately 31 and 89% less than those in control muscle, respectively.

Gain of body weight was significantly inhibited by feeding β-GPA, although the daily food consumption was identical for control and β-GPA groups. Body weight of the β-GPA group was approximately 13% less than of the controls when they were sacrificed ($p < 0.01$). Whole body fat content relative to the body weight was also less (approximately 22%, $p < 0.05$), suggesting that slower gain of body weight in creatine-depleted rats was partly due to reduced fat accumulation. However, the weight of interscapular BAT was significantly greater in β-GPA-fed than in control rats (67%, $p < 0.001$).

Rats fed β-GPA had a lower mean ± SEM rectal temperature (37.5 ± 0.1 °C) than control rats (38.3 ± 0.2 °C, $p < 0.01$). Tail skin temperature of β-GPA-fed rats was also less than in controls ($p < 0.001$), but that in calf and foot was not significantly different from controls.

Resting metabolic rate was elevated in rats fed β-GPA. The mean levels of $\dot{V}O_2$ expressed as ml/min and ml/min/kg in β-GPA rats were 15 and 31% greater than in control rats ($p < 0.001$), respectively. Those of $\dot{V}CO_2$ were elevated by 16 and 30% ($p < 0.01$), respectively. However, the $\dot{V}CO_2/\dot{V}O_2$ ratio was identical in the two groups. Resting heart rates tended to be less in β-GPA than control rats but insignificantly ($p > 0.05$).

The activities of citrate synthase, β-hydroxyacyl CoA dehydrogenase, and cytochrome oxidase in muscle were significantly greater in the β-GPA than the control group by 30, 40, and 35%, respectively. On the other hand, LDH activity in β-GPA muscle was approximately 33% less than the control level. Similar elevations of mitochondrial enzyme activities have been reported by Shoubridge et al. [17], and Lai and Booth [18] found an increase in cytochrome c mRNA in muscle of rats fed β-GPA.

Such increases in mitochondrial enzyme activities are known to occur following an intensive endurance training [1]. However, the rats in the current study did not perform any exercise training. Furthermore, the daily voluntary activity of the β-GPA group was less (approximately 35%, $p < 0.001$) than in control rats. Therefore, it is clear that the increase in mitochondrial enzyme activities in the β-GPA-fed rats was not stimulated by increased use of muscles.

Enhanced activities of mitochondrial enzymes [3] and mitochondrial size [19] have been found to occur in response to chronic cold exposure. Cold exposure increases the weight of BAT and stimulates nonshivering thermogenesis [20]. Elevation of BAT could be due to stress as it has been reported that BAT weight is increased in response to immobilization [21]. An increase in BAT weight was also found in rats fed β-GPA in the current study. However, the weight of the adrenals, which is an indicator of stress, remained normal, suggesting that the muscular responses may not be directly stress-related.

High-energy phospate contents in muscle of cold-exposed frog were also less than normal [3] as was found in rats fed β-GPA. However, the effects of these two conditions on muscle metabolism may be different, although both metabolic rate and mitochondrial protein synthesis are increased similarly by these two stimuli. In cold-exposed animals, mitochondrial biogenesis may be stimulated in order to increase heat production. Therefore, uncoupling might be stimulated and ATP content may be lowered as a result. On the other hand, mitochondrial biogenesis in rats fed β-GPA may be stimulated in order to increase ATP synthesis. Although the BAT volume and $\dot{V}O_2$, which are the measures of heat production, were increased, body temperature was less in β-GPA-fed rats than in controls. These results may indicate that uncoupling might be inhibited to increase ATP synthesis or that heat loss is enhanced due to the lower fat volume in the rats fed β-GPA. Activities of mitochondrial enzymes in skeletal muscles which are not dependent on iron are also elevated when iron deficiency anemia is severe enough to lower the high-energy phosphates [2]. But those levels were identical to controls if the high-energy phosphate levels are normal even with iron deficiency [22]. These results suggest that mitochondrial biogenesis, as well as protein synthesis, may be stimulated if high-energy phosphate contents are lowered chronically, regardless of the cause.

Elongated mitochondria are found in skeletal muscles of rats fed β-GPA [23]. Such mitochondria also contain inclusions. A bundle of inclusions consisted of four parallel arrays in which two outer arrays are continuous with cristal membranes. Similar intramitochondrial inclusions were also observed in diseased skeletal muscles [24].

Physical work capacity at mild, but not at high, intensity both in swimming and treadmill running is greater in rats fed β-GPA than in controls, even though the high-energy phosphate contents, especially PC, of skeletal muscles are depleted [25]. These results show that the elongated mitochondria associated with increased enzyme activity in muscles of β-GPA-fed rats are functional, and suggested that elevation of mitochondrial enzyme activities unrelated to exercise training can improve work capacity. Resistance to fatigue measured in situ was also improved even in fatigable fast-twitch extensor digitorum longus muscle in rats following β-GPA feeding [unpubl. observation]. It is also suggested that maintenance of high-energy phosphates at high levels is not necessary for endurance capacity of muscle, if the work intensity is submaximal.

In conclusion, effects of chronic depletion of high-energy phosphates by feeding the creatine analogue β-GPA on mitochondrial enzyme levels were studied in rats. High-energy phosphates, especially PC, in skeletal muscles were depleted in rats fed β-GPA. Resting $\dot{V}O_2$ and $\dot{V}CO_2$ were elevated by β-GPA feeding, as was the volume of BAT. However, the rectal and tail skin temperatures were less in the rats fed β-GPA than in controls. Body weight in β-GPA-fed rats was also less than controls partly due to lower fat content. Although the daily voluntary activity was significantly less, mitochondrial enzyme activities of muscles were greater in β-GPA-fed than in control rats. It is suggested that mitochondrial protein synthesis is stimulated if high-energy phosphates are depleted chronically.

References

1 Holloszy JO: Biochemical adaptations in muscle. Effects of exercise on mitochondrial oxygen uptake and respiratory enzyme activity in skeletal muscle. J Biol Chem 1967;242:2278–2282.

2 Ohira Y, Cartier L-J, Chen M, Holloszy JO: Induction of an increase in mitochondrial matrix enzymes in muscle of iron-deficient rats. Am J Physiol 1987;253:C639–C644.

3 Ohira M, Ohira Y: Effects of exposure to cold on metabolic characteristics in gastrocnemius muscle of frog *(Rana pipiens).* J Physiol (Lond) 1988;395:589–595.

4 Fitch CD, Jellinek M, Fitts RH, Baldwin KM, Holloszy JO: Phosphorylated β-guanidinopropionate as a substitute for phosphocreatine in rat muscle. Am J Physiol 1979;228:1123–1125.

5 Fitch CD, Jellinek M, Mueller EJ: Experimental depletion of creatine and phosphocreatine from skeletal muscle. J Biol Chem 1974;249:1060–1063.

6 Mainwood GW, Alward M, Eiselt B: The effect of metabolic inhibitors on the contraction of creatine-depleted muscle. Can J Physiol Pharmacol 1982;60:114–119.

7 Brooks GA, White TP: Determination of metabolic and heart rate responses of rats to treadmill exercise. J Appl Physiol 1978;45:1009–1015.

8 Ohira Y, Koziol BJ, Edgerton VR, Brooks GA: Oxygen consumption and work capacity in iron-deficient anemic rats. J Nutr 1981;111:17–25.
9 Lamprecht W, Trautschold I: ATP determination with hexokinase and glucose-6-phosphate dehydrogenase; in Bergmeyer HU (ed): Methods of Enzymatic Analysis, ed 2. New York, Academic Press, 1974, pp 2101–2110.
10 Lamprecht W, Stein P, Heinz F, Weisser H: Creatine phosphate; in Bergmeyer HU (ed): Methods of Enzymatic Analysis, ed 2. New York, Academic Press, 1974, pp 1777–1781.
11 Guynn RD, Veloso D, Veech RL: Enzymatic determination of inorganic phosphate in the presence of creatine phosphate. Anal Biochem 1972;45:277–285.
12 Srere PA: Citrate synthase. Methods Enzymol 1969;13:3–5.
13 Bass A, Brdiczka D, Eyer P, Hoffer S, Pette D: Metabolic differentiation of distinct muscle types at the level of enzymatic organization. Eur J Biochem 1969;10:198–206.
14 Wharton DC, Tzagoloff A: Cytochrome oxidase from beef heart mitochondria. Methods Enzymol 1967;10:245–250.
15 Pesce A, McKay RH, Stolzenbach F, Cahn RD, Kaplan NO: Comparative enzymology of LDH. J Biol Chem 1964;239:1753–1761.
16 Moerland TS, Wolf NG, Kushmerick MJ: Administration of a creatine analogue induces isomyosin transitions in muscle. Am J Physiol 1989;257:C810–C816.
17 Shoubridge EA, Challiss RAJ, Hayes DJ, Radda GK: Biochemical adaptation in the skeletal muscle of rats depleted of creatine with the substrate analogue β-guanidinopropionic acid. Biochem J 1985;232:125–131.
18 Lai MM, Booth FW: Cytochrome c mRNA and α-actin mRNA in muscles of rats fed β-GPA. J Appl Physiol 1990;69:843–848.
19 Yahata T, Kuroshima A: Changes in fine structure of rat skeletal muscle related to cold acclimation. Hokkaido J Med Sci 1977;52:63–67.
20 Cannon B, Nedergaard J: The biochemistry of inefficient tissue: Brown adipose tissue. Essays Biochem 1985;20:110–164.
21 Kuroshima A, Habara Y, Uehara A, Murazumi K, Yahata T, Ohno T: Cross adaption between stress and cold in rats. Pflügers Arch 1984;402:402–408.
22 Cartier L-J, Ohira Y, Chen M, Cuddihee RW, Holloszy JO: Perturbation of mitochondrial composition in muscle by iron deficiency. J Biol Chem 1986;261:13827–13832.
23 Ohira Y, Kanzaki M, Chen C-S: Intramitochondrial inclusions caused by depletion of creatine in rat skeletal muscles. Jpn J Physiol 1988;38:159–166.
24 Schochet SS Jr: Electron microscopy of skeletal muscle and peripheral nerve biopsy specimens; in MacKay B (ed): Introduction to Diagnostic Electron Microscopy. New York, Appleton-Century-Crofts, 1981, pp 131–169.
25 Inoue N, Ohira Y, Yamasaki K, Ishine S, Tabata I, Nishi I: Chronic depletion of creatine causes an enhanced metabolic activity in rats. 14th Int Congr Nutr, Seoul, 1989, p 340.

Y. Ohira, PhD, Department of Physiology and Biomechanics,
National Institute of Fitness and Sports, Kanoya City, Kagoshima 891–23 (Japan)

Sato Y, Poortmans J, Hashimoto I, Oshida Y (eds): Integration of Medical and Sports Sciences. Med Sport Sci. Basel, Karger, 1992, vol 37, pp 324–328

Mitochondrial Myopathy and Transcriptional Control

Yasuo Kagawa, Shigeo Ohta

Department of Biochemistry, Jichi Medical School, Minamikawachi, Tochigi-ken, Japan

Introduction

The velocity of ATP synthesis in muscle cells fluctuates greatly depending on exercise, and it is regulated by both rapid respiratory control [1] and slow synthesis of mitochondrial enzyme complexes [2]. ATP synthase (FoF↑) is the central enzyme complex of oxidative phosphorylation [3], utilizing the electrochemical potential of protons across the membrane [4]. Respiratory control is regulation of FoF↑ by the potential; and, in fact, the V_{max} of FoF↑ was shown to be controlled by the voltage across a planar lipid bilayer containing FoF↑ [5]. While much progress has been made in studies on respiratory control, little is known about the regulation of muscular energy metabolic capacity. It is known that the activities of several energy-transducing enzymes are coupled to exercise [6, 7], and it induces coordinate increases in mitochondrial and nuclear mRNAs of the cytochrome c oxidase subunits [8, 9]. However, it is not known how the contractile signal induces protein synthesis. More than 10 human genes encoding enzymes of energy metabolism have been sequenced in this laboratory [2, 10] and these have been used as probes to sequence both the enhancers coordinating transcription [11, 12] and mutated DNAs in patients with myopathy [13–18]. The analysis of muscle cells of mitochondrial myopathy is an approach for the expression of mitochondrial DNA (mtDNA). Here we report the regulation of the mitochondrial enzyme synthesis in relation to mitochondrial myopathy.

Materials and Methods

A muscle specimen was obtained by biopsy of biceps brachii with patients and their family's consent. In order to overcome the difficulty caused by the polymorphism of human DNA, both myopathic and normal myogenic cells from the same patient's muscle were compared. The myogenic cell lines were established by transfecting an SV40 virus DNA lacking its replication origin as reported previously [16]. The respiratory enzyme deficient cell line was selected by cytochemical staining for cytochrome c oxidase for sequencing the mtDNA. Total DNA was isolated from the myogenic cells by digestion with proteinase K. The mtDNA fragments were amplified by the polymerase chain reaction and purified by agarose gel electrophoresis. The nucleotide sequences of both strands were determined by using fluorescent dye dideoxy nucelotide terminators (Applied Biosystems, Inc.). All other procedures used have already been reported [13–19].

Results and Discussion

(1) Mitochondrial myopathy, encephalopathy, lactic acidosis and stroke-like episodes (MELAS) is a major group of heterogeneous mitochondrial disorders. On the modified Gomori-trichrome stained section, ragged-red fibers were seen. One of the substitutions was a transition of A to G in the tRNA-Leu(UUR) gene at Cambridge nucleotide number 3,243, in 8 patients from unrelated families (fig. 1a) [17]. This mutated tRNA may be inactive, because the protein synthesis was impaired in the myogenic cell. The gene for tRNA-Leu(UUR) is the binding site of the terminator that regulates the relative amounts of ribosomal RNA and mRNA for complexes of oxidative phosphorylation (fig. 1b). The 90% inhibition of transcription of mtDNA under low oxygen tension was lost in the MELAS myogenic cells.

(2) Multiple deletions of mitochondrial DNA (mtDNA) in the non-D region were found in the muscle of two brother patients (51 and 49 years old) with myopathic and neuropathic changes in extremities [15]. These multiple deletions (at least 12) were flanked by direct repeats, which were mainly localized to the non-D loop region. The lengths of the direct repeats were short (4–12 bases) (fig. 2). Thus, the mutational changes of this disease were different from those of an autosomal-dominant mitochondrial myopathy with multiple deletions starting in the D-loop region, and Kearns-Sayre syndrome and progressive external ophthalmoplegia with a single sporadic deletion of mtDNA. The deletions occur between origins of replication for the heavy and light chain. Many ragged-red fibers with mitochondria containing paracrystalline inclusions were found.

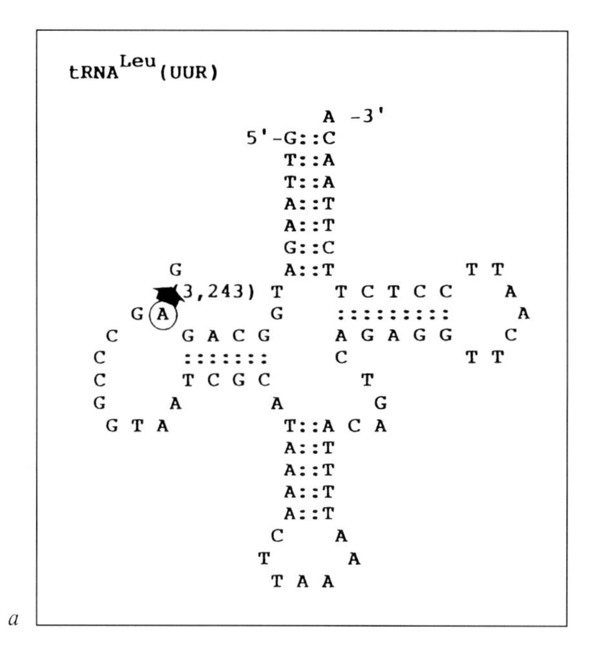

a

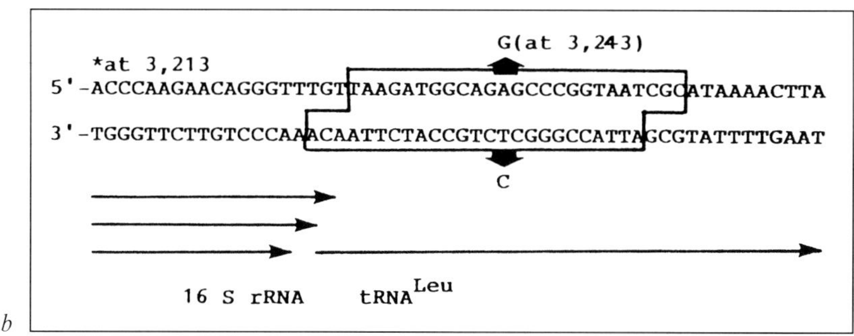

b

Fig. 1. a Secondary structure of mitochondrial tRNA-Leu(UUR) and the position of the substitution in MELAS. *b* The region where the terminator of RNA synthesis binds. The binding site is boxed.

In contrast to other cases of familial mitochondrial myopathy that are associated with a compensatory increase of mitochondria, a marked decrease (15% of the controls) of muscle mtDNA with multiple deletions in non-D region was found in another patient (54 years old, female) with familial myopathy similar to Kearns-Sayre syndrome.

In order to determine the effect of deletion on the transcription of nuclear genes for mitochondrial energy-transducing enzymes, mtDNA of mouse cells was removed by using ethidium bromide [19]. This treatment

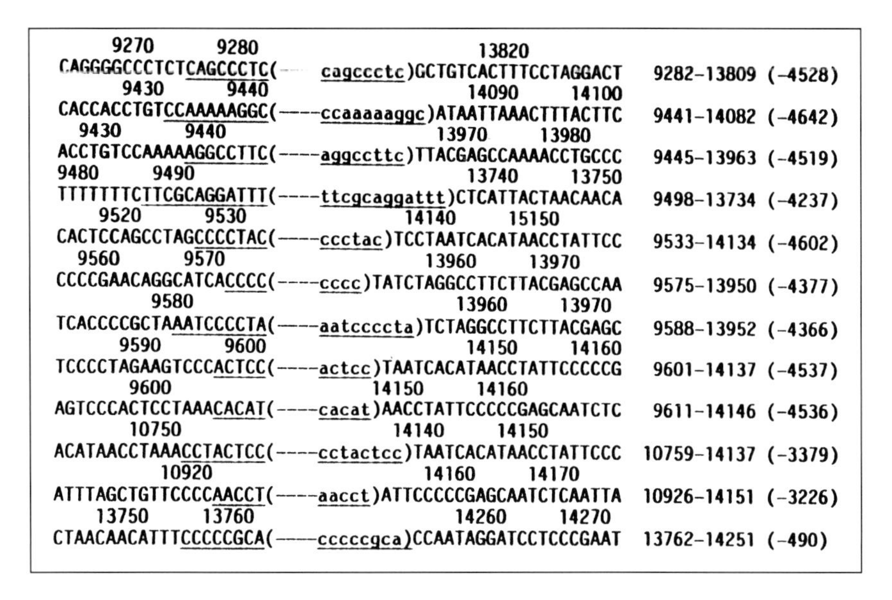

Fig. 2. Nucleotide sequences of the breakpoint regions in mtDNA from the patient with familial mitochondrial myopathy. Small letters indicate nucleotides within deletions. Regions of direct repeats are underlined.

did not affect expressions of the nuclear genes, but did affect their assembly [19]. These phenomena will be analyzed in the light of genetic regulation of the mitochondrial ATP synthesizing system that is under the control of a common enhancer [2, 11, 12].

References

1 Chance B: Electron transfer: Pathways mechanisms and controls. Ann Rev Biochem 1977;46:967–980.
2 Kagawa Y, Ohta S: Regulation of mitochondrial ATP synthesis in mammalian cells by transcriptional control. Int J Biochem 1989;22:219–229.
3 Kagawa Y: Proton motive ATP synthesis; in Ernster L (ed): Bioenergetics. Amsterdam, Elsevier, 1984, pp 149–186.
4 Mitchell P: Keilin's respiratory chain concept and its chemiosmotic consequences. Science 1979;206:1148–1159.
5 Muneyuki E, Kagawa Y, Hirata H: Steady state kinetics of proton translocation catalyzed by thermophilic FoF↑-ATPase reconstituted in planar lipid bilayer membranes. J Biol Chem 1989;264:6092–6096.

6 Schwerzmann K, Hoppeler H, Kayar SR, Weibel ER: Oxidative capacity of muscle and mitochondria: correlation of physiological biochemical, and morphometric characteristics. Proc Natl Acad Sci USA 1989;86:1583–1587.
7 Williams RS, Garcia-Moll M, Mellor J, Salmons S, Harlan W: Adaptation of skeletal muscle to increased contractile activity. Expression of nuclear genes encoding mitochondrial proteins. J Biol Chem 1987;262:2764–2767.
8 Hood DA, Zak R, Pette D: Chronic stimulation of rat skeletal muscle induces coordinate increases in mitochondrial nuclear mRNAs of cytochrome-c-oxidase subunits. Eur J Biochem 1989;179:275–280.
9 Hevner RF, Wong-Riley MTT: Neuronal expression of nuclear and mitochondrial genes for cytochrome oxidase (CO) subunits analyzed by in situ hybridization: Comparison with CO activity and protein. J Neurosci 1991;11:1942–1956.
10 Ohta S, Tomura H, Matsuda K, Kagawa Y: Gene structure of the human mitochondrial adenosine triphosphate synthase beta subunit. J Biol Chem 1988;263:11257–11262.
11 Tomura H, Endo H, Kagawa Y, Ohta S: Novel regulatory enhancer in the gene of the human mitochondrial ATP synthase beta subunit. J Biol Chem 1990;265:6525–6527.
12 Nagley P: Coordination of gene expression in the formation of mammalian mitochondria. Trends Genet 1991;7:1–4.
13 Endo H, Hasegawa K, Narisawa K, Tada K, Kagawa Y, Ohta S: Defective gene in lactic acidosis: abnormal pyruvate dehydrogenase E1 alpha subunit caused by a frame shift. Am J Hum Genet 1989;44:358–364.
14 Kobayashi Y, Momoi MY, Tominaga K, Momoi T, Nihei K, Yanagisawa M, Kagawa Y, Ohta S: A point mutation in the mitochondrial tRNA Leu(UUR) gene in MELAS (mitochondrial myopathy, encephalopathy, lactic acidosis and stroke-like episodes). Biochem Biophys Res Commun 1990;173;816–822.
15 Yuzaki M, Ohkoshi N, Kanazawa I, Kagawa Y, Ohta S: Multiple deletions in mitochondrial DNA at direct repeats of non-D-loop regions in cases of familial mitochondrial myopathy. Biochem Biophys Res Commun 1989;164:1352–1357.
16 Shimoizumi H, Yoshida-Momoi M, Ohta S, Kagawa Y, Momoi T, Yanagisawa M: Cytochrome c oxidase-deficient myogenic cell lines in mitochondrial myopathy. Ann Neurol 1989;25:615–621.
17 Kobayashi Y, Momoi MY, Tominaga K, Shimoizumi H, Nihei K, Yanagisawa M, Kagawa Y, Ohta S: Respiration-deficient cells are caused by a single point mutation in the mitochondrial tRNA-Leu(UUR) gene in mitochondrial myopathy, encephalopathy, lactic acidosis and strokelike episodes (MELAS). Am J Hum Genet 1991; 49:590–599.
18 Ohtsuka M, Niijima K, Mizuno Y, Yoshida M, Kagawa Y, Ohta S: Marked decrease of mitochondrial DNA with multiple deletions in patients with familial mitochondrial myopathy. Biochem Biophys Res Commun 1990;167:680–685.
19 Hayashi J, Tanaka M, Sato W, Ozawa T, Yonekawa H, Kagawa Y, Ohta S: Effects of ethidium bromide treatment of mouse cells on expression and assembly of nuclear-coded subunits of complexes involved in the oxidative phosphorylation. Biochem Biophys Res Commun 1990;167:216–221.

Yasuo Kagawa, MD, Department of Biochemistry, Jichi Medical School,
Minamikawachi, Tochigi-ken 329–04 (Japan)

Sato Y, Poortmans J, Hashimoto I, Oshida Y (eds): Integration of Medical and Sports Sciences. Med Sport Sci. Basel, Karger, 1992, vol 37, pp 329–335

Effects of Exercise and Glucose Ingestion on Adipose Tissue Metabolism

A Microdialysis Study

Jens Bülow[a], *Lene Simonsen*[a], *Joop Madsen*[b, 1]

[a] Department of Clinical Physiology, Bispebjerg Hospital, and

[b] Institute of Medical Physiology C, Panum Institute, Copenhagen, Denmark

Adipose tissue metabolism during exercise has mainly been studied indirectly by measurements of concentrations of the lipolytic products glycerol and free fatty acids in blood and by turnover studies of these substances, or studies have been performed in vitro either on isolated, perfused fat pads, on tissue biopsies or isolated fat cells. Another approach has been to use adipose tissue blood flow as an indirect indicator of metabolic rate in the tissue [1]. Adipose tissue is not easily accessible in vivo due to its dispersion throughout the body and the complicated blood supply which does not allow for catheterization of a selective venous drainage. Only few attempts have previously been made to assess adipose tissue metabolism in vivo in awake animals during rest and exercise [2].

However, in recent years two new techniques have been developed by which subcutaneous adipose tissue metabolism can be measured in vivo in man.

Techniques for Measurement of Human Adipose Tissue Metabolism in vivo

Frayn et al. [3] have developed a technique by which a superficial vein on the abdominal wall can be catheterized percutaneously. These veins

[1] Birgitte Kiærskov and Karen Klausen are thanked for skillful technical assistance. The study was supported by grants from The Novo Foundation and King Christian X's Foundation.

drain mainly subcutaneous adipose tissue, but to some extent also skin, implying that the metabolite concentrations measured in venous blood represent a mixture of the metabolic events in adipose tissue and skin.

The microdialysis technique, originally developed for measurements of brain metabolism in vivo, has been adapted for measurements of subcutaneous adipose tissue metabolism by Lönnroth et al. [4]. The principle of the technique is that a dialysis fiber is placed in the subcutaneous tissue and then perfused with known solutions of the metabolite of interest (e.g. glucose, glycerol, lactate). If the concentration of the metabolite is higher in the interstitial water than in the perfusate, an increase in concentration will occur in the dialysate sampled after the fiber and vice versa if the concentration is lower. When perfusate and tissue have the same concentration no net exchange will take place. If an in vivo calibration is performed before the system is pertubated (e.g. by food intake or exercise) the recovery fraction over the dialysis membrane can be determined and this allows for calculation of tissue concentrations during non-steady states.

The Effect of Glucose Ingestion on Subcutaneous Adipose Tissue Metabolism

The effect of glucose ingestion on adipose tissue metabolism during rest has been studied in man both by abdominal vein catheterization [3, 5] and by microdialysis [6–9]. However, these studies have suffered from the lack of simultaneous measurements of adipose tissue blood flow. Since blood flow increases significantly in adipose tissue after glucose ingestion [10, 11], it is necessary to measure this parameter simultaneously with the metabolic parameters if mobilization and deposition are to be calculated.

The main findings of the studies [5–9] have been that glucose administration inhibits the rate of glycerol mobilization (i.e. lipolysis) by at least 70%. Simultaneously there is a significant uptake of glucose in adipose tissue. An interesting finding is that adipose tissue even in the fasting state releases lactate [3, 9, 12], and that the release increases following glucose ingestion.

In our laboratory we have measured the lactate production from abdominal subcutaneous adipose tissue in resting man in the fasting state and after an oral 75-gram glucose load simultaneously by abdominal wall vein catheterization and microdialysis. Subcutaneous adipose tissue blood flow was measured by ^{133}Xe washout. Table 1 shows the lactate release and the glucose uptake from adipose tissue under these circumstances. Both techniques show a significant increase in lactate release after glucose; however, the increase measured by catheterization is modest, and, if correct, metabolically insignificant. In contrast, the microdialysis technique shows a significant lactate production before glucose with a doubling in excretion rate after glucose. Similarly, the glucose uptake measured by microdialysis seems to be greater than that measured by catheterization. The difference between the results obtained by these two techniques is probably due to a major contribution of blood derived from nonadipose tissue to the blood drawn from the abdominal wall vein. Another explanation can be the uncertainty in the determination of the recovery fraction over the dialysis membrane. However, the results obtained by both techniques emphasize that 'aerobic glycolysis' takes place in adipose tissue in spite of a high (70–90%) oxygen saturation in the blood derived from the tissue. Thus, adipose tissue may contribute significantly to the metabolic rate of the whole organism via the Cori cycle.

Table 1. Glucose uptake and lactate output in abdominal subcutaneous adipose tissue determined simultaneously after a 75-gram oral glucose load by vein catheterization and microdialysis

	Control	30 min	60 min	90 min	120 min
Abdominal wall vein					
Glucose uptake, µmol/(100 g·min)	0.70 ± 0.56	7.56 ± 3.18	5.18 ± 2.61	2.62 ± 1.55	4.18 ± 1.81
Lactate output, µmol/(100 g·min)	0.36 ± 0.57	0.46 ± 1.35	−0.08 ± −2.07	2.15 ± 1.53	
Microdialysis					
Glucose uptake, µmol/(100 g·min)	1.28 ± 2.56	13.36 ± 6.24	13.54 ± 11.19	5.24 ± 4.89	1.87 ± 5.65
Lactate output, µol/(100 g·min)	7.25 ± 3.05	18.00 ± 9.93	15.77 ± 9.64	17.12 ± 9.64	13.75 ± 5.48

Mean ± SD are given. For glucose uptake, n = 4. For lactate output, n = 5.

Table 2. Abdominal subcutaneous adipose tissue glucose uptake, lactate output and glycerol output during rest and at 1, 2 and 3 h of moderate exercise

	Rest	Exercise		
		1 h	2 h	3 h
Glucose uptake, μmol/(100 g·min)	1.08±0.37	3.30±1.46	2.78±1.17	5.26±2.94
Lactate output, μmol/(100 g·min)	1.76±2.02	5.17±4.53	7.12±6.75	7.87±5.96
Glycerol output, μmol/(100 g·min)	0.62±0.46	1.63±1.09	1.75±1.17	2.21±1.52

Mean ± SD are given. Glucose uptake, n = 4. Lactate output, n = 6. Glycerol output, n = 7.

Subcutaneous Adipose Tissue Metabolism during Prolonged Exercise

We have previously proposed a hypothesis with regard to the regulation of free fatty acid mobilization during exercise [13]. According to the hypothesis, this regulation takes place via the amount of fatty acids which is reesterified in the adipose tissue without being released to the blood, and not via a negative feedback mechanism regulating stimulation of lipolysis in relation to demand. Thus, it was proposed that lipolysis is stimulated in excess of demand, and evidence for this was obtained from our previous animal studies [2].

With the above-mentioned microdialysis method, it is now possible to examine subcutaneous adipose tissue metabolism in man during prolonged exercise.

Arner et al. [14] studied the adrenergic regulation of lipolysis in subcutaneous, abdominal adipose tissue during rest and 30 min of exercise at two-thirds of maximal work capacity by microdialysis. In this study it is demonstrated that lipolysis during rest is modulated by α-adrenergic inhibition while β-adrenergic mechanisms are not involved during these circumstances. On the other hand, β-adrenergic, but not α-adrenergic mechanisms are involved in the regulation of lipolysis during exercise. As estimated by the increase in glycerol concentration in adipose tissue intercellular water, the lipolytic rate increases from the beginning of the exercise bout and it continues to increase during the 30 min of exercise reaching a concentration level about 2.5-fold higher than the baseline level.

However, concentration changes are only indirect evidence of enhanced mobilization from the tissue. Thus, in our laboratory we performed the following experiments during prolonged exercise of moderate intensity in order to elucidate subcutaneous, abdominal adipose tissue metabolism during these circumstances in healthy man.

Seven male subjects were investigated after an overnight fast. Three microdialysis fibers (cuprophan, 20,000 D cut-off) were placed in the periumbilical subcutaneous tissue and in vivo calibration was then performed for measurements of glucose, glycerol and lactate concentrations. Adipose tissue blood flow in the region was measured continuously by ^{133}Xe washout. While the in vivo calibration was performed, the subjects rested supine. After this basal period the subjects worked for three 50-min periods on an ergometer bicycle at about 50% of maximal work capacity. Ten-minute periods were interspaced between the work periods. In this period arterialized blood was drawn for glucose, glycerol and lactate determination.

Table 2 shows the glucose uptake, the glycerol mobilization and the lactate mobilization from the tissue during the experiment. During rest there is a glucose uptake in the range of 1 μmol/(100 g·min). During exercise the glucose uptake in adipose tissue increases to 2–3 μmol/(100 g·min).

Glycerol release is on average 0.7 μmol/(100 g·min) during rest, increasing about 3-fold during exercise, and, as it appears, there is a relatively constant release of glycerol from the tissue during the 3 h of exercise, indicating a constant rate of lipolysis.

Lactate is released from adipose tissue during rest as well as during exercise. On average, the rate of lactate release increases 4-fold from rest to the end of the 3 h of exercise.

Discussion

The results of Arner et al. [14] and the present results confirm our previous hypothesis [13] that stimulation of lipolysis during exercise takes place very rapidly after the onset of exercise, and then is rather constant during prolonged exercise. However, the rate of lipolysis is much greater than needed to cover the need for mobilization of free fatty acids to be oxidized. Thus, this mobilization is regulated via reesterification, and the degree of reesterification is determined by the amount of free plasma binding sites for fatty acids presented to the adipose tissue. This last parameter is determined by the plasma free fatty acid/albumin ratio and the adipose tissue blood flow.

A key process in the reesterification is the presence of glucose in adipose tissue as precursor of α-glycerophosphate. In exercising dogs we have previously shown [6] that about 70% of the released fatty acids are reesterified. If this figure holds true for human adipose tissue, about 25–50% of the glucose uptake in adipose tissue can be explained by reesterification. On the other hand, the glucose uptake can approximately account for the lactate production, leaving the glucose used for reesterification to be derived from glycogen stored in the adipocytes.

Thus, two futile metabolic cycles are taking place in adipose tissue during prolonged exercise. The first lipolysis/reesterification participates as a key process in the regulation of free fatty acid mobilization. The second, the adipose tissue/liver glucose/lactate cycle, seems more futile but may contribute significantly to glucose homeostasis during exercise.

References

1 Bülow J: Adipose tissue blood flow during exercise. Dan Med Bull 1983:30:85–100.
2 Bülow J: Subcutaneous adipose tissue blood flow and triacylglycerol mobilization during prolonged exercise in dogs. Pflügers Arch 1982;392:230–234.
3 Frayn KN, Coppack SW, Humphreys SM, Whyte PL: Metabolic characteristics of human adipose tissue in vivo. Clin Sci 1989:76:509–516.
4 Lönnroth P, Jansson P-A, Smith U: A microdialysis method allowing characterization of intercellular water space in humans. Am J Physiol 1987:253:E228–E231.
5 Coppack SW, Fisher RM, Gibbons GF, Humphreys SM, McDonough MJ, Potts JL, Frayn KN: Postprandial substrate deposition in human forearm and adipose tissues in vivo. Clin Sci 1990:79:339–348.
6 Hagström C, Arner P, Engfeldt P, Rörsner S, Bolinder J: In vivo subcutaneous adipose tissue glucose kinetics after glucose ingestion in obesity and fasting. Scand J Clin Lab Invest 1990:50:129–136.
7 Jansson P-A, Favelin J, Smith U, Lönnroth P: Characterization by microdialysis of intercellular glucose level in subcutaneous tissue in humans. Am J Physiol 1988; 255:E218–E220.
8 Jansson P-A, Smith U, Lönnroth P: Interstitial glycerol concentration measured by microdialysis in two subcutaneous regions in humans. Am J Physiol 1990: 258:E918–E922.

9 Jansson P-A, Smith U, Lönnroth P: Evidence for lactate production by human adipose tissue in vivo. Diabetologia 1990:33:253–256.
10 Bülow J, Astrup A, Christensen NJ, Kastrup J: Blood flow in skin, subcutaneous adipose tissue and skeletal muscle in the forearm of normal man during an oral glucose load. Acta Physiol Scand 1987:130:657–661.
11 Simonsen L, Bülow J, Astrup A, Madsen J, Christensen NJ: Diet induced changes in subcutaneous adipose tissue blood flow in man: β-Adrenoceptor inhibition. Acta Physiol Scand 1990:139:341–346.
12 Hagström E, Arner P, Ungerstedt U, Bolinder J: Subcutaneous adipose tissue: A source of lactate production after glucose ingestion in humans. Am J Physiol 1990: 258:E888–E893.
13 Bülow J: Lipid mobilization and utilization; in Poortmans JR (ed): Principles of Exercise Biochemistry. Med Sport Sci, vol 27, Basel, Karger 1988, pp 140–163.
14 Arner P, Kriegholm E, Engfeldt P, Bolinder J: Adrenergic regulation of lipolysis in situ at rest and during exercise. J Clin Invest 1990:85:893–898.

Dr. Jens Bülow, Department of Clinical Physiology and Nuclear Medicine,
Bispebjerg Hospital, Bispebjerg Bakke 23, DK-2400 Copenhagen NV (Denmark)

Sato Y, Poortmans J, Hashimoto I, Oshida Y (eds): Integration of Medical and Sports Sciences. Med Sport Sci. Basel, Karger, 1992, vol 37, pp 336–341

Recommended Dietary Allowances for Chinese Athletes

Suggestions and Illustrations

J.D. Chen, J.F. Wang, S.W. Wang, K.J. Li, Z.M. Chen

Institute of Sports Medicine, Beijing Medical University, Beijing, China

Establishment of the recommended dietary allowances (RDA) of athletes has become an urgent need for evaluating both the diet quality and quantity, and as a guide for balancing the diet of athletes.

On the basis of the RDA of normal healthy people in China [1], information of athletes' RDA of some other countries [2, 3], and a series of studies on nutrition metabolism, requirements and systematic nutritional surveys [4–12], in 1990 we organized and requested a group of experts to examine the material from nutritional surveys and research on athletes and to set up the first RDA for athletes through attentive discussion and analysis. We hope to revise this RDA after practicing and extending research in this area.

RDA of Energy Requirement for Athletes

The energy requirement of athletes depends on a series of factors such as exercise intensity and duration, body weight, sport events, age, sex, and ambient temperature. On the basis of the assumption that the energy intakes of normal healthy people usually adapt to energy consumption, in adults keeping a balanced energy intake body weight would be stable, and, based on a series of surveys of energy intake of a large number of athletes [4, 5] and some information of RDA for athletes of other countries [2, 3, 13], we set the suggestive RDA of energy for athletes into five ranks according to their energy consumption. We presented the data as calories per kilogram of body weight. Considering the individual and some other

Table 1. Energy RDA for Chinese athletes

Sport events	Energy	
	kcal	kcal/kg
Weiqi (M, F)	2,400 (2,000–2,800)	45 ± 5
Diving (M, F), gymnastics (F)	2,800 (2,200–3,200)	50 ± 5
Gymnastics (M), sprint (F) Table tennis, badminton (M,F) Wt. lifting, swimming (M, F)	3,500 (2,700–4,200)	55 ± 5
Sprint (M), soccer (M) Distance running (M, F) Basketball, volleyball (M, F) Throwing (F), swimming (M, F)	4,200 (3,700–4,700)	60 ± 5
Soccer (M), throwing (M), Swimming (M, F), marathon (M) Weight lifting (M, > 75 kg)	> 4,700	> 65

Weiqi is a game played with black and white pieces on a board of 361 crosses.
M = Male; F = female.

variables, we extended the RDA of energy within ± 10% (table 1). We have provided the energy RDA for male and female athletes separately, but we can also use a factor of 0.85 times the data of males for female athletes.

RDA of Calorie Sources for Athletes

The proportion of calorie source of energy was based on the specific nutritional requirement of athletes [2, 3, 13]; calories from protein was set at 12–15% of the total, which is about 1.2–2.0 g/kg body weight. The RDA of protein is mainly based on the data of estimated nitrogen balance on athletes [7]. A nutrition survey of elite athletes showed that there had been a problem concerning overnutrition of protein in elite athletes, the majority of whom had a protein intake of around 20% of the total calories which is about 2 g/kg body weight or even more. Protein intakes have risen significantly in recent years, which has drawn our attention as overnutrition of protein brings about problems such as acid metabolite accumulation, strain on the liver and renal system, disturbances of water and mineral metabolism, the possibility of renal stone formation, and the risk factor of chronic disease in later life. The amount of calories of fat was set at 30%, but should not exceed 35% of the total. A nutrition survey showed that the dietary fat intake of the elite athletes investigated reached about 40% (38–49%) of the total calories [6], and that the cholesterol intake was also too high (418 ± 42 to 3,003 ± 607 mg/day). The incidence of hypercholesterolemia was 13.3% which needed some consideration.

Regarding the fatty acid ratio, we suggested that FA:PUFA:MUFA be in the order of 1:1:1–1.5 [14].

Carbohydrate is an important fuel for exercise. The RDA of carbohydrate for athletes has been set at 50–60% of the total calories. A nutrition survey showed that dietary carbohydrate intake has decreased in recent years, which may cause adverse effects, especially on endurance.

RDA of Minerals for Athletes

There was no significant difference between the mineral requirements of normal people and athletes training at normal temperatures, but the RDA of minerals for athletes training in hot environments should be increased. Na^+ and K^+ loss could be around 2.5 and 1.5 g per training period [8, 9]. Besides, attention should be paid to mineral balance for athletes who often reduce their weight and have saunas. On the other hand, excess of Na^+ intake is related to the incidence of high blood pressure, it has been accepted that Na^+ intake should be limited to < 5 g/day, but for the intake of K^+, 3–6 g/day is safe.

Although the FAO/WHO suggested that dietary calcium should be balanced to a daily intake of 400–500 mg and considering that Ca absorption is affected by a series of factors and that milk Ca content is low in the Chinese diet, we set the RDA for Ca at 800–1,200 mg/day for athletes. Since adequate intake of Mg^{2+} protects heart function, and we found that the serum Mg of athletes decreased significantly after exercising in hot environments [15], the RDA for Mg^{2+} has been set at 400–500 mg/day (table 2).

The incidence of iron deficiency was high. Fe absorption was affected by dietary factors and training. Performance ability would be affected even

Table 2. Mineral RDA for Chinese athletes

Training condition	K^+, g	Na^+, g	Ca^{2+}, mg	Mg^{2+}, mg
Normal temperature	3–4	<4	800–1,200	400–500
Hot environment	4–6	<5	800–1,200	400–500

Table 3. Trace element RDA for Chinese athletes

Training condition	Zn^{2+}, mg	Cu^{2+}, mg	Fe^{2+}, mg
Normal temperature	20–25	3–3.5	20–25
Hot environments	20–25	3–3.5	25–30

with a mild degree of iron deficiency. Fe deficiency affects Hb levels and thus limits the oxygen-carrier capacity. Besides, iron proteins in blood and muscle such as myoglobin, SDH and cytochrome C are important components in the citric cycle and electron-transfer chain, and are involved in ATP formation. Based on these manifestations, the RDA of Fe was set at 20–25 mg/day for athletes. Since the iron concentration in sweat is rather high and frequent heavy sweating may become a risk factor inducing Fe deficiency, the RDA of Fe was set at 25–30 mg/day for athletes training in hot environments.

Average Zn levels of elite athletes was significantly low as compared with those of normal healthy students (96 ± 19 vs. 109 ± 17 μg/dl). Although the dietary intakes of Zn and Cu of athletes were twice the amount of the RDA for normal people, 9.1% (male) and 12.5% (female) of the 360 athletes investigated still had low serum Zn levels [11]. Zn deficiency affects muscle contraction function and endurance performance. Both Zn and Cu deficiency in mice and humans increased free radical generation and impaired the defence system, as shown by the markedly low SOD activity in blood and liver [16]. Based on a series of studies on Zn and Cu metabolism, the RDA of Zn and Cu have been suggested to be 20–25 mg/day for athletes, which is about 5–10 mg/day more than that of normal persons (table 3).

RDA of Vitamins

Vitamin B_1 insufficiency would cause an increase in blood pyruvate and lactate concentrations of athletes at exercise. Studies from the USSR suggested that the RDA of vitamin B_1 should be 3–5 mg/day in the general population, and it should be 5–10 mg/day for athletes undergoing endurance exercise training. Nutrition surveys indicated that as the dietary intakes were 0.37–0.48 mg/1,000 kcal, 25% of the athletes investigated were in a state of vitamin B_1 insufficiency. Systematic nutrition investigation showed that there has been a trend towards a decrease in vitamin B_1 intake because cereal (especially whole grains) intake decreased and animal food intake increased. The RDA of vitamin B_1 has been set at 3–6 mg/day. A nutrition survey showed that there has been no significant vitamin B_2 deficiency in athletes as their dietary intake was 2.5 mg/day. Vitamin B_2 intake was only low in a small portion of athletes because the intake of animal food had increased. The RDA of vitamin B_2 was set at 2.5 mg/day. The RDA of vitamin C has been set at 140 mg/day and that of niacin at 25 mg/day which is ten times that of vitamin B_2.

Vitamin A intake was generally sufficient for elite athletes. Serum vitamin A levels showed that of the 182 athletes investigated 5.1% has values of < 30 μg/dl. Besides, there had been a tendency that vitamin A was negatively correlated with serum malondialdehyde. The RDA of vitamin A was primarily fixed at 1,500 RE/day which corresponds to 5,000 IU/day for athletes; and the RDA of vitamin A was set at 2,400 RE/day (8,000 IU) for those athletes who need an alert vision response (table 4).

Table 4. Vitamin RDA for Chinese athletes

Condition	Vit. A RE	Vit. B_1 mg	Vit. B_2 mg	Niacin mg	Vit. C mg
Normal training	1,500	3–6	2.5	25	140
Special condition	2,400	5–10	2.5	25	200

Special condition refers to intensive vision for vitamin A, endurance training for vitamin B_1, competition period for vitamin C.
RE = Retinol equivalent.

References

1 Chinese Nutrition Society: Illustration of the recommended dietary allowances of the Chinese. Acta Nutr Sin 1990;12:1–9.
2 Sports Science Committee of Japanese Sports Association: In Suzuki (ed): Diet Manipulation of Sportsmen. Japan Amateur Sports Association, 1977, pp 1–34.
3 Grandjean AC: Macronutrient intake of US athletes compared with the general population and recommendation made for athletes. Am J Clin Nutr 1989;49:1070–1076.
4 Chen JD, et al: The investigation on energy expenditure of athletes. J Beijing Med Coll 1965;3:204–208.
5 Chen JD: Some Sports Nutrition Researches in China; in Qu MY, Yu CL (eds): China Sports Medicine. Med Sport Sci, vol 28. Basel, Karger, 1988, pp 94–113.
6 Chen JD, et al: Nutritional problems and measures in elite and amateur athletes. Am J Clin Nutr 1989;49:1084–1089.
7 Chen JD, Chen ZM, Yang ZY, et al: Research of the protein metabolism and requirements of athletes. Sports Sci 1982;2:49–57.
8 Chen JD, et al: Primary research of water and mineral metabolism of long distance runners. J Beijing Med Coll 1962;1:72.
9 Chen JD, Yang ZY, et al: Nutrition and metabolism of athletes training in hot environment. Chin J Sports Med 1987;6:65.
10 Li KJ, Chen JD, et al: Investigation on iron status of adolescent athletes. Basic Med Clin 1990;10:49.
11 Chen JD, Wu YZ, Bai RY, et al: Study on serum zine and copper levels of elite athletes. Chin J Sports Med 1987;6:194–199.
12 Chen JD, Yu XX, Nie FE: The study of vitamin C status and requirement of the gymnasts and distance runners. Chin Med J 1962;7:454–457.
13 Yakovlev NN: Sports nutrition. Nutr Problems 1957;16:58.
14 Brown LM: Present knowledge in Nutrition. Washington, Int Life Sci Nutr Foundation, 1990, pp 1–65.
15 Chen JD, et al: The effects of exercise load on element metabolism of organism. Can J Sports Sci 1988;13:8.
16 Cao GH, Chen JD: Free radical generation and elimination in athletes and their relation to zinc and copper status. Chin J Sports Med 1992;10:132–134.

Prof. J.D. Chen, Institute of Sports Medicine, Beijing Medical University, Beijing, 100083 (China)

Sato Y, Poortmans J, Hashimoto I, Oshida Y (eds): Integration of Medical and Sports Sciences. Med Sport Sci. Basel, Karger, 1992, vol 37, pp 342–348

Effects of Exercise and Nutrition on Branched-Chain Amino Acid Metabolism

Activation of Branched-Chain Alpha-Keto Acid Dehydrogenase Complex by Exercise and Effect of High-Fat Diet Intake on the Activation of the Enzyme Complex

Yoshiharu Shimomura, Hisao Kotsuka, Shinichi Saitoh, Masashige Suzuki

Laboratory of Biochemistry of Exercise and Nutrition,
Institute of Health and Sport Sciences, University of Tsukuba, Tsukuba,
Ibaraki, Japan

Introduction

The branched-chain amino acids (BCAAs), leucine, isoleucine and valine, are indispensable amino acids for animals. The BCAAs comprise about 35% of the indispensable amino acids in muscle proteins and about 40% of the amino acids required preformed by mammals [1]. Exercise enhances protein metabolism, and the BCAAs produced by protein degradation could supply a significant fraction of the increased energy expenditure during exercise [2].

In the catabolism of BCAAs, the first step is a reversible transamination catalyzed by BCAA transaminase, and the second is an irreversible oxidative decarboxylation of the branched-chain α-keto acids by the branched-chain α-keto acid dehydrogenase (BCKDH) complex. The regulation of BCAA catabolism is achieved in large part at the second step [3]. The BCKDH complex is subject to regulation by covalent modification [3]. Phosphorylation causes inactivation and dephosphorylation causes activation of the enzyme complex. A specific kinase [4] and a specific phosphatase [5] are responsible for the phosphorylation and dephosphorylation, respectively. Thus, the in vivo activity of the enzyme complex depends on the amount of complex in the dephosphorylated form.

High-fat diet intake has great effects on energy metabolism [6, 7]. However, such an effect on amino acid metabolism has not yet been examined, although the amino acids appear to contribute significantly to energy metabolism during exercise. In this paper, we summarize the effects of exercise and high-fat diet intake on the activity of the BCKDH complex.

Effects of Running Exercise and High-Fat Diet Intake on the Activity of BCKDH Complex in Rat Liver and Skeletal Muscle

Since liver has a high activity of BCKDH complex compared to other organs, it is an important tissue for the catabolism of BCAAs [1]. In our study [8], running exercise elevated the activity state of BCKDH complex to approximately 100% in liver of rats fed a high-fat diet, whereas the exercise did not significantly affect the activity state (65–75%) of the complex in liver of rats fed a high-carbohydrate diet, resulting in a significantly higher activity state of the complex in the former than in the latter during exercise (fig. 1). The total activity of the enzyme complex was not altered by the exercise or the high-fat diet intake. These findings suggest that BCAA catabolism in rat liver during exercise is enhanced by high-fat diet intake.

Although skeletal muscle has a high activity of BCAA transaminase, the activity of BCKDH complex in the muscle is low due to a low percentage of the enzyme complex in active state and to a low total activity of the enzyme complex compared to that in liver [1]. However, the muscle should be an important tissue for BCAA metabolism because of the high content of muscle tissue compared to other tissues in animals. Running exercise greatly activated the enzyme complex in our study [8] (fig. 1). This explains the increased leucine oxidation in skeletal muscle during exercise as previously reported [9–12]. However, the high-fat diet intake had little effect on the activation of the enzyme complex caused by exercise.

The mechanism of BCKDH complex activation by exercise in rat skeletal muscle has been examined by Kasperek [13]. In his experiment, nontrained rats were used and it was found that activation of the BCKDH complex was associated with a decrease in the ATP content in the muscle. From these results, he suggested that the ATP level was a regulator of the activity of the BCKDH complex in skeletal muscle, since ATP is required for BCKDH kinase to inactivate the BCKDH complex by phosphorylation. On the other hand, we found activation of the BCKDH complex

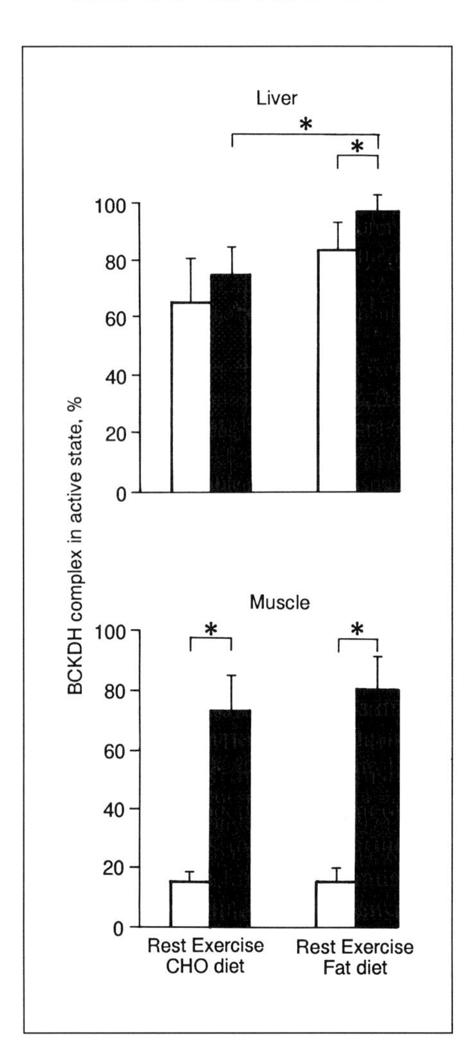

Fig. 1. Effect of running exercise on percentage of BCKDH complex in active state. Values are means ± SD. * $p < 0.05$. The trained rats fed either high-carbohydrate (CHO) or high-fat (Fat) diets were forced to run for 2 h at a speed of 30 m/min, up an 8° incline, and then the gastrocnemius muscles and livers were immediately removed and frozen by the freeze-clamp method. The total activities of the BCKDH complex in muscle and liver were not altered by 2-hour exercise bout or diet.

without a decrease in the ATP content in the muscle of trained rats [8], suggesting that the ATP content is not the regulator of the activity of the BCKDH complex. BCAA concentrations are considered as other candidates for the regulators of the activity of the BCKDH complex, because the BCAA levels in the muscle were enhanced by exercise [8] and α-keto acids derived from BCAAs, especially leucine, are potent inhibitors of BCKDH kinase [14].

Activation of BCKDH Complex in Skeletal Muscle by Tetanic Contractions Caused by Electrical Stimulation

Activation of the BCKDH complex in skeletal muscle was further examined using a model of tetanic contractions caused by electrical stimulation, since it has been demonstrated that this model of muscle contractions increased leucine oxidation in the muscle [10]. Nontrained rats were used in this experiment. The muscle contractions caused by electrical stimulation for 5–60 min significantly increased the activity state of the BCKDH complex at all of the time points measured (fig. 2), explaining an increase in leucine oxidation during the muscle contractions as reported previously [10]. The total activity of the enzyme complex in the muscle was not altered by the contractions. There was a trend that the concentrations of BCAAs in the muscle were increased by the contractions (fig. 2), and the concentrations of BCAAs were significantly correlated with the activity state of the BCKDH complex, suggesting that the concentrations of BCAAs are activation factors of the enzyme complex. On the other hand, the ATP content in the muscle was markedly decreased by initiating the muscle contractions (at 5–15 min), recovered to the original level at 30–45 min, and again decreased at 60 min of the muscle contractions (fig. 2). This alteration pattern is obviously different from the profile of the activation of BCKDH complex during the contractions. It might be reasonable to conclude that the ATP level in the muscle is ruled out from the regulation factors of the activity of the BCKDH complex.

It was interesting to examine the effects of exercise training and high-fat diet intake on the activity state of the BCKDH complex in skeletal muscle using this model of muscle contractions. As given in table 1, the training elevated the activity state of the enzyme complex in both the rested and the stimulated muscles, but high-fat diet intake had little effect on the activity state of the enzyme complex, suggesting that training, but not high-fat diet intake, enhances the catabolism of BCAAs in rat skeletal muscle.

Conclusion

The running exercise elevated the activity state of BCKDH complex in liver of rats fed the high-fat diet, but not in liver of rats fed the high-carbohydrate diet, suggesting that high-fat diet intake enhances the catab-

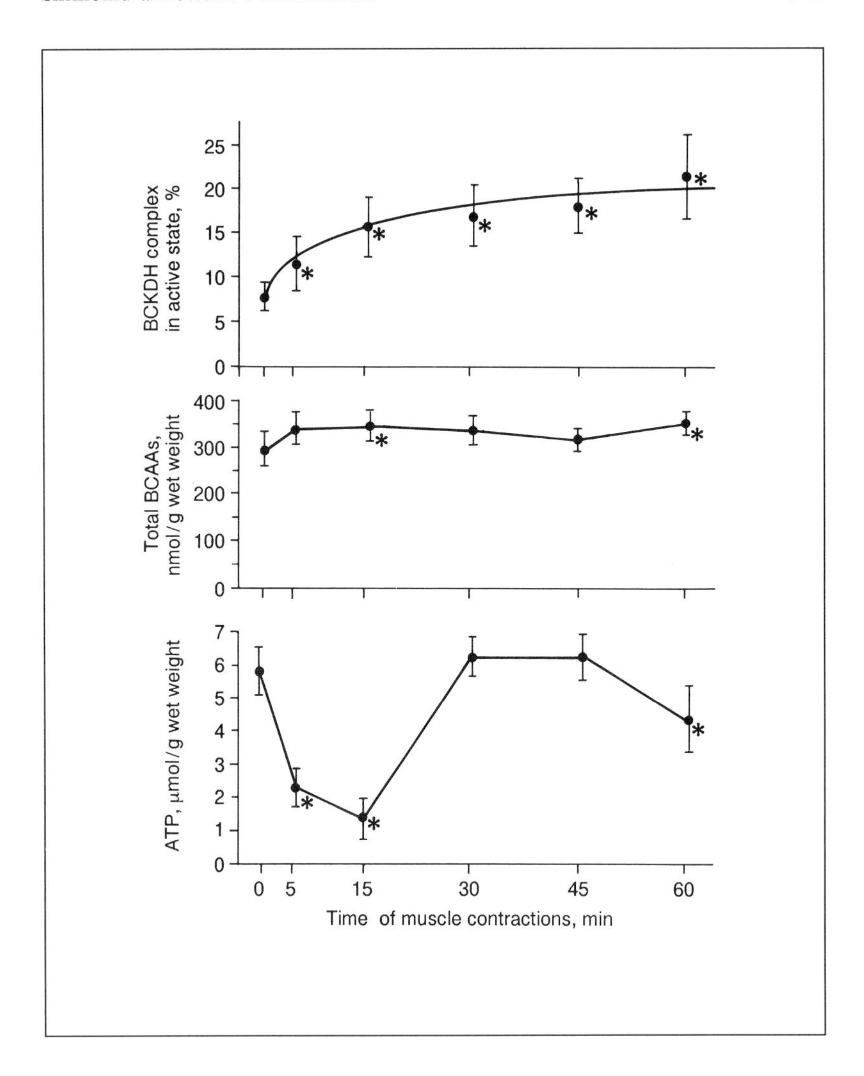

Fig. 2. The time course of BCKDH complex activation, the concentrations of total BCAAs and ATP in muscle. Values are means ± SD. * Significantly different from the 0-min time point ($p < 0.05$). Nontrained rats were anesthetized with sodium pentobarbital, and the muscles of one hindlimb were contracted via supramaximal square wave pulses (6 V, 0.05 ms duration) at a tetanic frequency of 60 tetani/min (100 ms train, 100 Hz). At the indicated time point, gastrocnemius muscle was removed and immediately frozen by the freeze-clamp method. The activity of BCKDH complex was expressed as % active state of the total activity. The total activity was not altered during the contractions.

Table 1. Effect of electrical stimulation on percentage of BCKDH complex in the active state in muscle

Diet	BCKDH complex in active state, %	
	sedentary	trained
Before stimulation		
CHO	8.1 ± 0.6	14.7 ± 2.3*
Fat	9.9 ± 1.8	15.9 ± 1.6*
After 20 min stimulation		
CHO	15.7 ± 2.4	21.0 ± 2.8*
Fat	15.1 ± 3.8	21.2 ± 2.4*

Values are means ± SD.
* Significantly different from sedentary rats ($p < 0.05$). Sedentary and trained rats fed either high-carbohydrate (CHO) or high-fat (Fat) diet were used in this experiment. The muscle contractions were caused for 20 min by electrical stimulation as described in figure 2. Then, gastrocnemius muscle was removed and frozen by the freeze-clamp method. Activity of the BCKDH complex was expressed as % active state of the total activity. Activity state of the enzyme complex was significantly elevated by the muscle contractions in all of the groups of rats, but the total activity was not altered by either training or diet.

olism of BCAAs in rat liver during exercise. The running exercise and the muscle contractions caused by electrical stimulation elevated the activity state of BCKDH complex in rat skeletal muscle, suggesting that the exercise enhances the catabolism of BCAAs in skeletal muscle. ATP and BCAAs had been considered as regulators of the activity of BCKDH complex, and both concentrations in the muscle were measured. Then, the level of BCAA concentrations, but not ATP, were associated with the activity state of the BCKDH complex, suggesting that the BCAAs (especially leucine) are the regulators of the BCKDH complex in rat skeletal muscle.

References

1 Harper AE, Miller RH, Block KP: Branched-chain amino acid metabolism. Ann Rev Nutr 1984;4:409–454.
2 Poortmans JR: Protein metabolism; in Poortmans JR (ed): Principles of Exercise Biochemistry. Med Sport Sci, vol 27, Basel, Karger, 1988, pp 164–193.

3 Harris RA, Paxton R, Powell SM, Goodwin GW, Kuntz MJ, Han AC: Regulation of branched-chain α-keto acid dehydrogenase complex by covalent modification; in Weber G (ed): Advance in Enzyme Regulation, vol 25, New York, Pergamon Press, 1986, pp 219–237.
4 Shimomura Y, Nanaumi N, Suzuki M, Popov KM, Harris RA: Purification of branched-chain α-ketoacid dehydrogenase kinase from rat liver and rat heart. Arch Biochem Biophys 1991;283:293–299.
5 Damuni Z, Reed LJ: Purification and properties of the catalytic subunit of the branched-chain α-keto acid dehydrogenase phosphatase from bovine kidney mitochondria. J Biol Chem 1987;262:5129–5132.
6 Jansson E, Hjemdahl P, Kaijser L: Diet induced changes in sympatho-adrenal activity during submaximal exercise in relation to substrate utilization in man. Acta Physiol Scand 1982;114:171–178.
7 Miller WC, Bryce GR, Conlee RK: Adaptations to a high-fat diet that increase exercise endurance in male rats. J Appl Physiol 1984;56:78–83.
8 Shimomura Y, Suzuki T, Saitoh S, Tasaki Y, Harris RA, Suzuki M: Activation of branched-chain α-keto acid dehydrogenase complex by exercise: Effect of high-fat diet intake. J Appl Physiol 1990;68:161–165.
9 Dohm GL, Hecker AL, Brown WE, Klain GJ, Puente FR, Askew EW, Beecher GR: Adaptation of protein metabolism to endurance training: increased amino acid oxidation in response to training. Biochem J 1977;164:705–708.
10 Hood DA, Terjung RL: Leucine metabolism in perfused rat skeletal muscle during contractions. Am J Physiol 1987;253:E636–E647.
11 Lemon PWR, Nagle FJ, Mullin JP, Benevenga NJ: In vivo leucine oxidation at rest and during two intensities of exercise. J Appl Physiol 1982;53:947–954.
12 Wolfe RR, Goodenough RD, Wolfe MH, Royle GT, Nadel ER: Isotopic analysis of leucine and urea metabolism in exercising humans. J Appl Physiol 1982;52:458–466.
13 Kasperek GJ: Regulation of branched-chain 2-oxo acid dehydrogenase activity during exercise. Am J Physiol 1989;256:E186–E190.
14 Paxton R, Harris RA: Regulation of branched-chain α-ketoacid dehydrogenase kinase. Arch Biochem Biophys 1984;231:48–57.

Dr. Yoshiharu Shimomura, Department of Bioscience, Nagoya Institute of Technology, Showa-ku, Nagoya 466 (Japan)

Sato Y, Poortmans J, Hashimoto I, Oshida Y (eds): Integration of Medical and Sports Sciences. Med Sport Sci. Basel, Karger, 1992, vol 37, pp 349–355

Sustained Exercise Endurance Capacity after Depletion of Liver Glycogen Levels

Stephanus E. Terblanche[a], *Kishor Gohil*[b], *Johanna K. Lang*[b], *Lester Packer*[b]

[a] Department of Biochemistry, University of Zululand, KwaDlangezwa, South Africa; [b] Department of Molecular and Cell Biology, University of California, Berkeley, Calif., USA

It is well documented that depletion of glycogen and/or development of hypoglycemia during prolonged strenuous exercise are major factors in the development of exhaustion [1–3]. It has also been reported that liver glycogen remains low in fasting rats, showing that the liver is enzymatically geared for glucose production rather than glycogen synthesis during fasting [4]. The aim of this study was to ascertain the effects of fasting and exhaustive exercice on certain variables involved in oxidative stress (first study phase) and to provide a possible explanation for sustained exercise endurance capacity after depletion of liver glycogen levels as a result of prolonged fasting (second study phase).

First Study Phase

Materials and Methods

Animal Care and Running to Exhaustion Protocol. Female Sprague-Dawley rats were randomly divided into 4 groups of 16 rats each, viz. (1) normally fed; (2) fasted 24 h; (3) fasted 48 h, and (4) fasted 72 h.

All 4 groups were subdivided into 2 subgroups of 8 rats each, the one subgroup to be sacrificed at rest and the other after a bout of exercise to exhaustion. Endurance capacity was measured by running animals at $26\ m \cdot min^{-1}$ up a 15% gradient as previously described [5].

Tissue Preparation. Rats either at rest or exhaustion were anesthetized with pentobarbital sodium, $6\ mg \cdot 100\ g^{-1}$ body mass, injected intraperitoneally. A plantaris muscle and a liver lobe were excised and stored on ice in aluminum foil for determination of the levels of activity of cytochrome oxidase. A liver lobe and the other plantaris muscle were

Table 1. Plasma levels of lactate dehydrogenase activity ($nmol \cdot ml^{-1} \cdot min^{-1}$)

Treatment	Rested	Exhausted
Normally fed	55.4 ± 21.0	188.7 ± 22.0
Fasted 24 h	17.6 ± 5.7	195.6 ± 6.9
Fasted 48 h	12.8 ± 3.3	161.8 ± 15.2
Fasted 72 h	53.6 ± 22.1	223.4 ± 27.1

also freeze-clamped for the determination of vitamin E, ubiquinones and ubiquinols. The protein and glycogen content of the liver were also determined. Blood was drawn from the abdominal aorta into a heparinized syringe for the determination of the levels of activity of plasma lactate dehydrogenase.

Biochemical Analysis. The activity levels of plasma lactate dehydrogenase were determined according to the method described by Vassault [6].

Muscle and liver cytochrome oxidase activity were measured by a method similar to that of Cooperstein and Lazarow [7]. Vitamin E, ubiquinones and ubiquinols were determined by HPLC after a one-step lipid extraction according to the procedure described by Lang and Packer [8]. Liver protein content was determined according to the method described by Lowry et al. [9] and liver glycogen content according to the method reported by Keppler and Decker [10]. Liver DNA levels were determined according to a modified method described by Ceriotti [11].

Results and Discussion

Body Mass. The body masses of the normally fed group and of the groups fasted for 24, 48 and 72 h were 240.9 ± 3.2 , 210.7 ± 1.5, 205.4 ± 1.8 and 194.2 ± 1.6 g, respectively ($p < 0.05$). Exhaustive exercise did not exhibit any significant effect on body mass.

Endurance Capacity. Endurance capacity was not significantly affected ($p < 0.05$). Mean run times to exhaustion were 89.3 ± 6.8, 67.4 ± 2.2, 67.3 ± 6.2 and 75.7 ± 7.9 min for 0, 1, 2 and 3 days of fasting, respectively.

Liver Glycogen. It is well known that liver glycogen is depleted during periods of extended fasting. Liver glycogen levels in fed rats were 135.7 ± 9.9 $\mu mol \cdot mg\ DNA^{-1}$. After 1 day of fasting, liver glycogen was almost totally depleted to a residual concentration of 1.8 ± 0.6 $\mu mol \cdot mg\ DNA^{-1}$. Mean liver glycogen on day 3 of fasting was 12.2 ± 9.1 $\mu mol \cdot mg\ DNA^{-1}$ and suggests that glycogen synthesis had occurred. Almost complete glycogen depletion also resulted from a single run to exhaustion (135.7 ± 9.9 to 3.8 ± 1.2 $\mu mol \cdot mg\ DNA^{-1}$) in fed rats.

Table 2. Liver protein and DNA levels

Treatment		Rested	Exhausted
Protein	Normally fed	177.8 ± 4.1	164.1 ± 9.3
mg·mg DNA^{-1}	Fasted 24 h	142.2 ± 7.8	133.0 ± 5.1
	Fasted 48 h	140.4 ± 6.5	144.1 ± 5.9
	Fasted 72 h	121.0 ± 8.6	119.1 ± 5.2
DNA	Normally fed	1.251 ± 0.020	1.622 ± 0.097
mg·g wet weight^{-1}	Fasted 24 h	1.663 ± 0.088	1.820 ± 0.064
	Fasted 48 h	2.081 ± 0.196	1.647 ± 0.085
	Fasted 72 h	2.076 ± 0.096	2.051 ± 0.072

Plasma Lactate Dehydrogenase. Exhaustive exercise resulted in a 3- to 12-fold increase in plasma lactate dehydrogenase (table 1). This is probably due to a leakage from erythrocytes and muscle cells.

Liver Protein and DNA Levels. When calculating liver protein per milligram DNA it was apparent that liver protein was continually lost during fasting and that there was no significant protein loss with a single bout of exercise (table 2). On the contrary, liver protein content expressed as milligrams per gram wet weight shows a steady increase which is misleading because it is important to take the changes in total liver weight into account. It was therefore decided to determine liver DNA concentrations as the DNA content of liver can be expected to remain constant over a 3-day fasting period. The decrease in liver weight over the 3 days of fasting is reflected in liver DNA concentration expressed as milligrams DNA per gram liver wet weight (table 2).

Liver and Muscle Cytochrome Oxidase. The levels of activity of cytochrome oxidase in liver were significantly higher in sedentary fed animals than in any fasted or exercised groups. These levels were also significantly decreased with exercise in animals which had been fasted for 3 days (table 3). In plantaris muscle the levels of activity of cytochrome oxidase did not show a definite trend but in exhausted animals an increase in the activity levels of cytochrome oxidase with prolonged fasting was observed. This is probably indicative of an adaptive mechanism in this respect (table 3).

Liver and Muscle Ubiquinones and Vitamin E. In liver, vitamin E was almost exclusively present as α-tocopherol, ubiquinones were mainly ubiquinone 9 homologue and minor amounts of ubiquinone 10. Vitamin E

Table 3. Liver and muscle (plantaris) levels of cytochrome oxidase activity (μmol · g^{-1} · min^{-1})

Treatment		Rested	Exhausted
Liver cytochrome oxidase	Normally fed	43.1 ± 7.0	27.7 ± 5.4
	Fasted 24 h	21.4 ± 2.9	23.9 ± 3.1
	Fasted 48 h	30.9 ± 4.2	28.9 ± 5.3
	Fasted 72 h	26.9 ± 6.1	11.2 ± 2.8
Muscle cytochrome oxidase	Normally fed	16.3 ± 2.6	10.6 ± 2.5
	Fasted 24 h	13.3 ± 2.6	13.7 ± 2.8
	Fasted 48 h	11.9 ± 1.7	14.5 ± 2.7
	Fasted 72 h	16.0 ± 3.7	18.9 ± 3.9

Table 4. Liver ubiquinone and vitamin E levels (nmol · mg DNA^{-1})

Treatment		Rested	Exhausted
Ubiquinones	Normally fed	103.2 ± 4.5	95.6 ± 4.7
	Fasted 24 h	118.1 ± 6.3	120.4 ± 9.1
	Fasted 48 h	103.8 ± 16.6	117.0 ± 3.5
	Fasted 72 h	92.0 ± 6.3	92.3 ± 6.8
Vitamin E	Normally fed	40.1 ± 1.5	39.4 ± 2.5
	Fasted 24 h	37.3 ± 1.9	37.0 ± 2.1
	Fasted 48 h	30.1 ± 1.8	38.0 ± 1.5
	Fasted 72 h	29.0 ± 1.8	31.1 ± 1.2

Table 5. Muscle (plantaris) ubiquinone and vitamin E levels (nmol · g^{-1})

Treatment		Rested	Exhausted
Ubiquinones	Normally fed	48.8 ± 2.1	43.7 ± 3.5
	Fasted 24 h	49.6 ± 1.4	52.4 ± 2.3
	Fasted 48 h	48.5 ± 3.6	48.1 ± 2.4
	Fasted 72 h	50.3 ± 4.2	54.2 ± 3.6
Vitamin E	Normally fed	45.5 ± 0.8	43.6 ± 1.4
	Fasted 24 h	48.0 ± 1.7	43.9 ± 1.5
	Fasted 48 h	43.1 ± 1.4	46.4 ± 1.5
	Fasted 72 h	46.6 ± 1.7	46.0 ± 1.4

values expressed as nanomoles per milligram DNA revealed a small but continuous loss of the vitamin during the fasting period, whereas ubiquinones, which can be synthesized endogenously, do not show this trend (table 4). In muscle, approximately 7% of the vitamin E was present as γ-tocopherol, the remainder was α-tocopherol. Both parameters measured were unaffected by fasting or exercise, the percentage of oxidized ubiquinones to total ubiquinones was between 75.7 and 90.5%; the highest values were determined in sedentary fed rats, the lowest in sedentary rats after 3 days of fasting (table 5). Exercise did not have a consistent effect on the ubiquinone redox state. In contrast to muscle tissue, the percentage of oxidized ubiquinones in liver was only between 24.3 and 30.4%. It was increased in 3-day-fasted animals, but differences to fed controls were not statistically significant.

Second Study Phase

In an attempt to find a possible explanation for the sustained exercise endurance capacity after depletion of liver glycogen levels as a result of prolonged fasting, a second phase of the investigation was conducted.

Materials and Methods

Rats were divided into two groups. One group had access to food ad libitum and the other group was fasted for 72 h. Blood was drawn from the abdominal aorta, and the gastrocnemius muscle and a liver lobe were excised for the determination of the activity levels of lactate dehydrogenase and the levels of β-hydroxybutyrate and acetoacetate.

The activity levels of lactate dehydrogenase in liver, gastrocnemius muscle and blood were determined by the method described by Vassault [6]. The levels of β-hydroxybutyrate and acetoacetate were quantified according to the method described by Kientsch-Engel and Siess [12].

Results and Discussion

Fasting resulted in significantly ($p < 0.05$) increased levels of activity of lactate dehydrogenase, β-hydroxybutyrate and acetoacetate in liver, gastrocnemius muscle and serum (table 6).

Conclusions

Based on the first phase, it is strongly recommended that results can be expressed on a DNA basis to compensate for possible positive or negative concentration effects or compositional changes. This is especially true for

Table 6. Levels of lactate dehydrogenase activity, β-hydroxybutyrate and acetoacetate in tissues and serum

Specimen	Fed	Fasted
Lactate dehydrogenase		
Liver, $\mu mol \cdot g^{-1} \cdot min^{-1}$	307.2 ± 18.4	552.4 ± 23.9
Gastrocnemius, $\mu mol \cdot g^{-1} \cdot min^{-1}$	513.0 ± 54.7	947.6 ± 63.6
Serum, $\mu mol \cdot l^{-1} \cdot min^{-1}$	140.7 ± 13.5	522.6 ± 37.7
β-Hydroxybutyrate		
Liver, $\mu mol \cdot g^{-1}$	0.205 ± 0.027	1.058 ± 0.109
Gastrocnemius, $\mu mol \cdot g^{-1}$	0.107 ± 0.039	0.673 ± 0.175
Serum, $\mu mol \cdot ml^{-1}$	0.052 ± 0.004	0.650 ± 0.065
Acetoacetate		
Liver, $\mu mol \cdot g^{-1}$	0.008 ± 0.001	0.066 ± 0.005
Gastrocnemius, $\mu mol \cdot g^{-1}$	0.013 ± 0.006	0.036 ± 0.003
Serum, $\mu mol \cdot ml^{-1}$	0.041 ± 0.008	0.347 ± 0.043

tissue such as liver which is very susceptible to change. In the present study decreases of the order of 40% were observed in liver wet weight.

Based on the observations during the second phase, it is concluded that the sustained exercise endurance capacity even after prolonged fasting might be attributable (at least in part) to the increased levels of ketone bodies which could serve as an alternative source of energy.

Acknowledgements

This research was supported by the Office of Naval Research, The National Foundation for Cancer Research, Hoffmann-LaRoche, National Institutes of Ageing and Health grants AG-04 818 and AM-19577 (USA), as well as the University of Zululand and the South African Sugar Association.

References

1 Ahlborg B, Bergstrom J, Ekelund L-G, Hultman E: Muscle glycogen and muscle electrolytes during prolonged physical exercise. Acta Physiol Scand 1967;70:129–142.

2 Christensen EH, Hansen O: Hypoglykämie, Arbeitsfähigkeit und Ermüdung. Scand Arch Physiol 1939;81:172–179.

3 Hermansen L, Hultman E, Saltin B: Muscle glycogen during prolonged severe exercise. Acta Physiol Scand 1967;71:129–139.
4 Hers HG: The control of glycogen metabolism in liver. Annu Rev Biochem 1976;45: 167–189.
5 Gohil K, Henderson S, Terblanche SE, Brooks GA, Packer L: Effects of training and exhaustive exercise on the mitochondrial oxidative capacity of brown adipose tissue. Biosci Rep 1984;4:987–993.
6 Vassault A: Lactate dehydrogenase: UV-method with pyruvate and NADH; in Bergmeyer HU (ed): Methods of Enzymatic Analysis, ed 3. Weinheim, Verlag Chemie, 1983, vol VIII, pp 118–126.
7 Cooperstein SJ, Lazarow A: A microspectrophotometric method for the determination of cytochrome oxidase. J Biol Chem 1951;189:665–670.
8 Lang JK, Packer L: Quantitative determination of vitamin E and oxidized and reduced coenzyme Q by high-performance liquid chromatography with in-line ultraviolet and electrochemical detection. J Chromat 1987;385:109–117.
9 Lowry OH, Rosebrough NJ, Farr AL, Randall RJ: Protein measurement with the Folin phenol reagent. J Biol Chem 1951;193:265–275.
10 Keppler D, Decker K: Glycogen: Determination with amyloglucosidase; in Bergmeyer HU (ed): Methods of Enzymatic Analysis, ed 2. New York, Academic Press, 1974, pp 1127–1131.
11 Ceriotti G: A microchemical determination of deoxyribonucleic acid. J Biol Chem 1952;198:297–303.
12 Kientsch-Engel RI, Siess EA: *D*-(–)-3-Hydroxybutyrate and acetoacetate; in Bergmeyer HU (ed): Methods of Enzymatic Analysis, ed 3. Weinheim, Verlag Chemie, 1985, vol VIII, pp 60–69.

Prof. S.E. Terblanche, Department of Biochemistry, University of Zululand,
Private Bag X1001, KwaDlangezwa 3880 (South Africa)

Sato Y, Poortmans J, Hashimoto I, Oshida Y (eds): Integration of Medical and Sports Sciences. Med Sport Sci. Basel, Karger, 1992, vol 37, pp 356–363

Carbohydrate Supplementation during Exercise

Edward F. Coyle

Human Performance Laboratory, Department of Kinesiology and Health, The University of Texas at Austin, Tex., USA

Introduction

The relative contribution of carbohydrate and fat to energy depends upon the intensity and duration of exercise as well as upon the preexercise diet and the state of physical training [1–4]. Carbohydrate metabolism during exercise has typically been manipulated by altering diet either before or during exercise. In a study of a dog named Joe who ran for longer period on a laboratory treadmill when fed glucose, Dill et al. [5] concluded that the limiting factor in the performance of prolonged exercise 'seems to be merely the quantity of easily available fuel' in the form of blood-borne glucose.

In contrast to these observations in the dog, early studies of humans focused upon the effects of altering blood glucose concentration upon the central nervous system [6, 7]. Carbohydrate ingestion during exercise was therefore thought to improve performance primarily by preventing symptoms of neuroglucopenia. This concept was reinforced by the classic studies of Christensen and Hansen [8–10].

Carbohydrate Energy and Strenuous Exercise: Importance of Muscle Glycogen

The introduction of the muscle biopsy technique in the 1960s demonstrated that muscle glycogen levels become depleted after intense exercise performed to fatigue [11]. Furthermore, the duration that strenuous exercise could be maintained was altered by raising and lowering preexercise muscle glycogen levels through manipulation of diet and exercise [12]. This has led to the concept that muscle glycogen is the primary carbohydrate source during exercise. Although this is not incorrect, this oversimplified concept indirectly minimizes the potential of blood glucose to serve as a fuel for carbohydrate oxidation during exercise at 60–80% VO_2max.

Carbohydrate Energy and Strenuous Exercise: Contribution of Blood Glucose

Data from studies conducted during the 1970s, which directly measured muscle glucose uptake, indicated that blood glucose could make substantial contributions to energy metabolism. Wahren et al. [13] observed that leg glucose uptake increases during exercise of increasing duration and intensity. Ahlborg et al. [14] also reported that the progressive increase in leg glucose uptake with increasing exercise duration was halted due to a decline in blood glucose concentration secondary to reduced liver glucose output. This was assumed to result from a depletion of liver glycogen stores. Ahlborg and Felig [15] subsequently demonstrated that when blood glucose concentration was maintained throughout exercise by glucose ingestion, leg glucose uptake was maintained at high levels. Additionally, Gollnick et al. [16] observed that the continued increase in leg glucose uptake was related to the number of muscle fibers low in muscle glycogen. These studies of leg glucose uptake did not typically employ exercise of sufficient intensity or duration and with exogenous blood glucose supplementation and therefore the association between blood glucose concentration, oxidation and endurance was not established.

Evidence that Carbohydrate Feedings Delay Fatigue

Although it was recognized that muscle glucose uptake can increase to high levels especially when ingesting carbohydrate [15], in 1982 Felig et al. [17] concluded that glucose feedings do not delay fatigue during bicycle ergometer exercise at 60–65% of VO_2max. An indication that carbohydrate ingestion may improve endurance performance in people was provided by Ivy et al. [18] yet this concept was not generally accepted [19].

Coyle et al. [20] demonstrated that carbohydrate feedings delay fatigue and improve endurance in people by preventing blood glucose from declining to levels which, in most of the subjects, causes local muscle fatigue during the latter stages of prolonged exercise.

Carbohydrate Feedings Do Not Affect Muscle Glycogen Use during Continuous Strenuous Exercise

In a subsequent study we measured the pattern of decline in muscle glycogen concentration in the vastus lateralis during exercise to fatigue at 74% VO_2max on two occasions, with and without carbohydrate feedings [21]. Fatigue occurred after 3.0 ± 0.2 h when fed a placebo, whereas fatigue was delayed until 4.0 ± 0.3 h when fed carbohydrate (i.e. approximately 70 g of maltodextrins in a 50% solution at 20 min followed by 28 g in a 10% solution every 20 min thereafter). As shown in figure 1C, the pattern of decline in muscle glycogen concentration was similar during the first 3 h of exercise with and without carbohydrate feeding. Remarkably, the additional hour of exercise made possible by carbohydrate feedings occurred without a further decline in muscle glycogen concentration.

We interpret these observations to indicate that the lowering of blood glucose during the latter stages of prolonged strenuous exercise (fig. 1A) play a major role in the development of muscular fatigue by not allowing leg glucose uptake to increase sufficiently to offset reduced muscle glycogen availability. When plasma glucose was maintained at 4–5 m*M*, through carbohydrate ingestion, it was theorized that blood glucose can largely replace muscle glycogen in providing carbohydrate for oxidation during the latter stages of prolonged strenuous exercise. These concepts are summarized in figure 2. An interesting question regards the cause of fatigue during exercise with the carbohydrate feedings.

Carbohydrate Supplementation at Fatigue

In order to more directly test the hypothesis that carbohydrate feedings improve exercise performance by preventing the decline in blood glucose concentration and oxidation late in exercise, we reasoned that it should be possible to reverse the decline in carbohydrate oxidation as well

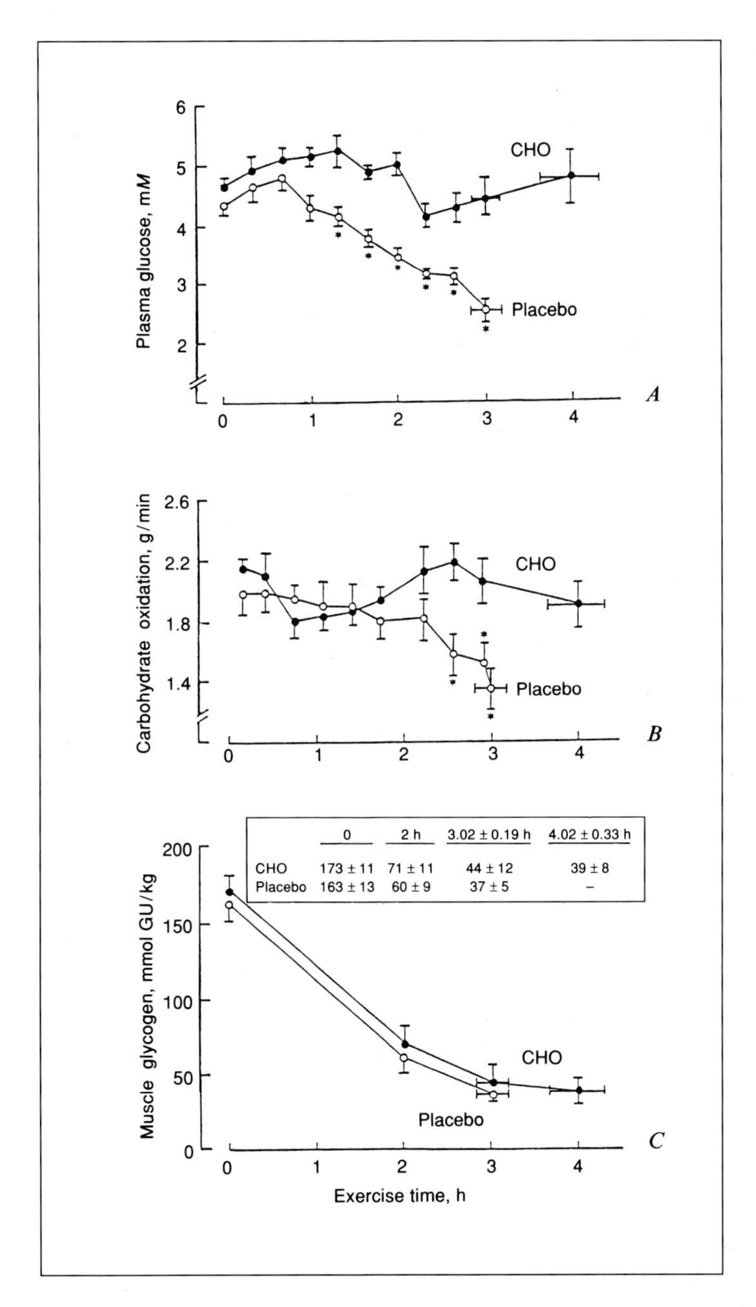

Fig. 1. Responses when cycling at 74% VO_2max with a placebo or when ingesting carbohydrate every 20 min (CHO). Placebo different from carbohydrate: * $p < 0.05$.

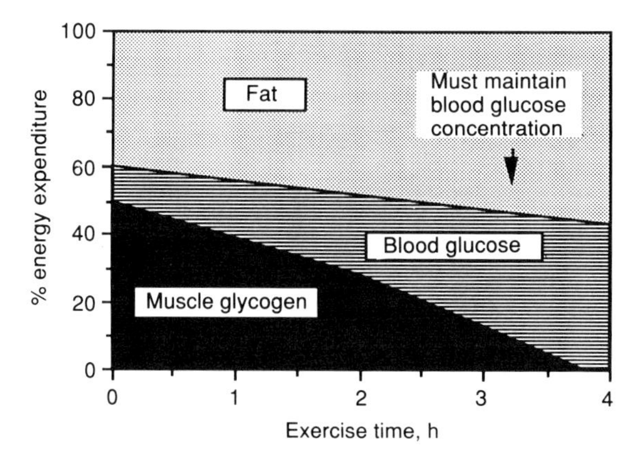

Fig. 2. Various sources of energy during prolonged exercise at 70% VO_2max. Note that blood glucose becomes the predominant source of carbohydrate energy during the latter stages of exercise and thus it is important to maintain glucose concentration by eating carbohydrate.

as fatigue during exercise when fasted by restoring euglycemia [22]. Therefore, the study displayed in figure 3 was performed. On three separate occasions subjects first exercised at 70% VO_2max to the point of fatigue when fasted (exercise bout 1, fig. 3) displaying a decline in plasma glucose concentration (i.e. to 3.1 m*M*) and RER (i.e. to 0.81) prior to fatigue. After a 20-min rest, the subjects were encouraged to perform further exercise (i.e. exercise bout 2, fig. 3) with three different treatments. When they received a placebo solution to drink during the rest, the subjects tolerated only an additional 10 ± 1 min of exercise. During a second trial, 200 g of a 50% maltodextrins solution was ingested during the rest period. As a result, during exercise bout 2, plasma glucose concentration and RER were initially increased above levels at fatigue of exercise bout 1, but they could not be maintained, declining progressively to the point of fatigue, which occurred after 26 ± 4 min. During a third trial, glucose was infused (via a pump) intravenously at the beginning of exercise bout 2 at the rate required to maintain plasma glucose concentration at 5 m*M*. This maintained RER above the levels observed at fatigue during exercise bout 1 and the subjects completed an additional 43 ± 5 min of exercise (fig. 3). It was also observed that muscle glycogen use was minimal during this additional exercise, suggesting that blood glucose was the primary energy source for carbohydrate oxidation, which was occurring at 1.6 g/min.

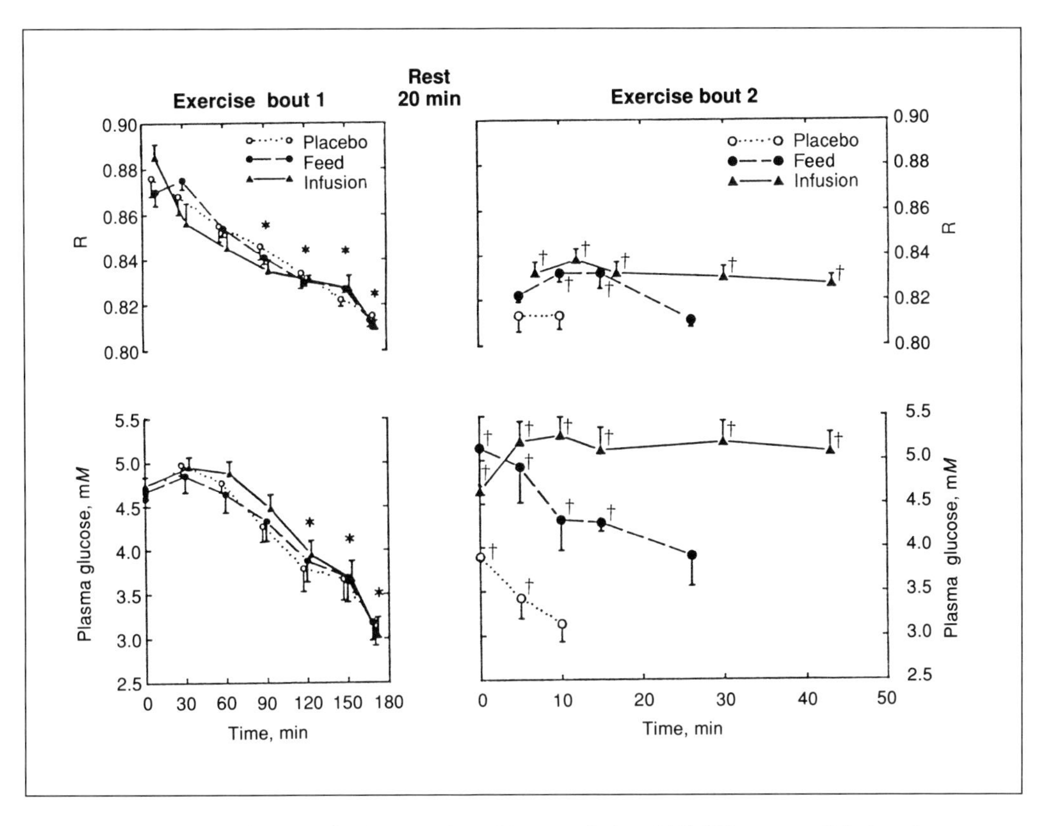

Fig. 3. During exercise bout 1, the subjects cycled at 74% VO_2max until fatigued. After a 20-min rest, they continued to exercise (bout 2) with 3 different treatments. Placebo = After drinking flavored water and being infused with saline; infusion = receiving intravenous glucose infusion at a rate which maintained plasma glucose at 5 m*M*; feeding = ingestion of 200 g of a 50% maltodextrin solution.

Significant decline during exercise bout 1: † $p < 0.05$; values during exercise bout 2 which were significantly higher than at the end of exercise bout 1: * $p < 0.05$.

An important finding of this study was that a glucose infusion rate of over 1.1 g/min was required to maintain euglycemia, suggesting that this exogenous glucose was being oxidized at a high rate under these conditions (i.e. low muscle glycogen, hypoinsulinemia). Since muscle glycogen contributed little to energy, it is likely that endogenous glucose production supplemented the exogenous glucose infusion in providing the carbohydrate needs of exercise at this intensity. We have recently observed that in order to keep blood glucose concentration at 10 m*M* during exercise, the rate of exogenous intravenous glucose infusion must be increased progressively during the second hour of exercise to a value of 2.6 g/min which is approximately equal to the rate of total carbohydrate oxidation [23]. However, these data cannot provide information as to whether glucose disposal is oxidized in the exercising muscle or taken-up by other tissues.

References

1 Krogh A, Lindhard J: Relative value of fat and carbohydrate as a source of muscular energy. Biochem J 1920;14:290–298.

2 Hultman E: Physiological role of muscle glycogen in man, with special reference to exercise. Circ Res 1967;20(suppl 1):I99–I112.

3 Hermansen L, Hultman E, Saltin B: Muscle glycogen during prolonged severe exercise. Acta Physiol Scand 1967;71:129–139.

4 Holloszy J, Coyle E: Adaptations of skeletal muscle to endurance exercise and their metabolic consequences. J Appl Physiol 1984;56:831–838.

5 Dill D, Edwards H, Talbott J: Studies in muscular activity. VII. Factors limiting the capacity for work. J Physiol (Lond) 1932;77:49–62.

6 Levine S, Gordon B, Derick C: Some changes in the chemical constituents of the blood following a marathon race. J Am Med Assoc 1924;82:1778–1779.

7 Gordon B, Kohn L, Levine S, Matton M, Scriver W de M, Whiting W: Sugar content of the blood in runners following a marathon race. J Am Med Assoc 1925;85:508–509.

8 Christensen E, Hansen O: Teil II. Untersuchungen über die Verbrennungsvorgänge bei langdauernder, schwerer Muskelarbeit. Scand Arch Physiol 1939;81:152–161.

9 Christensen E, Hansen O: Teil III. Arbeitsfähigkeit und Ernährung. Scand Arch Physiol 1939;81:161–172.

10 Christensen E, Hansen O: Teil IV. Hypoglykamie, Arbeitfähigkeit und Ernährung. Scand Arch Physiol 1939;81:172–179.

11 Bergstrom J, Hultman E: The effect of exercise on muscle glycogen and electrolytes in normals. Scand J Clin Invest 1966;18:16–20.

12 Bergstrom J, Hermansen L, Hultman E, Saltin B: Diet, muscle glycogen, and physical performance. Acta Physiol Scand 1967;71:140–150.

13 Wahren J, Felig P, Ahlborg G, Jorfeldt L: Glucose metabolism during leg exercise in man. J Clin Invest 1971;50:2715–2725.

14 Ahlborg F, Felig P, Hagenfeldt L, Hendler R, Wahren J: Substrate turnover during prolonged exercise in man: Splanchnic and leg metabolism of glucose, free fatty acids, and amino acids. J Clin Invest 1974;53:1080–1090.

15 Ahlborg G, Felig P: Influence of glucose ingestion on fuel-hormone response during prolonged exercise. J Appl Physiol 1976;41:683–688.

16 Gollnick P, Pernow P, Essen B, Jansson E, Saltin B: Availability of glycogen and plasma FFA for substrate utilization in leg muscle of man during exercise. Clin Physiol 1981;1:27–42.

17 Felig P, Cherif A, Minagawa A, Wahren J: Hypoglycemia during prolonged exercise in normal men. New Engl J Med 1982;306:895–900.
18 Ivy J, Costill D, Fink W, Lower R: Influence of caffeine and carbohydrate feedings on endurance performance. Med Sci Sports 1979;11:6–11.
19 Costill D, Miller J: Nutrition for endurance sport: carbohydrate and fluid balance. Int J Sports Med 1980;1:2–14.
20 Coyle E, Hagberg J, Hurley B, Martin W, Ehsani A, Holloszy J: Carbohydrate feedings during prolonged strenuous exercise can delay fatigue. J Appl Physiol 1983;55: 230–235.
21 Coyle E, Coggan A, Hemmert M, Ivy J: Muscle glycogen utilization during prolonged strenuous exercise when fed carbohydrate. J Appl Physiol 1986;61:165–172.
22 Coggan A, Coyle E: Reversal of fatigue during prolonged exercise by carbohydrate infusion or ingestion. J Appl Physiol 1987;63:2388–2395.
23 Coyle E, Hamilton M, Gonzalez Alonso J, Montain S, Ivy J: Carbohydrate metabolism during intense exercise when hyperglycemic. J Appl Physiol 1991;70:834–840.

Edward F. Coyle, PhD,
Human Performance Laboratory, Department of Kinesiology and Health,
The University of Texas at Austin, Bellmont Hall 222, Austin, TX 78712 (USA)

Sato Y, Poortmans J, Hashimoto I, Oshida Y (eds): Integration of Medical and Sports Sciences. Med Sport Sci. Basel, Karger, 1992, vol 37, pp 364–373

Noninvasive and Continuous Determination of Energy Metabolism during Muscular Contraction and Recovery

Takayoshi Yoshida[a], *Hiroshi Watari*[b,1]

[a]Exercise Physiology Laboratory, Faculty of Health and Sport Sciences, Osaka University, Toyonaka, Osaka; [b]Department of Molecular Physiology, National Institute for Physiological Sciences, Okazaki, Aichi, Japan

Introduction

That adenosine triphosphate (ATP) is the immediate energy source for muscular contraction has been well documented. While the amount of energy released from ATP is estimated to be 12–15 kcal in the body per mole, the total muscular store of ATP is very small, which would be estimated as a resting ATP of 5.5 mmol·kg^{-1} wet muscle. To sustain the exercise, ATP must be resynthesized at the same rate as ATP is broken down.

The following three biochemical pathways are available for ATP resynthesis: (1) phosphocreatine (PCr) breakdown; (2) anaerobic glycolysis, and (3) aerobic metabolism in the mitochondria. In other words, the maximal rate of ATP resynthesis is dependent on the combined rates of PCr breakdown, anaerobic glycolysis and oxidative phosphorylation. Phosphocreatine is stored in the muscles as an energy-rich phosphate closely related to ATP. The muscle content of PCr is three to four times higher than that of ATP. When PCr is broken down, a large amount of energy is released for ATP resynthesis.

Currently, muscle metabolites such as PCr and ATP in human muscle during exercise have been obtained by the invasive needle biopsy tech-

[1] The authors wish to acknowledge Mr. O. Ichikawa, and H. Ogasawara for their technical assistance. This study was performed at the Laboratory for Magnetic Resonance Imaging and Spectroscopy, National Institute for Physiological Science, Okazaki, Japan.

nique [Bergström, 1962]. However, by using the needle biopsy method, muscle samples must be repeatedly obtained from different muscle sites to determine the change in muscle metabolites at the transient phase of exercise or the repeated measurements during exercise. Furthermore, the exercise needs to bc stopped during the muscle sampling which needs to be rapid. The sample must then be frozen quickly to inhibit metabolic activity. The number of samples were limited in the biopsy studies.

^{31}P-MRS for Determination of Muscle Metabolites during Exercise

Recently, ^{31}P-nuclear magnetic resonance spectroscopy (^{31}P-MRS) has been used as a tool for a noninvasive means of detecting relative changes in muscle metabolites in vivo human skeletal muscle at rest and during exercise [Chance et al., 1986; Inch et al., 1985; Sapega et al., 1986; Wilikie, 1986]. However, the most obvious shortcoming of ^{31}P-MRS is the restriction in the small muscle mass such as hand grip or foot extension activity by the gastrocnemius that can be studied and type of exercise employed, due to the limited size of the clear bore of the magnet. Furthermore, the time resolution for the determination of ^{31}P-MRS is restricted to obtain PCr kinetics during exercise, because of the relatively small muscle mass involved in the exercise in the magnet. Thus, only a weak signal results in a low signal-to-noise ratio.

To avoid these problems, a larger bore apparatus and a higher magnetic field are needed, in which larger amounts of muscle would be able to exercise with a high signal-to-noise ratio. We used a large-bore magnet with a 67-cm diameter to determine muscle metabolites during exercise and recovery (fig. 1).

Changes in Muscular Metabolites and Intracellular pH by Time-Resolved ^{31}P-MRS

In our study, femoral biceps can be laid on the magnet and the surface coil is slid into the center of the magnet where the field is homogeneous. Computer control emits the ratio frequency to the coil, at the same time receiving the free induction decays, and Fourier transforms gives a spectrum of the principal chemicals. During this time, data acquisition was repeated using an ensemble-averaging method, allowing collection of a ^{31}P-

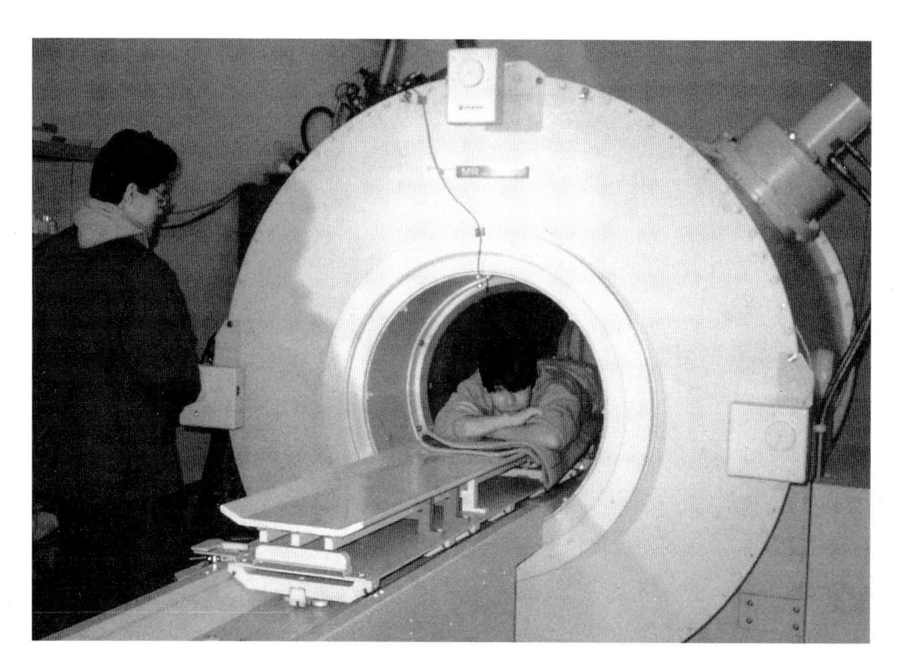

Fig. 1. NMR apparatus for whole body (2.1 Tesla, 67 cm of bore diameter).

MRS signal with a high S/N ratio within a couple of seconds. ^{31}P-MRS was accumulated using 32 scans per spectrum requiring 12.8 s per spectrum.

Six normal male subjects performed femoral flexion exercise at '0' (loadless), 10, 20 and 30 kg m/min for 4 min and the successful recovery in the 2.1 Tesla superconducting magnet.

Phosphorus Metabolism

Although it has been observed that muscle contraction in vivo results in a rapid decrease in PCr and a rapid increase in Pi (fig. 2), few studies are concerned with the time course of such metabolites at the start and end of exercise in humans and in rats. A continuous Pi time course at the start of exercise and at recovery is shown in figure 3. The Pi increased exponentially during exercise and returned during recovery. Recovery and exercise Pi kinetics appear to be symmetrized. However, during recovery from the low exercise intensity, an apparent reduction in Pi peak to an almost undetectable level is observed, which makes pH estimation difficult.

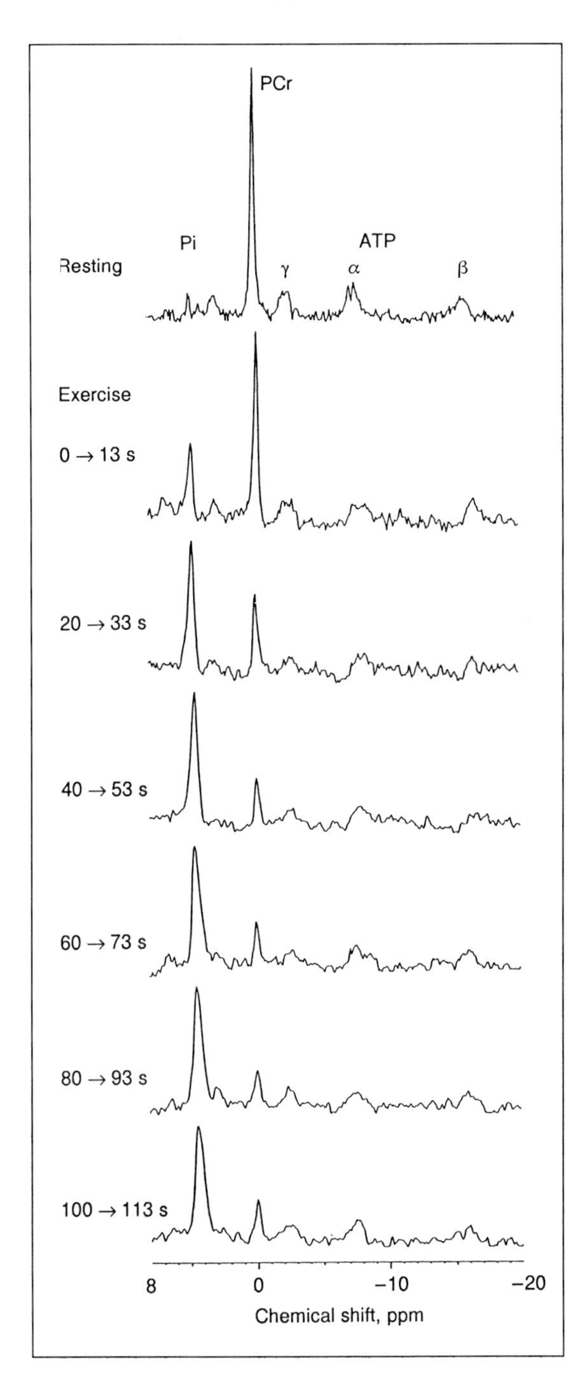

Fig. 2. An example of the changes in ^{31}P-MRS at the start of exercise from the femoral flexion exercise at 30 kg m/min work rate.

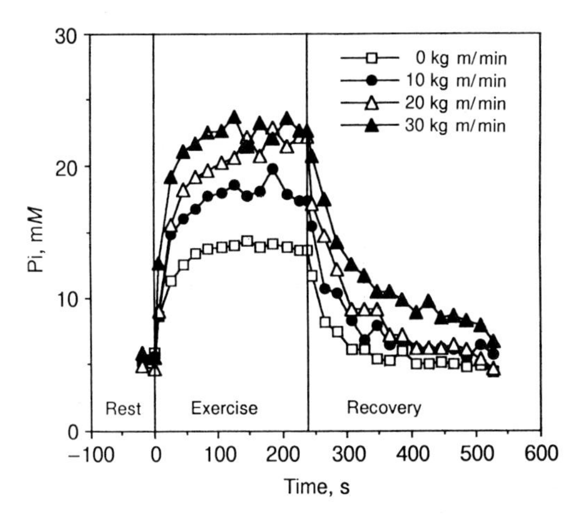

Fig. 3. Group mean responses in Pi at the start and end of the different exercise intensities.

The PCr time course at the start and end of exercise in the magnet is shown in figure 4. At the start of exercise PCr decreased exponentially and increased during recovery. PCr decreased in response to exercise with a time constant of about 30 s. The rate of PCr decrease was independent of work rate. At the end of exercise the magnitude of PCr decrease was more marked, the higher the work rate. At the end of exercise, the PCr values attained at loadless work rate and at other work rate were significantly different. During recovery, the time constant for PCr changes was about 30–40 s, and the values did not differ regardless of work rate employed. At the higher exercise intensity, PCr did not return to the pre-exercise level. Recovery and exercise PCr kinetics appear to be symmetrized.

The rate of PCr depletion with a time constant of about 30 s in vivo human muscle supports the result reported by Mole et al. [1985], who described that the observation of right forearm muscle during a constant bulb squeezing exercise revealed a time constant for PCr of about 30 s. Such a similar first-order kinetics for PCr was observed in various samples of animal muscle. A time constant for PCr in in situ rat gastrocnemius muscle is about 1.4 min [Meyer, 1988] and is about 3.8 min in frog sartorius muscle at 20 °C [Mahler, 1985].

PCr hydrolysis is controlled by myosin ATPase and sacromere creatine kinase and the PCr resynthesis rate during recovery is controlled by

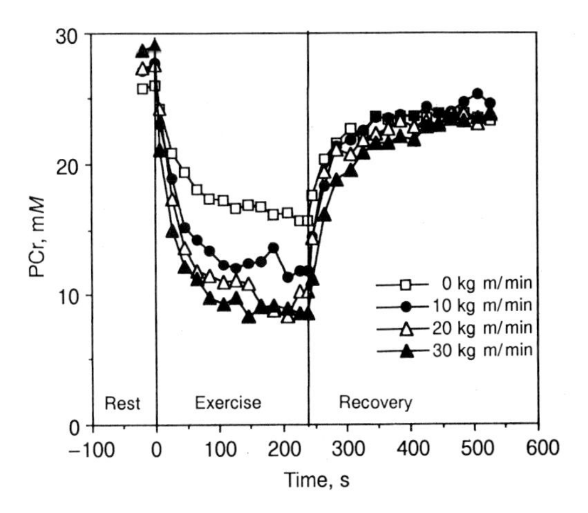

Fig. 4. Group mean responses in PCr at the start and end of the different exercise intensities.

aerobic metabolism and mitochondrial creatine kinase. In other words, PCr resynthesis at the end of exercise would reflect the mitochondrial oxidative phosphorylation at that time. It is of interest to note that the PCr time constant in recovery will evaluate the capacity of oxidative phosphorylation in the musculature. Figure 5 shows the PCr time course at light (fig. 5, left) and at heavy exercise (fig. 5, right) in long-distance runners and controls. At a given work rate the consumed PCr is significantly small in long-distance runners compared to normal subjects. In addition, a faster PCr recovery rate in long-distance runners is observed than that of normal subjects, suggesting a greater oxidative capacity in long-distance runners. In this way, ^{31}P-MRS would provide some useful information concerning muscle metabolites for athletes' performance and muscle fatigue [McCully et al., 1988].

It is well documented that oxygen uptake ($\dot{V}O_2$) following the onset of a step function exercise increases exponentially to a steady state. The quantitative evaluation for this $\dot{V}O_2$ exponential function can be expressed in the time constant, which is about 30 s at the exercise intensity below the lactate threshold.

Hickson et al. [1978], Hagberg et al. [1979] and, more recently, Yoshida et al. [1992] demonstrated more rapid $\dot{V}O_2$ kinetics following training or in endurance runners than in control subjects. This faster $\dot{V}O_2$

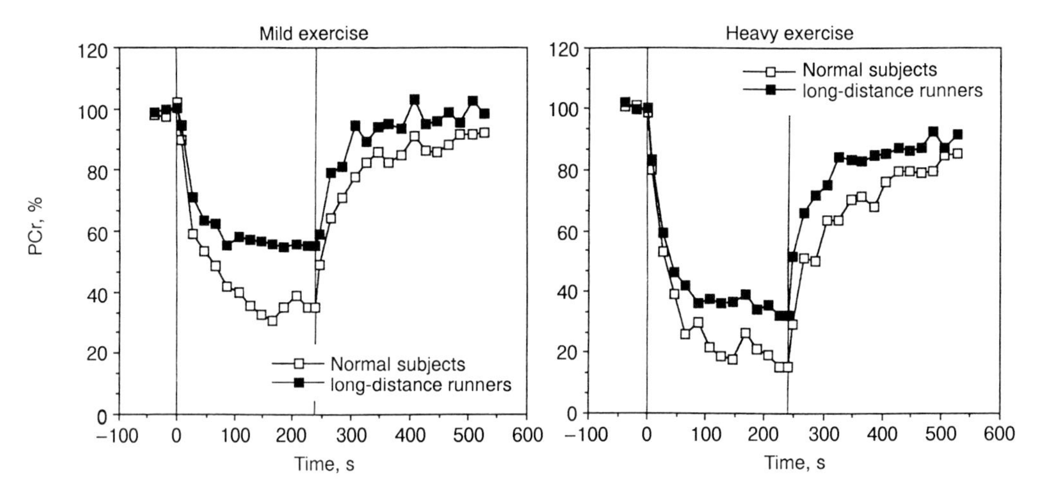

Fig. 5. The relative changes in PCr time course for long-distance runners (■) and normal subjects (□) at a given exercise intensity. The data are expressed as 100% at the resting value for both groups (left: mild exercise intensity; right: heavy exercise intensity).

kinetic response is dependent on the rate of substrate utilization in aerobic ATP synthesis, and is a more complex process for oxygen utilization in mitochondria or muscle blood flow. Although it is an over-simplification to link $\dot{V}O_2$ kinetics and PCr kinetics, on a one-to-one basis the control mechanisms for $\dot{V}O_2$ kinetics might be reflected by the rate of PCr resynthesis.

Intracellular pH with Exercise

Another interesting point from the ^{31}P-MRS study is the noninvasive determination of intracellular pH during exercise and recovery. Using ^{31}P-MRS, a continuous change in intracellular pH can be estimated during exercise, based on the chemical shift for Pi and PCr. Figure 6 shows the continuous determinations of intracellular pH at the start and end of different exercise intensities. There is a transient alkalosis at the early phase of exercise regardless of exercise intensity (see arrow in fig. 6). A similar observation has been certified in animal studies using a glass electrode technique [Gebert and Friedman, 1973], and by ^{31}P-MRS [Tanuoka and Yamada, 1984] and during forearm exercise in human studies [Arnold et al., 1984; Mole et al., 1985]. A transient alkalosis at the onset of exercise may be attributed to the utilization of hydrogen ions that accompanies PCr

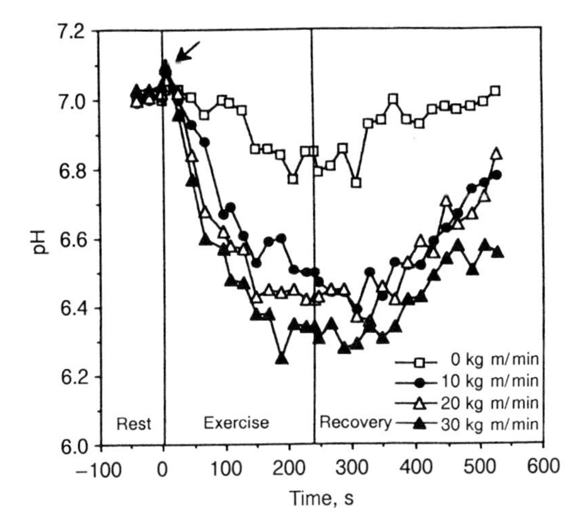

Fig. 6. Group mean responses of the time course in intracellular pH at different exercise intensities (arrow indicates a transient alkalosis phase at the start of exercise).

hydrolysis. Thereafter, intracellular pH fell gradually towards the acidification. At the end of exercise, the degree of the pH value was dependent on the exercise intensity. Furthermore, during recovery from exercise, the Pi peak was shifted to a more acidosis site, in which the pH value was found to go increasingly acid even during the recovery phase. A similar phenomenon has been observed by Pan et al. [1988] using ^{1}H-NMR. This acidification is attributed to the formation of lactic acid.

Thereafter, the return of pH to the resting level was significantly slower than that at the start of exercise. As shown in figure 4, PCr recovery was relatively rapid ($\tau = 30$ s), while pH recovery was relatively slower. The slow pH recovery likely reflects the rate of HCO_3^- regeneration or lactate removal from the cell (a process independent of PCr regeneration but a determinant of the return of Pi peak to the control).

During recovery, it is occupationally observed that the Pi peak reduced to almost undetectable levels, which makes it difficult to estimate the pH. Then after a couple of minutes the Pi peak again increases to a detectable level. The reason why the Pi peak reduces at the end of exercise is considered to be as follows [Newman and Cady 1990]: (1) inhomogeneous pH distribution causing a spread in chemical shift of Pi, and/or (2) pH changing rapidly with time causing a drift of the Pi peak during data acquisition. However, as shown in the present study, pH change after exercise is not so rapid. The precise reason why the undetectable Pi reduction after exercise occurs is still unclear.

Conclusion

The time-resolved measurement of muscle metabolites and intracellular pH during exercise and the successful recovery by the ^{31}P-MRS method provides that: (1) PCr and Pi changed in first-order kinetics with a time constant of about 30 s during exercise and recovery regardless of the exercise intensities employed; (2) at the early phase of the start of exercise, a transient alkalosis in intracellular pH was observed, which may be attributed to the utilization of hydrogen ions that accompanies the PCr hydrolysis; and (3) PCr recovery is relatively rapid, while pH recovery is relatively slower. The slow pH recovery likely reflects the rate of HCO_3^- regeneration or lactate removed from the cell.

References

Arnold DL, Matthews PM, Radda GK: Metabolic recovery after exercise and the assessment of mitochondria function in vivo in human skeletal muscle by means of ^{31}P NMR. Magn Reson Med 1984;1:307–315.

Bergström J: Muscle electrolytes in man. Determined by neuron activation analysis in needle biopsy specimens. A study on normal subjects, kidney patients, and patients with chronic diarrhoea. Scand J Clin Lab Invest 1962;(suppl 68).

Chance B, Leigh JS Jr, Kent J, McCully K: Metabolic control principles and ^{31}P NMR. Fed Proc 1986;45:2915–2920.

Gebert G, Friedman SM: An implantable glass electrode used for pH measurement in working skeletal muscle. J Appl Physiol 1973;34:122–124.

Hagberg JM, Nagel FJ, Carson JL: Transient O_2 uptake response at the onset of exercise. J Appl Physiol 1978;44:90–92.

Hickson RC, Bonze HA, Holloszy JO: Faster adjustment of O_2 uptake to the energy requirement of exercise in the trained state. J Appl Physiol 1979;44:877–881.

Inch WR, Serebrin B, Taylor AW, Thomson RT: Exercise muscle metabolism measured by magnetic resonacne spectroscopy. Can J Sports Sci 1986;11:60–65.

Mahler M: First-order kinetics of muscle oxygen consumption, and an equivalent proportionality between QO_2 and phosphorylcreatine level. J Gen Physiol 1985;86:135–165.

McCully KK, Kent JA, Chance B: Application of ^{31}P magnetic resonance spectroscopy to the study of athletic performance. Sports Med 1988;5:312–321.

Meyer RA: A linear model of muscle respiration explains monoexponential phosphocreatine changes. Am J Physiol 1988;254:C548 C553.

Mole PA, Coulson RL, Caton JR, Nichols BG, Barstow T: In vivo ^{31}P-NMR in human muscle: Transient patterns with exercise. J Appl Physiol 1985;59:101–104.

Newman DJ, Cady EB: A ^{31}P study of fatigue and metabolism in human skeletal muscle with voluntary, intermittent contraction at different forces. NMR Biomed 1990;3: 211–219.

Pan JW, Hamm JR, Rothman DL, Shulman RD: Intracellular pH in human skeletal muscle by ^{1}H NMR. Proc Natl Acad Sci USA 1988;85:7836–7839.

Sapega AA, Sokolow DP, Graham TJ, Chance B: Phosphorus nuclear magnetic resonance: A non-invasive technique for the study of muscle bioenergetics during exercise. Med Sci Sports Exerc 1987;19:410–420.

Tanoura M, Yamada K: Changes in intracellular pH and inorganic phosphate concentration during and after muscle contraction as studied by time-resolved ^{31}P-NMR: Alkalinization by contraction. FEBS Lett 1984;171:165–168.

Wilkie DR: Muscular fatigue: Effect of hydrogen ions and inorganic phosphate. Fed Proc 1986;45:2921–2923.

Yoshida T, Udo M, Ohmori T, Matsumoto Y, Uramoto T, Yamamoto K: Day-to-day changes in $\dot{V}O_2$ kinetics at the onset of exercise during strenuous endurance training. Eur J Appl Physiol 1992;64:78–83.

T. Yoshida, PhD, Exercise Physiology Laboratory, Faculty of Health and Sport Sciences, Osaka University, Toyonaka, Osaka 560 (Japan)

Sato Y, Poortmans J, Hashimoto I, Oshida Y (eds): Integration of Medical and Sports Sciences. Med Sport Sci. Basel, Karger, 1992, vol 37, pp 374–380

Oxygen Uptake: Work Rate Relationship in Patients with Heart Disease[1]

Haruki Itoh

The Cardiovascular Institute, Minato-ku, Tokyo, Japan

In patients with heart disease, restricted oxygen delivery to peripheral tissue due to impaired cardiac function causes characteristic symptoms such as exercise intolerance. Oxygen uptake ($\dot{V}O_2$) during steady state exercise reflects oxygen consumption, including not only oxygen consumption of the working muscle but of other organs as well. Therefore, $\dot{V}O_2$ for the certain work load may change when blood redistribution occurs during exercise, even though the work efficiency of the working muscle itself is unchanged. On the other hand, the plasma norepinephrine level in heart failure patients is elevated [1]. This is recognized as one of the compensatory mechanisms via the myocardial beta-adrenoceptor for impaired left ventricular contractility, but it also affects blood distribution during exercise through the vascular alpha-adrenoceptor. We conducted this study to investigate the $\dot{V}O_2$-work rate relationship in patients with heart disease from the point of view of compensatory mechanism for restricted oxygen transport.

Method

Eighteen normal subjects and 29 cardiac patients entered this study. Regarding the New York Heart Association (NYHA) functional classifications of the patients, 10 patients were in class I, 11 in class II, and 8 in class III. The underlying diseases of the patients were valvular disease in 12, ischemic heart disease in 7, hypertensive heart disease in 5, dilated cardiomyopathy in 2, and others in 3. The study excluded patients with

[1] This work was partly presented at the 55th Meeting of the Japanese Circulation Society in Nagoya, Japan, March, 1991.

angina pectoris who stopped exercise testing due to chest pain. Also excluded were patients with respiratory dysfunction, obesity (30% more than ideal weight) or emaciation (20% less than ideal weight), or pacemakers. None of them had obvious edema or ascitis.

This study employed a symptom-limited exercise testing with a ramp protocol using electromagnetically controlled cycle ergometer. After a 4-min rest on the ergometer, exercise began with a 4-min warm-up at 20 W, followed by ramping of work rate at the rate of 10 W/min. The ECG and heart rate were monitored throughout the testing. Cuff blood pressure was also measured every minute with an automatic indirect manometer. Blood samples for norepinephrine were obtained at rest, at the end of the 4-min 20-watt warm-up. Thyroid-stimulating hormone, triiodothyronine, and thyroxine were also measured at rest. Cardiac output at rest and warm-up were measured by the dye dilution method with indocyanine green using an ear photoelectric transducer, the output of which was analyzed by a computer [2]. Respiratory gas analysis was performed throughout testings using a breath-by-breath respiromonitor consisting of a zirconia oxygen analyzer, infrared carbon dioxide analyzer, and hot wire spirometer. The system was carefully calibrated before each study.

Oxygen uptake ($\dot{V}O_2$), CO_2 output ($\dot{V}CO_2$), and expired tidal volume (TV) were measured continuously on a breath-by-breath basis. Derived parameters such as minute ventilation ($\dot{V}E$), $\dot{V}E/\dot{V}CO_2$, $\dot{V}E/\dot{V}O_2$, respiratory exchange ratio (R; $\dot{V}CO_2/\dot{V}O_2$), end-tidal O_2, and end-tidal CO_2 were computed simultaneously and displayed with heart rate and $\dot{V}O_2$ on a monitor during measurement.

Anaerobic threshold (AT) was determined from gas exchange data by the V slope method [3] using our software, which determines the turning point of the $\dot{V}O_2$–$\dot{V}CO_2$ relation curve by 2 linear regression lines. Peak $\dot{V}O_2$ was defined as the average of $\dot{V}O_2$ over the last 30 s of each testing.

The ratio of increase in $\dot{V}O_2$ to increase in work rate ($\Delta\dot{V}O_2/\Delta WR$) was also determined by a linear regression of $\dot{V}O_2$ plots from 1 min after the start of ramping up to the AT point using a personal computer with our software.

Results

The peak $\dot{V}O_2$ of normal subjects was 34.2 ± 6.2 ml/min/kg. In cardiac patients, peak $\dot{V}O_2$ was 21.1 ± 3.5 in class I, 17.4 ± 2.6 in class II, and 13.8 ± 4.8 in class III. The index of %AT, which is the percentage of predicted AT for each subject, is calculated in order to take into consideration sex, body weight and age [4]. The %AT decreased with increasing severity of disease, from 99.3 ± 23.3% in normal subjects to 61.8 ± 9.4% of NYHA class III patients.

Figure 1 showed the relationship between %AT and $\dot{V}O_2$ at rest and 20 W. AT appeared after the 4-minute warm-up period except in 1 case in class III. This case was excluded from the data at 20-watt warm-up in order to ensure that all the subjects were in steady state at the end of the warm-

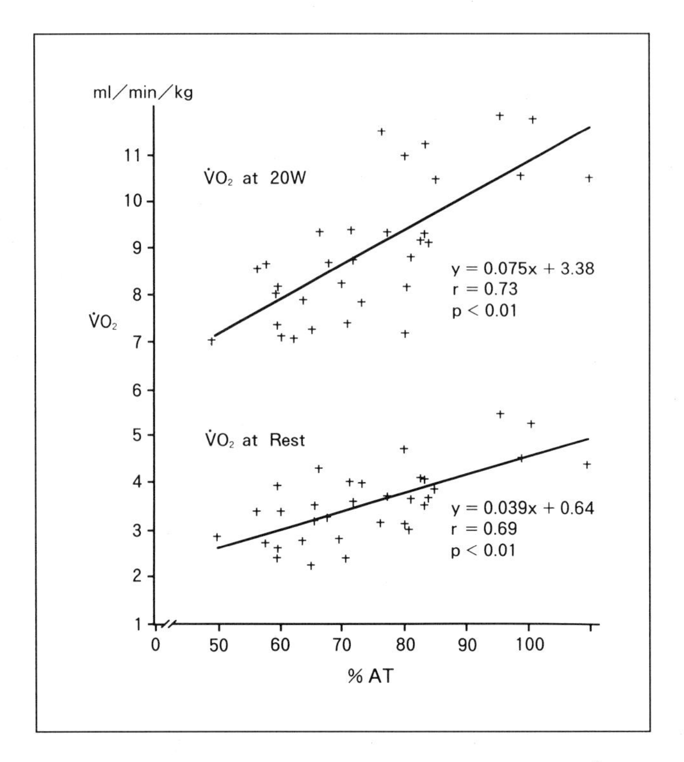

Fig. 1. Relationship between anaerobic threshold (AT) and $\dot{V}O_2$ at rest and at 20 W. AT is expressed as a percentage of predicted AT with respect to age, sex, and body weight (%AT). $\dot{V}O_2$ was also corrected by body weight. Both the $\dot{V}O_2$ at rest and 20-watt steady state loading showed a significant positive correlation to exercise capacity.

up. $\dot{V}O_2$ at 20 W and $\dot{V}O_2$ at rest and 20 W decreased significantly as the severity of exercise intolerance rose. These findings demonstrate that $\dot{V}O_2$ requirements for the same work load decreased in patients with exercise intolerance.

The average $\Delta\dot{V}O_2/\Delta WR$ in a normal subject was 10.0 ± 1.1 ml/min/W and it decreased in accordance with the severity of heart disease, 9.3 ± 1.1 in class I, 8.0 ± 1.1 in class II, and 6.7 ± 2.6 in class III patients. $\Delta\dot{V}O_2/\Delta WR$ also showed a significant relationship to the exercise capacity.

There was no significant difference among the groups of subjects regarding thyroid function tests.

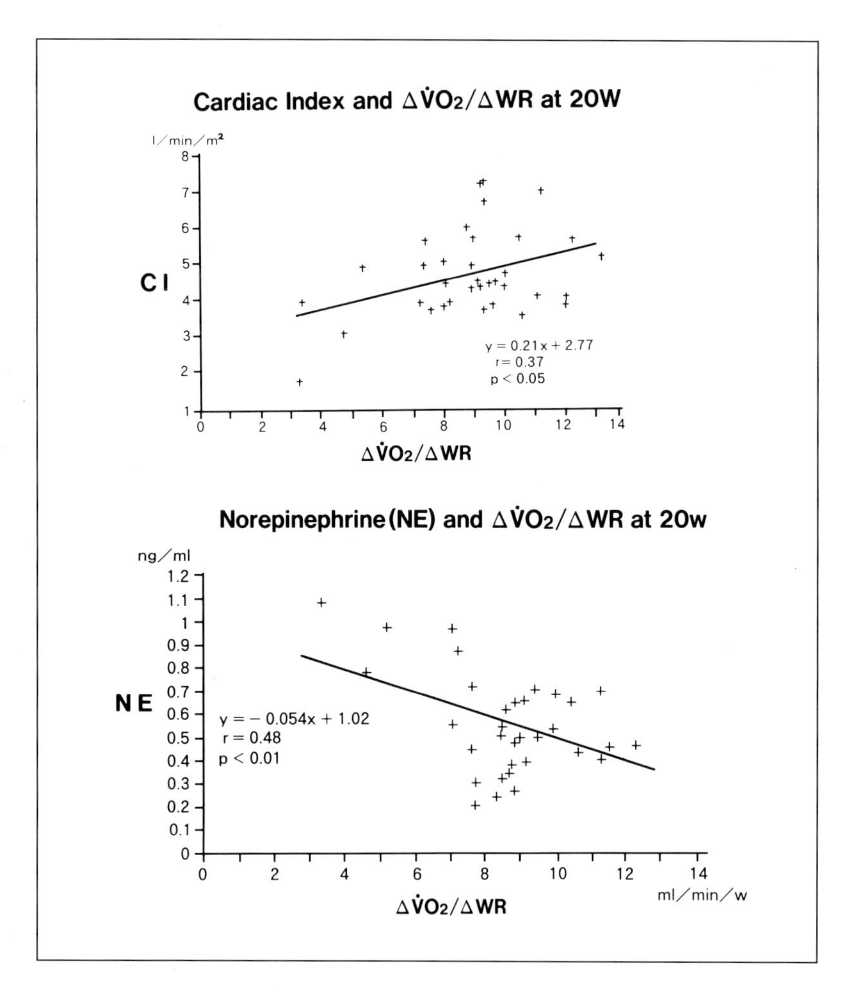

Fig. 2. The relation of $\Delta\dot{V}O_2/\Delta WR$ to cardiac index (CI) and norepinephrine concentration (NE) at 20 W steady state. CI and NE showed a very weak correlation to $\Delta\dot{V}O_2/\Delta WR$.

There was no significant relationship between $\Delta\dot{V}O_2/\Delta WR$ and cardiac index and norepinephrine level at rest. Figure 2 demonstrates the relation of $\Delta\dot{V}O_2/\Delta WR$ to those two parameters at 20-watt steady state exercise. There was a significant but very weak positive correlation between $\Delta\dot{V}O_2/\Delta WR$ and cardiac index ($r = 0.37$, $p < 0.05$), and a negative correlation to norepinephrine level ($r = 0.48$, $p < 0.01$) at 20 W.

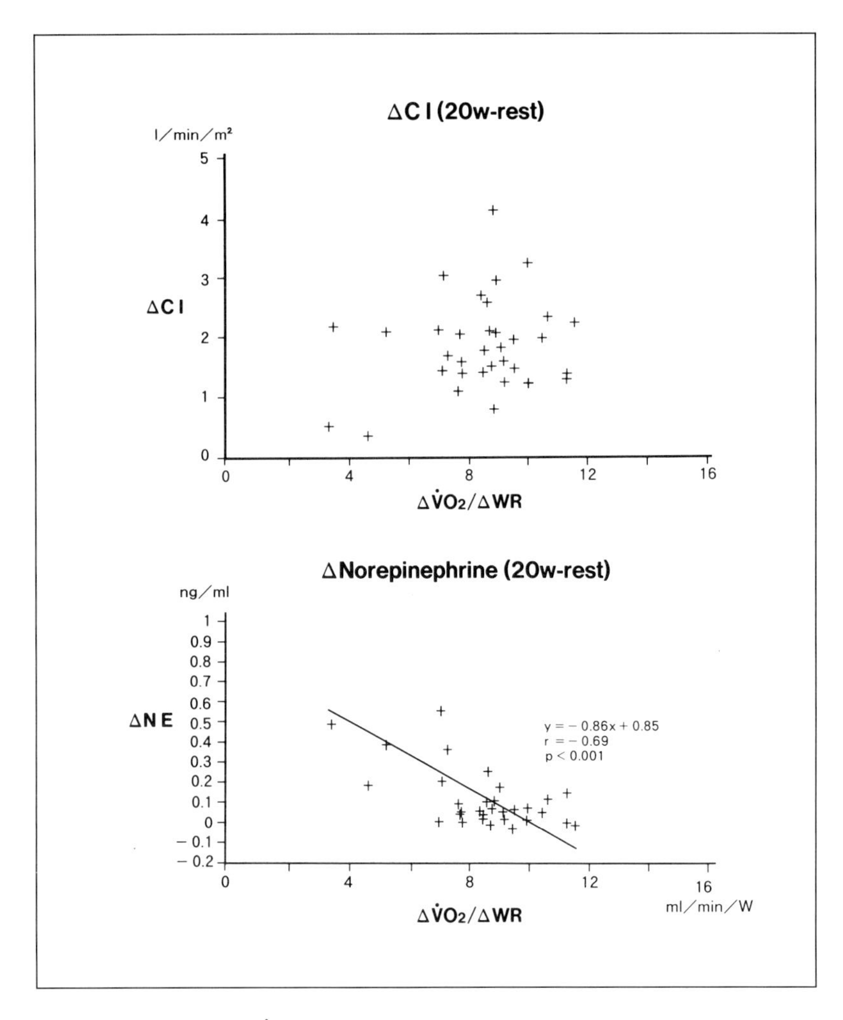

Fig. 3. The relation of $\Delta\dot{V}O_2/\Delta WR$ to the delta values from rest to 20-watt warm-up of cardiac index (ΔCI) and norepinephrine concentration (ΔNE).

To clarify the contribution of changes in blood flow and sympathetic nerve response to oxygen transportation to the peripheral tissue, delta values of these parameters were compared with $\Delta\dot{V}O_2/\Delta WR$ (fig. 3). Only delta norepinephrine showed a significant negative correlation with $\Delta\dot{V}O_2/\Delta WR$ ($r = -0.69$, $p < 0.001$).

Discussion

A decrease in the rate of rise in $\dot{V}O_2$ at high work rate was reported in patients with impaired O_2 transport, which is interpreted as an insufficiency of oxygen transport to peripheral tissue [5]. However, $\Delta\dot{V}O_2/\Delta WR$ below the AT level was believed not to decrease even in patients with heart disease. Our data disclosed that $\Delta\dot{V}O_2/\Delta WR$ up to the AT point did decrease in accordance with the severity of heart disease concordant with the decrease in $\dot{V}O_2$ at rest and at steady state in constant work rate of 20 W. This means that $\dot{V}O_2$/work rate decreased in heart failure patients.

There should be some mechanism affecting $\dot{V}O_2$ work rate relationship such as work efficiency of working muscles or mitochondrial function. We recently reported that a single oral administration of enoximone, a new phosphodiesterase inhibitor, increases the anaerobic threshold and $\Delta\dot{V}O_2/\Delta WR$ acutely [6]. This finding and the fact that the rise in norepinephrine concentration is closely related to $\Delta\dot{V}O_2/\Delta WR$ strongly suggest the important role of blood flow distribution on the $\dot{V}O_2$-work rate relationship. Namely, the enhanced sympathetic tone and simultaneous elevation in serum norepinephrine level during exercise should cause blood flow redistribution in order to increase the working muscle blood flow at the sacrifice of other organs. Since the $\dot{V}O_2$ reflects the oxygen consumption of both working muscle and other organs, this compensatory mechanism can reduce the total oxygen requirement of the body in patients with heart failure. From these data, the values of $\dot{V}O_2$ for physical activity obtained from normal subjects such as the table of energy cost [7] and METs should be applied carefully to patients with heart disease.

In summary, $\dot{V}O_2$ for a certain work load decreases in heart failure patients with increasing severity of the disease. This is probably due to blood redistribution during exercise as a compensatory mechanism for exercise intolerance.

References

1 Cohn JN, Levine TB, Olivali MT, Garberg V, Lura D, Francis G, Simon AB, Rector T: Plasma norepinephrine as a guide to prognosis in patients with chronic congestive heart failure. Engl J Med 1984;311:819–823.

2 Kisman JM, Moore JW, Hamilton WF: Studies on the circulation: I. Injection method. Physical and mathematical considerations. Am J Physiol 1929;89:322–330.

3 Beaver WL, Wasserman K, Whipp BJ: A new method for detecting anaerobic threshold by gas exchange. J Appl Physiol 1986;60:2020–2027.
4 Itoh H, Koike A, Taniguchi K, Marumo F: Severity and pathophysiology of heart failure on the basis of anaerobic threshold (AT) and related parameters. Jpn Circ J 1989;53:146–154.
5 Wasserman K, Hansen JE, Sue DY, Whipp BJ: Principles of Exercise Testing and Interpretation. Philadelphia, Lea & Febiger, 1987, pp 30–32.
6 Itoh H, Taniguchi K, Doi M, Koike A, Sakuma A: Effects of enoximone on exercise tolerance in mild to moderate heart failure patients. Am J Cardiol 1991;68:360–364.
7 Fox SM, Naughton JP, Gorman PA: Physical activity and cardiovascular health. The exercise prescription: Intensity and duration. Mod Concepts Cardiovasc Dis 1972;41:21–25.

Haruki Itoh, MD, The Cardiovascular Institute, 7–3–10 Roppongi, Minato-ku, Tokyo 106 (Japan)

Sato Y, Poortmans J, Hashimoto I, Oshida Y (eds): Integration of Medical and Sports Sciences. Med Sport Sci. Basel, Karger, 1992, vol 37, pp 381–395

Characteristics of Cardiorespiratory Responses during Exercise in Patients with Chronic Airflow Obstruction

Kouhei Makiguchi

Kasumigaura Branch Hospital, Tokyo Medical and Dental University, Miho, Ibaraki, Japan

Introduction

The number of patients with chronic airflow obstruction (CAO) has been increased over the last decade. Once exertional dyspnea develops in patients with CAO, a continuous downhill cycle of deconditioning often develops. But many commonly encountered medical conditions such as lung disease, heart disease, obesity and so on can be a cause of exertional dyspnea.

Therefore, an exercise test is important to identify and distinguish between the various causes of exercise limitation. So, I would like to review about the characteristics of cardiorespiratory responses during exercise in CAO patients.

The Definitions and Epidemiology of CAO

Figure 1 [1] shows the basic concepts that define the mechanisms of airflow and the causes of airflow limitation. Airflow limitation is determined by a loss of elastic recoil from damage to the alveolar tissue as in emphysema, and airway narrowing as in asthma and chronic bronchitis.

Figure 2 [2] shows subsets of patients with chronic bronchitis, emphysema, and asthma in three overlapping circles. Subsets of patients lying within the rectangle have airflow obstruction. But subset 9, that is patients with asthma, is defined as having completely reversible airflow obstruc-

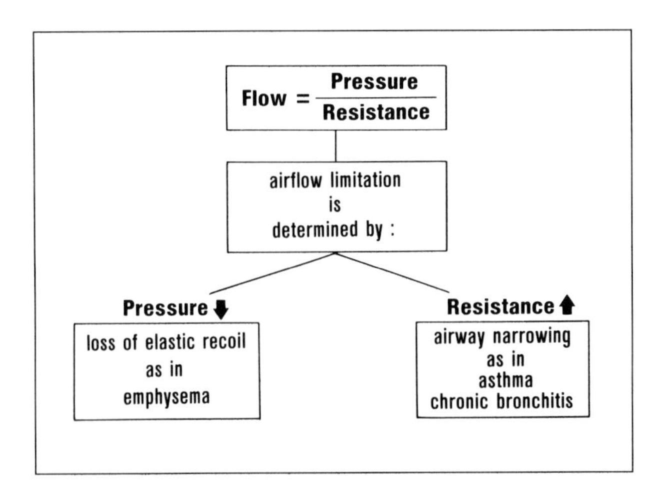

Fig. 1. Mechanisms of airflow and causes of airflow limitation [from ref. 1, with permission].

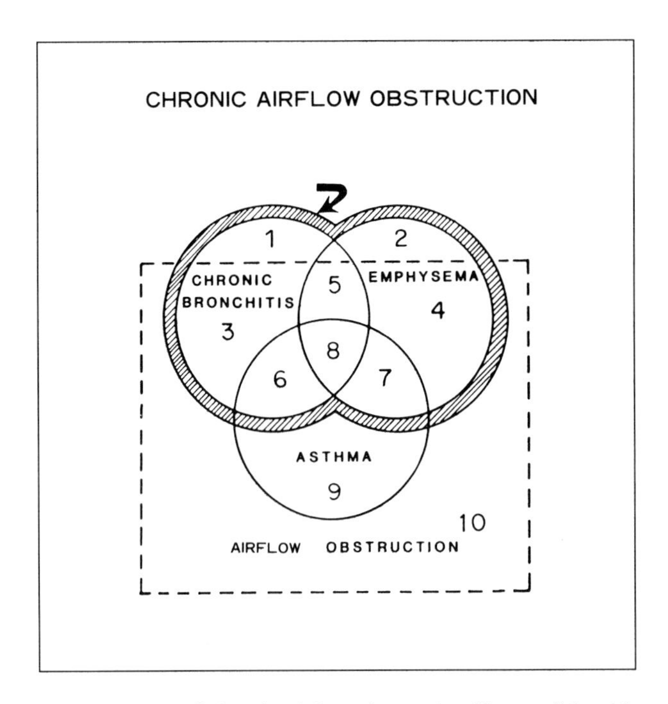

Fig. 2. Scheme of chronic airflow obstruction [from ref. 2, with permission].

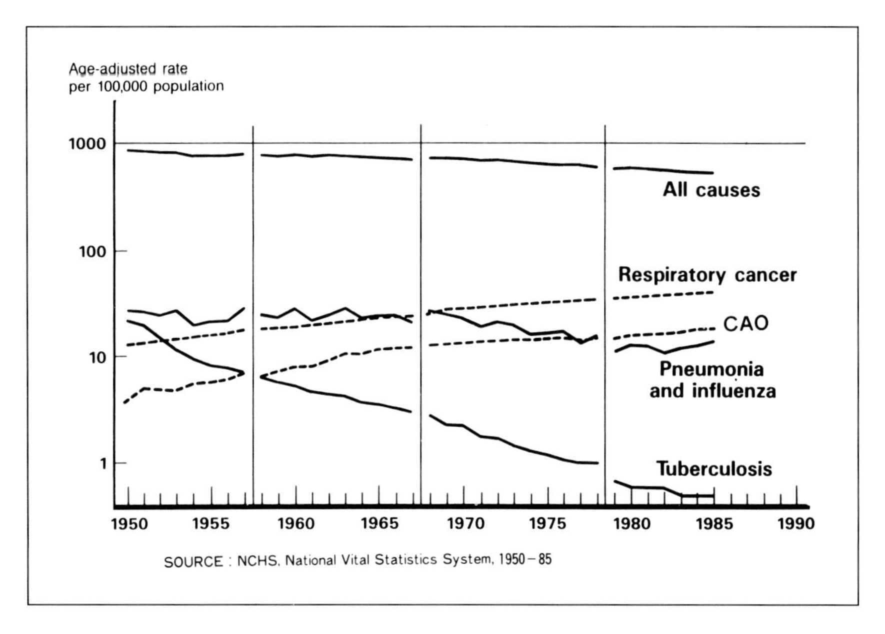

Fig. 3. Mortality from selected causes [from ref. 3, with permission].

tion. And subsets 1 and 2 do not have airflow obstruction as determined by the $FEV_{1.0}$, but clinically have features of CAO. Patients 1–8, included within the area outlined by a shaded band, denote CAO.

Figure 3 [3] shows the mortality from selected causes based on data of the National Center for Health Statistics in USA. In 1981, there were at least 60,000 deaths from CAO and related conditions. CAO has been the fifth leading cause of death since the late 1970s, and is one of the leading causes that has shown a steady increase since 1950. In Japan, although the absolute value was smaller, a similar tendency has been reported.

Figure 4 [4] shows that data from the Tecumseh Community Health Survey of over 9,000 men and women in all age groups revealed that approximately 14% of adult males and 8% of adult females had obstructive airway disease. At least 7.5 million Americans have chronic bronchitis, more than 2 million have emphysema, and at least 6.5 million have some form of asthma. In Japan, although the value is lower, a similar tendency has been reported. Therefore, there is no question of the importance of this disease spectrum as a major health problem.

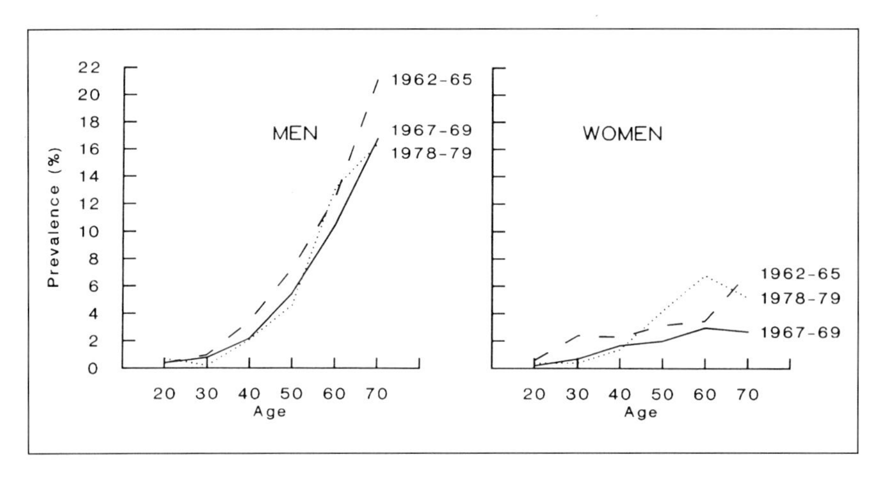

Fig. 4. Prevalence of obstructive airway disease [from ref. 4. with permission].

Characteristics of Cardiorespiratory Responses during Exercise in CAO

Ventilation and Gas Exchange

Figure 5 [5] shows the distribution of ventilation-perfusion ratio ($\dot{V}_A/\dot{Q}$ ratio) in CAO patients. CAO patients commonly have maldistribution of ventilation, resulting in a wide dispersion of $\dot{V}_A/\dot{Q}$ ratios. Figure 5A shows an example of $\dot{V}_A/\dot{Q}$ distribution of a patient with severe emphysema. A considerable part of ventilation distributes to the regions of high $\dot{V}_A/\dot{Q}$ ratio. This has the effect of increasing the wasted ventilation fraction. On the other hand, figure 5B shows the example of $\dot{V}_A/\dot{Q}$ distribution of a patient with chronic bronchitis, this shows a considerable part of blood flow distributes to the regions of low $\dot{V}_A/\dot{Q}$ ratio. This contributes to hypoxemia particularly during exercise.

Figure 6 shows the changes in the wasted fraction of the breath (V_D/V_T) during incremental exercise in CAO patients. The high $\dot{V}_A/\dot{Q}$ ratio lung units cause the calculated dead space to increase. In CAO patients, V_D/V_T may be high at rest and remain unchanged or even increase further with exercise, instead of decreasing normally. The failure of V_D/V_T to decrease normally in CAO reduces ventilatory efficiency and compounds the problem for these patients, since they already have reduced ventilatory capacity.

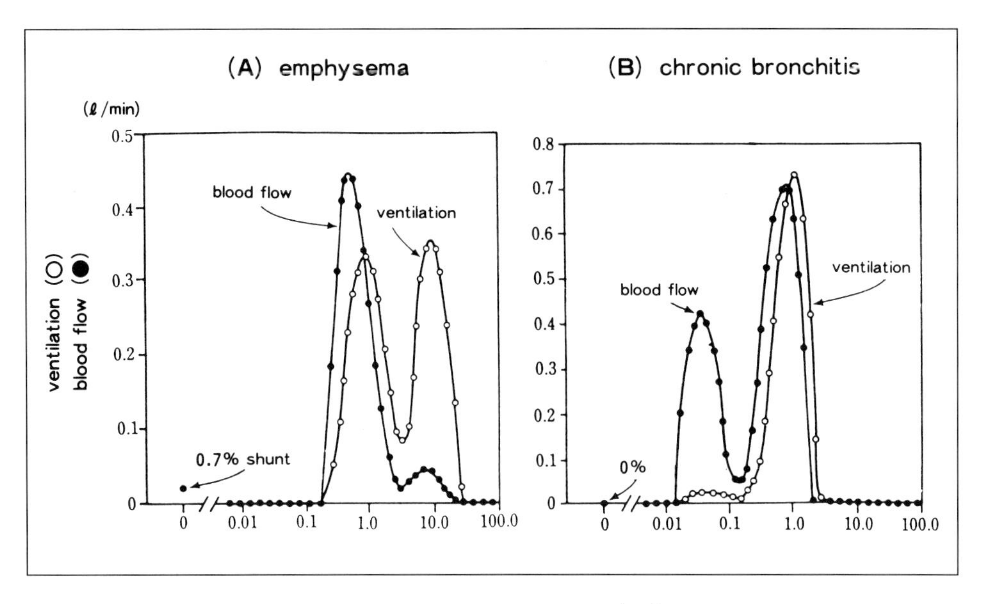

Fig. 5. Distributions of ventilation-perfusion ratio ($\dot{V}_A/\dot{Q}$) in patients with CAO [from ref. 5, with permission].

Figure 7 shows the changes in alveolar to arterial P_{O_2} difference (A-aD_{O_2}) during incremental exercise in CAO patients. A-aD_{O_2} may be high at rest and increase disproportionately with exercise compared with normals. Increased A-aD_{O_2} during incremental exercise can result from $\dot{V}_A/\dot{Q}$ mismatching and/or failure of complete transfer of oxygen by diffusion from alveolus to pulmonary capillary. However, in CAO patients, it is typically a result from underventilated regions of lung relative to their perfusion.

Figure 8 shows the changes in Pa_{O_2} during incremental exercise in CAO patients. During exercise as a result of the increase of AaD_{O_2} arterial hypoxemia increases. It may be a consequence of increased blood flow or reduced ventilation to lung units with a low $\dot{V}_A/\dot{Q}$ ratio.

Figure 9 shows the dyspnea index, that is, exercise minute ventilation ($\dot{V}_E$) divided by maximum voluntary ventilation (MVV), at exhaustion in CAO patients. CAO patients commonly have reductions in MVV, and therefore have reductions in ventilatory capacity. So, maximum exercise ventilation (peak $\dot{V}_E$) commonly meets or exceeds the MVV, reflecting complete utilization of ventilatory capacity.

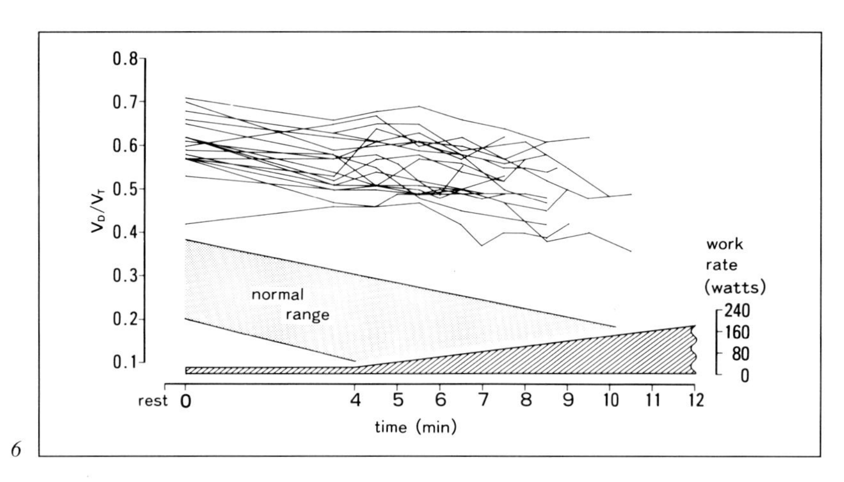
0.8
0.7
0.6
0.5
0.4
0.3
0.2
0.1
V_D/V_T
normal
range
work
rate
(watts)
240
160
80
0
rest 0
4 5 6 7 8 9 10 11 12
time (min)
6

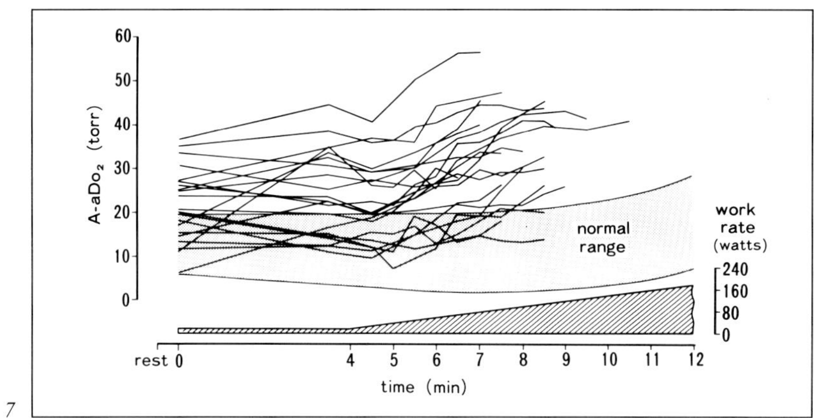
60
50
40
30
20
10
0
A-aD_{O_2} (torr)
normal
range
work
rate
(watts)
240
160
80
0
rest 0
4 5 6 7 8 9 10 11 12
time (min)
7

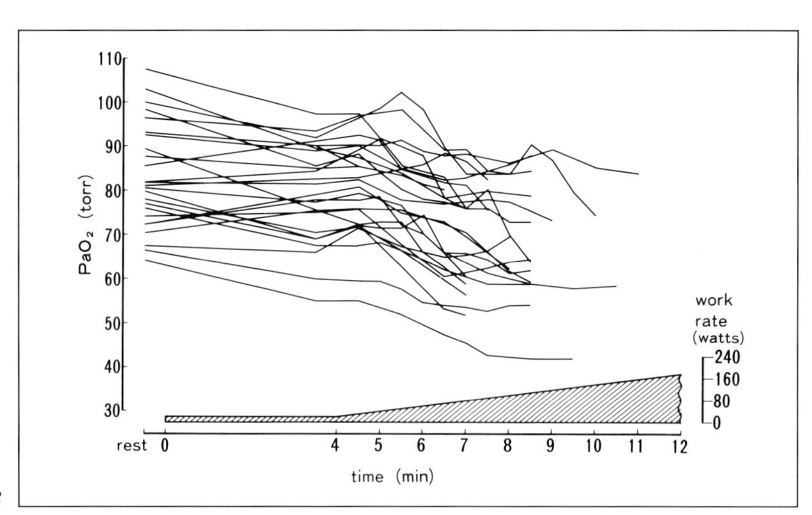
110
100
90
80
70
60
50
40
30
PaO_2 (torr)
work
rate
(watts)
240
160
80
0
rest 0
4 5 6 7 8 9 10 11 12
time (min)
8

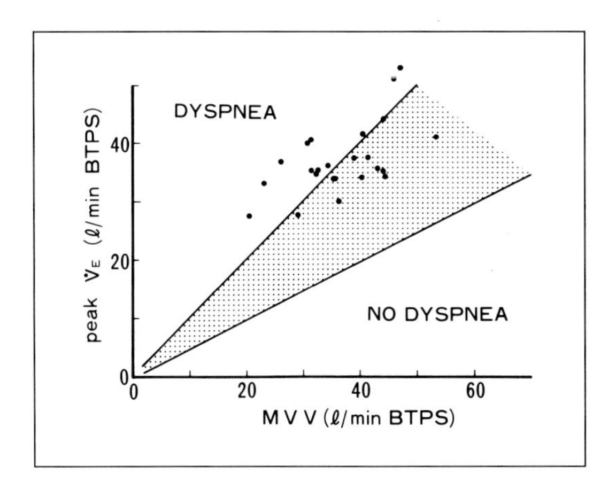

Fig. 9. Dyspnea index at exhaustion in CAO.

Table 1 shows the respiratory and cardiovascular responses in CAO at anaerobic threshold and maximum exercise. At maximum exercise, the dyspnea index exceeds 1.0, but the ratio of maximum heart rate to the predicted maximum heart rate is only 0.7. With increasing exercise, CAO patients are unable to reach their predicted maximum heart rate because of an early achievement of their maximum ventilation. The dyspneic CAO patients are not limited in their exercise by hemodynamic factors but by ventilatory limitation.

Oxygen Cost of Breathing

Figure 10 [6] shows the relationship between ventilation and total oxygen consumption and its respiratory and nonrespiratory components during exercise in normals and CAO patients. This shows that the total oxygen consumption at any ventilation is lower in CAO patients. In addition, at each level of ventilation the amount of oxygen consumed by the respiratory muscles is higher. And considerably less oxygen is available for the

---

Fig. 6. Changes in V_D/V_T during exercise in CAO.
Fig. 7. Changes in $A\text{-}aD_{O_2}$ during exercise in CAO.
Fig. 8. Changes in PaO_2 during exercise in CAO.

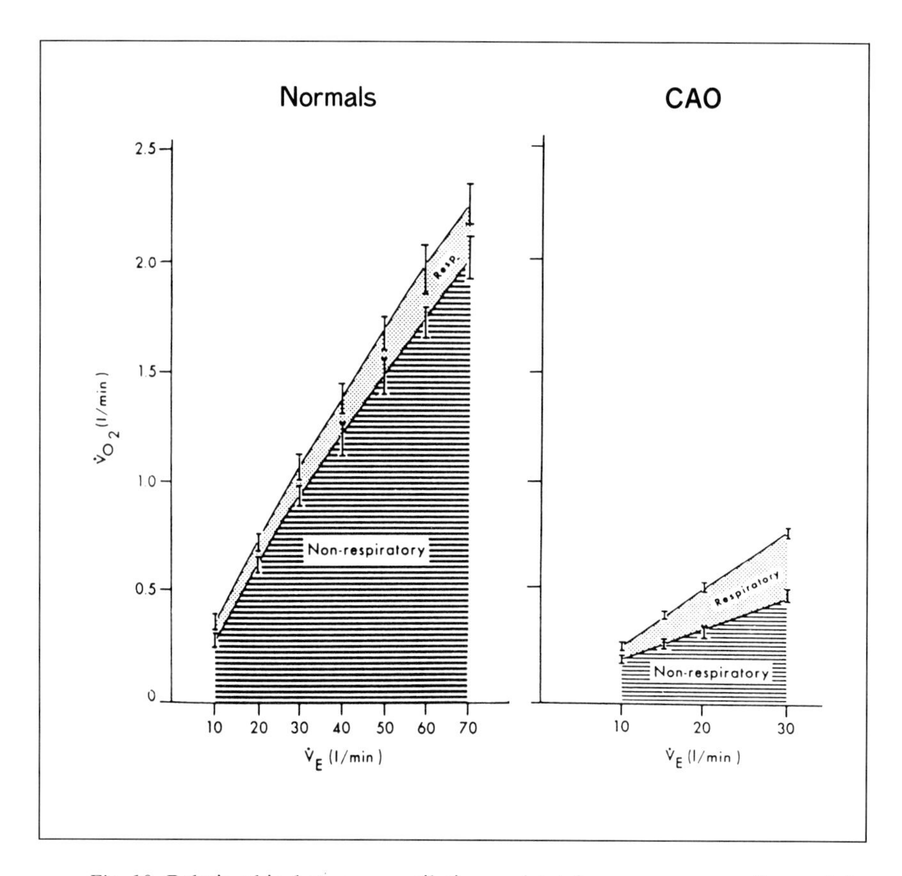

Fig. 10. Relationship between ventilation and total oxygen consumption and its respiratory and nonrespiratory components during exercise [from ref. 6, with permission].

Table 1. Respiratory and cardiovascular responses in CAO

	AT	max
$\dot{V}_E$/MVV	0.70 ± 0.20	1.06 ± 0.33
HR/$HR_{pred\ max}$	0.58 ± 0.09	0.72 ± 0.10

Mean ± SD.
$HR_{pred\ max}$ = The maximum heart rate predicted for age.

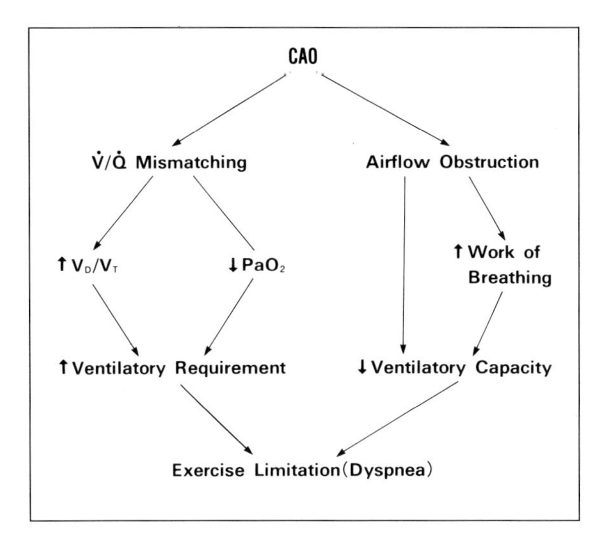

Fig. 11. Pathogenesis of ventilatory limitation in CAO [from ref. 7, with permission].

nonrespiratory muscles during exercise compared to normals. This may be an important factor that limits the exercise performance in CAO patients.

To review briefly, the pathogenesis of ventilatory limitation in CAO patients is conceptualized in figure 11 [7]. The increased ventilatory requirement may be due primarily to mismatching of ventilation to perfusion. And the decreased ventilatory capacity may be due to increased airflow obstruction.

Cardiovascular Performance

Figure 12 [8] shows the comparison of right ventricular ejection fraction at rest and during exercise in normals and CAO patients, evaluated with radionuclide angiocardiography. Of 30 patients 23 demonstrated abnormal right ventricular ejection fraction responses to exercise. Right ventricular dysfunction may be common in CAO patients, even in patients with mild airflow obstruction. Altered right ventricular afterload due to pulmonary artery hypertension is likely due to the major factor which causes right ventricular dysfunction rather than abnormalities in myocardial contractility.

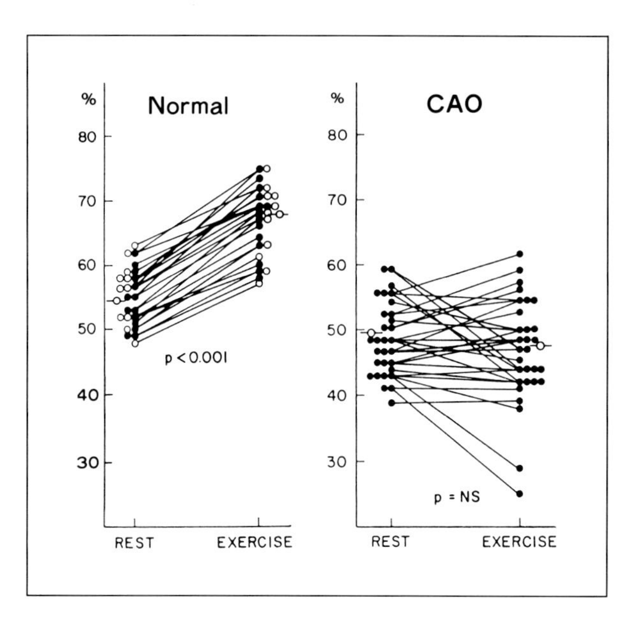

Fig. 12. Comparison of right ventricular ejection fraction at rest and during exercise in normals and patients with CAO [from ref. 8, with permission].

The primary and very important cardiovascular complication in CAO patients may be the development of pulmonary artery hypertension.

Table 2 [8] shows the causes of pulmonary artery hypertension in CAO patients. The most common cause may be increased pulmonary vascular resistance due to hypoxia-induced structural narrowing of the pulmonary artery, acidemia which augments hypoxic vasoconstriction, and increased alveolar and transthoracic pressure. Other causes include increased cardiac output which increases pulmonary artery pressure in patients with a compromised vascular bed, increased pulmonary blood volume, increased viscosity of blood, and raised pulmonary venous pressure.

Oxygen Uptake Kinetics

Gas exchange of phase 1 is postulated to be caused by the immediate increase in cardiac output at the start of exercise. Table 3 [9] shows the average phase 1 responses of respiratory and cardiovascular variables during 40 W constant load. The CAO patients had smaller phase 1 increases

Table 2. Causes of pulmonary artery hypertension in CAO [from ref. 8, with permission]

Increased pulmonary vascular resistance
  Hypoxia-induced structural narrowing of pulmonary arteries
  Acidemia augmenting hypoxic vasoconstriction
  Alterations in intrathoracic pressures
Increased cardiac output
Increased pulmonary blood volume
Increased viscosity of blood
Raised pulmonary venous pressure

Table 3. Average phase 1 responses of respiratory and cardiovascular variables [modified from ref. 9, with permission]

	Control	CAO
$\dot{V}_E$, liters/min BTPS	6.8 ± 1.05	3.4 ± 0.89*
$\dot{V}_{CO_2}$, liters/min STPD	0.22 ± 0.03	0.10 ± 0.03*
$\dot{V}_{O_2}$, liters/min STPD	0.24 ± 0.04	0.10 ± 0.03*
O_2-P, ml/beat	2.2 ± 0.45	0.93 ± 0.21*
HR, beats/min	16 ± 1.4	6 ± 0.9*

Air breathing, exercise at 40 W. * $p < 0.05$.

than controls for all variables. This indicates, in CAO patients, that an increase in pulmonary blood flow may be smaller than in controls. Several factors such as high pulmonary vascular resistance and changes in intrathoracic pressure have been reported to reduce the phase 1 responses in CAO patients, but further studies are needed.

Figure 13 [9] shows the pattern of $\dot{V}_{O_2}$ in a CAO patient and a matched normal subject during the performance of 40 W ergometer exercise. $\dot{V}_{O_2}$ during phase 1 is less in this CAO patient, and the rate of rise of $\dot{V}_{O_2}$ to its asymptote during phase 2 is also reduced, as shown by the longer time constant. Table 4 [9] shows the ventilatory and gas-exchange kinetics during moderate exercise in controls and CAO patients. CAO patients had, on average, significantly longer time constant for $\dot{V}_E$, $\dot{V}_{CO_2}$, and $\dot{V}_{O_2}$ than the controls. But $\tau\dot{V}_{CO_2}/\tau\dot{V}_E$ and $\tau\dot{V}_{O_2}/\tau\dot{V}_E$ were similar. This suggests that the

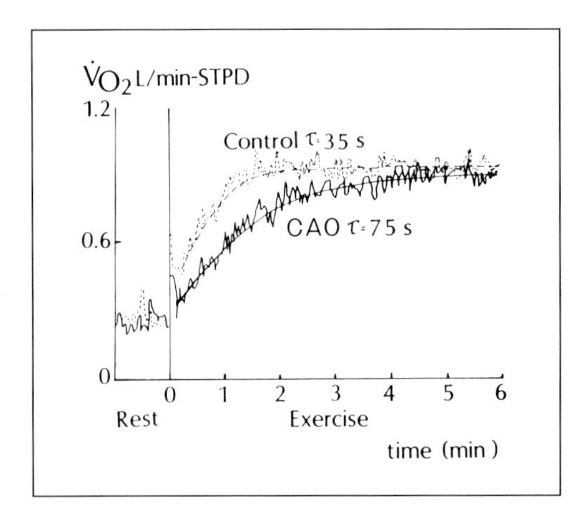

Fig. 13. Pattern of $\dot{V}_{O_2}$ in a patient with CAO and a matched normal subject during 40 W ergometer exercise [modified from ref. 9, with permission].

Table 4. Ventilatory and gas-exchange kinetics during moderate exercise [modified from ref. 9, with permission]

	Control	CAO		
$\tau\dot{V}_E$	65 ± 3	87 ± 7*		$p < 0.05$
$\tau\dot{V}_{CO_2}$	63 ± 3	79 ± 6*		$p < 0.05$
$\tau\dot{V}_{O_2}$	39 ± 2	56 ± 5*		$p < 0.05$
		($\tau\dot{V}_E > 80$ s)	($\tau\dot{V}_E < 80$ s)	
$\tau\dot{V}_{CO_2}/\tau\dot{V}_E$	0.96 ± 0.03	0.90 ± 0.03	0.94 ± 0.04	NS
$\tau\dot{V}_{O_2}/\tau\dot{V}_E$	0.60 ± 0.03	0.64 ± 0.04	0.67 ± 0.04	NS

Air breathing. Exercise at 40 W.

ventilatory response dynamics may be proportionally matched to the metabolic stimulus reaching the lungs in both groups of subjects. So, longer time constants for $\dot{V}_E$, $\dot{V}_{CO_2}$ and $\dot{V}_{O_2}$ in CAO patients are postulated to be caused by slowed cardiovascular kinetics. But the cause of slowed cardiovascular kinetics is not clear, and further examinations may be necessary.

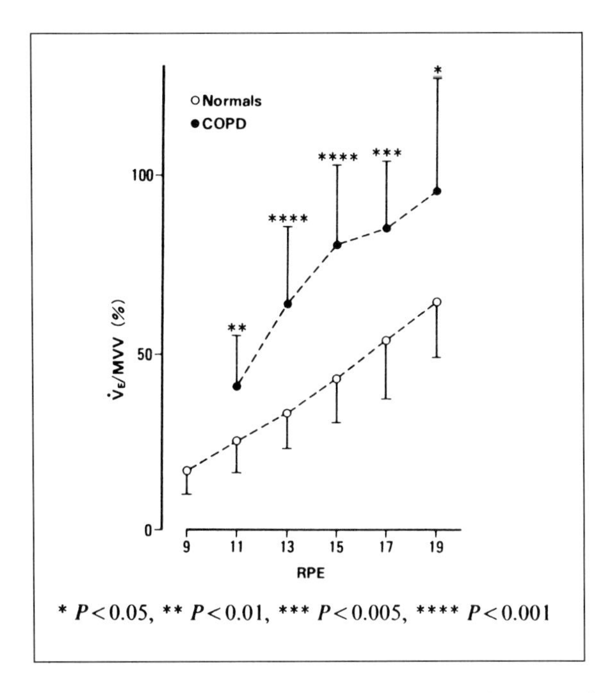

Fig. 14. Relationship between ratings of perceived exertion (RPE) and dyspnea index in patients with COPD and healthy subjects [from ref. 10, with permission].

Ratings of Perceived Exertion

Figure 14 [10] shows the relationship between ratings of perceived exertion (RPE) and dyspnea index in CAO patients and normals. At a given RPE during exercise, patients had a higher dyspnea index than normals. This indicates that patients are using a greater proportion of their MVV during exercise, and reduction of ventilatory capacity is thought to be a major factor contributory to limited exercise capacity.

Figure 15 [10] shows the relationship between RPE and the absolute and relative oxygen uptake in CAO patients and normals. At a given RPE, $\dot{V}_{O_2}$ was significantly greater in normals than patients, but $\%\dot{V}_{O_{2max}}$ at any RPE did not differ between the two groups. A clear difference in exercise capacity was found between two groups, but the relationship between $\%\dot{V}_{O_{2max}}$ and RPE was identical in both groups. These findings suggest that RPE could be a good indicator of relative exercise intensity in CAO patients.

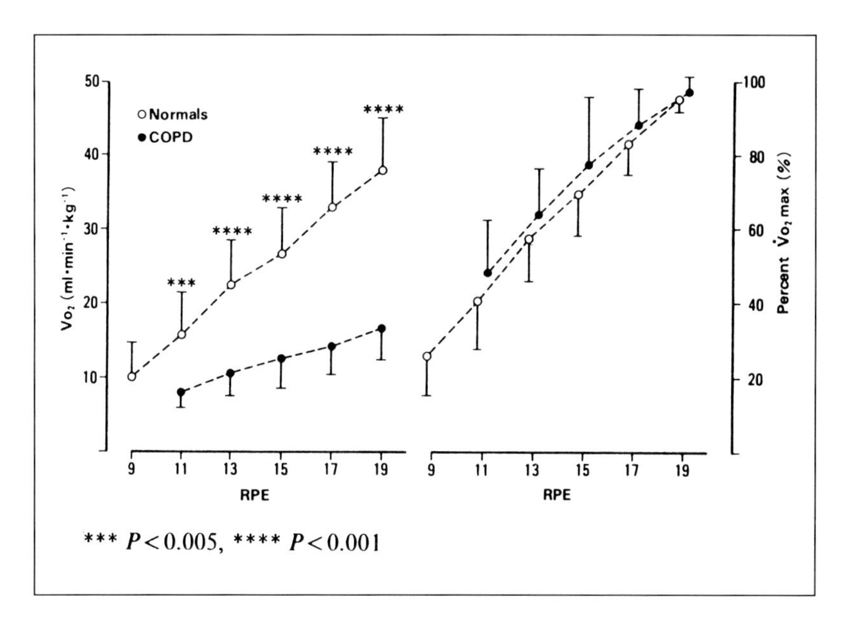

Fig. 15. Relationship between ratings of perceived exertion (RPE) and the absolute ($\dot{V}_{O_2}$, left) and the relative ($\%\dot{V}_{O_{2max}}$, right) oxygen uptake in patients with COPD and healthy subjects [from ref. 10, with permission].

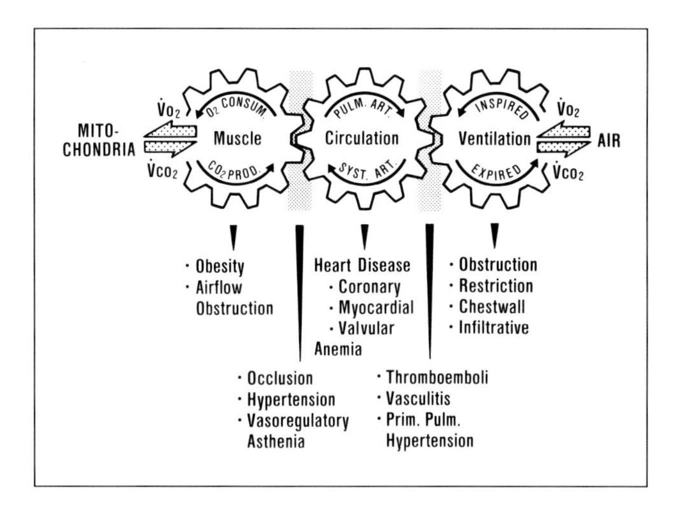

Fig. 16. Sites of interference in the metabolic-cardiovascular-ventilatory coupling for various disease states [from ref. 11, with permission].

Conclusion

Figure 16 [11] is a well-known figure by Dr. Wasserman which shows the site of interference in the metabolic-cardiovascular-ventilatory coupling for various disease states.

Many CAO patients have multiple medical problems other than their pulmonary condition. So, although exercise testing can be expensive, time-consuming and difficult to carry out in certain subjects, it can be of significant value to determine the etiology of a patient's symptom. When we perform exercise testing in patients and analyze the obtained information, we must always keep this figure in mind.

References

1 Hodgkin JE, Petty TL: In Manke D (ed): Chronic Obstructive Pulmonary Disease: Current Concepts. Philadelphia, Saunders, 1987.
2 Snider GL: Chronic bronchitis and emphysema; in Murray JF, Nadel JA (eds): Textbook of Respiratory Medicine. Boston, Saunders, 1988.
3 Feinleib M, Rosenberg HM, Collins JG, Delozier JE, Pokras R, Chevarley FM: Trends in COPD morbidity and mortality in the United States. Am Rev Respir Dis 1989;140:s9–s18.
4 Higgins MW, Keller JB: Trends in COPD morbidity and mortality in Tecumseh, Michigan. Am Rev Respir Dis 1989;140:s42–s48.
5 Wagner PD, Dantzker DR, Dueck R, Clausen JL, West JB: Ventilation-perfusion inequality in chronic pulmonary disease. J Clin Invest 1977;59:203–206.
6 Levison H, Cherniack RM: Ventilatory cost of exercise in chronic obstructive pulmonary disease. J Appl Physiol 1968;25:21–27.
7 Brown HV, Wasserman K: Exercise performance in chronic obstructive pulmonary diseases. Med Clin North Am 1981;65:525–547.
8 Matthay RA, Berger HJ: Cardiovascular performance in chronic obstructive pulmonary diseases. Med Clin North Am 1981;65:489–524.
9 Nery LE, Wasserman K, Andrews JD, Huntsman DJ, Hansen JE, Whipp BJ: Ventilatory and gas exchange kinetics during exercise in chronic airways obstruction. J Appl Physiol 1982;53:1594–1602.
10 Chida M, Inase N, Ichioka M, Miyazato I, Marumo F: Ratings of perceived exertion in chronic obstructive pulmonary disease. A possible indicator for exercise training in patients with this disease. Eur J Appl Physiol 1991;62:390–393.
11 Wasserman K, Hansen JE, Sue DY, Whipp BJ: Principles of Exercise Testing and Interpretation. Philadelphia, Lea & Febiger, 1987.

Dr. Kouhei Makiguchi, Division of Internal Medicine, Kasumigaura Branch Hospital, Tokyo Medical and Dental University, Miho Inashiki-gun, Ibaraki 300–04 (Japan)

Sato Y, Poortmans J, Hashimoto I, Oshida Y (eds): Integration of Medical and Sports Sciences. Med Sport Sci. Basel, Karger, 1992, vol 37, pp 396–404

Cardiopulmonary Responses and Blood Flow Distribution during Exercise in Patients with Left Ventricular Dysfunction

Takao Yoshioka, Toshikazu Hashizume, Tatuo Fujii, Yoshiaki Okano, Norifumi Nakanishi, Kazuo Haze, Katuro Shimomura

National Cardiovascular Center, Division of Cardiology, Department of Internal Medicine, Osaka, Japan

Introduction

Patients with left ventricular (LV) dysfunction due to ischemic heart disease have limited exercise capacity [1, 2], but this limitation is not always correlated with their LV functions at rest [3]. Cardiac output during exercise in patients with myocardial infarction (MI) has been studied, but its relation with O_2 utilization is not well understood. Recent advances in medical equipment have made it possible to monitor the mixed venous O_2 saturation (SvO_2), the arterial blood O_2 saturation (SaO_2) and the ventilatory parameters such as oxygen uptake ($\dot{V}_{O_2}$) or minute ventilation ($\dot{V}_E$) continuously. In this study, we examined the relationship between O_2 delivery to the exercising limbs and the cardiac output during exercise in patients with LV dysfunction by continuous monitoring of these parameters. We also analyzed the distribution of the systemic flow to the exercise limbs in patients with various LV functions by a thermodilution method [4, 5]. The purpose of this study is to clarify the mechanisms which maintain the exercise capacity of patients with LV dysfunction.

Subjects and Methods

We evaluated the exercise capacity in 37 male patients with MI (aged from 26 to 62 years old, mean 50 years). The patients with ischemic events manifested during exercise were excluded. All patients were hospitalized for more than 3 weeks before the exercise test. To classify the grade of the LV dysfunction, the patients were divided according to their LV ejection fraction (EF) which had been measured by LV angiograms into three groups; EF $< 30\%$ (group L), $30\% \leq$ EF $< 55\%$ (group M) and EF $\geq 55\%$ (group H), consisting of 12, 17 and 8 patients, respectively. To analyze the distribution of the blood to the exercising legs by the thermodilution method, cardiac output and leg flow were measured in 13 male patients with MI and without angina pectoris.

Pulmonary arterial pressure and SvO_2 were measured using the Oximetric III opticatheter system (Oximetrix Inc. Mountain View, Calif.), which was placed into the pulmonary trunk. We used a Biox III pulse oximeter (Omeda, Louisville, Ky.) to measure SaO_2 continuously. $\dot{V}_E$, $\dot{V}_{O_2}$, and carbon dioxide production ($\dot{V}_{CO_2}$) were determined using respiromonitor RM-300 (Minato Products, Osaka, Japan) connected to a massspectrometer (Perkin Elmer, Pomona, Calif.). Heart rates (HR) were monitored with an ECG tachometer, and systemic blood pressures (SBP) were determined using a sphygmomanometer at rest and at 1-minute intervals during exercise.

O_2 extraction was computer-calculated by the equation as: O_2 extraction in % = $(SaO_2 - SvO_2)/SaO_2$; cardiac output was also computer-calculated continuously by the Fick equation as described previously [6].

A cycle ergometer was used and a progressive, 1-min, 15-watt incremental workload test from 0 W to the patients' symptoms-limited maximum was performed in the supine position after a 2-min cycling warm-up. In the leg flow study, the same incremental workload from 0 W was limited to 60 W. In these patients, their peak $\dot{V}_{O_2}$ was measured before the examination. The flow to one exercising leg was measured by the thermodilution method described by Sorlie and Myhre [4] and Sullivan et al. [5] using a 5-French size Swan-Ganz catheter inserted from the femoral vein to the external iliac vein. Total leg flow (Q1) was determined as twice that of one leg flow.

Results

A timed record during exercise of the observed variables in a group H patient was shown in figure 1. Total exercise durations in group H, M and L were 392 ± 45 (mean ± 1 SD), 430 ± 100 and 400 ± 102 s, and peak exercise loads were 72.3 ± 11.3, 81.0 ± 25.9 and 70.7 ± 25.6 W, respectively. Mean peak $\dot{V}_{O_2}$ in each group was 19.3 ± 4.2 ml/min/kg in group H, 18.5 ± 3.6 ml/min/kg in group M and 16.9 ± 3.6 ml/min/kg in group L, respectively. There were no significant differences among the groups. SBP, HR and SaO_2 in each group at rest, at every 30 W and at peak exercise were similar among the groups. However, there were significant differences in pulmonary arterial mean pressure (PAm) at each exercise load between the groups L and H or M, and in SvO_2 in groups L and H. Calculated O_2 extraction and cardiac output (CO(F)) in all patients in each group are shown in figure 2. O_2 extraction and cardiac index in each group were also

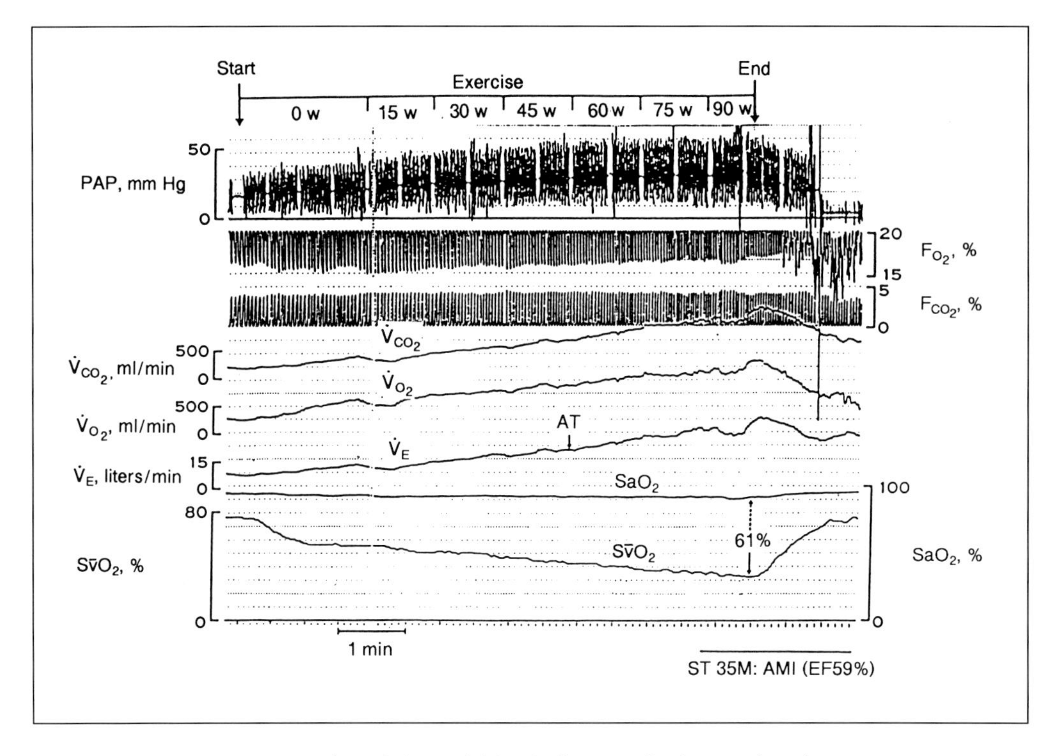

Fig. 1. A recording of the variables during exercise in a patient in group H: The pulmonary arterial pressure (PAP) and mean pressures every 30 s during exercise are shown at the top of this figure. The fractional concentrations of O_2 (F_{O_2}) and carbon dioxide (F_{CO_2}) are shown under PAP. The five curves under these records show from top to bottom, $\dot{V}_{CO_2}$, $\dot{V}_{O_2}$, $\dot{V}_E$, SaO_2 and SvO_2.

significantly different during exercise between group H or M and group L (fig. 3).

To make clear the relation between CO(F) and O_2 extractions, we plotted one against the other. Figure 4 shows the correlations between CO(F) and O_2 extraction in the subject with good LV function and with LV dysfunction. In subjects with good LV function, the slope was gentle, while in patients with poor LV function, the slope was very steep. We used these slopes of the linear regression lines of the CO(F)–O_2 extraction curves to compare the function of LV during exercise. The mean slope of group L was significantly steeper than the slopes of groups H and M.

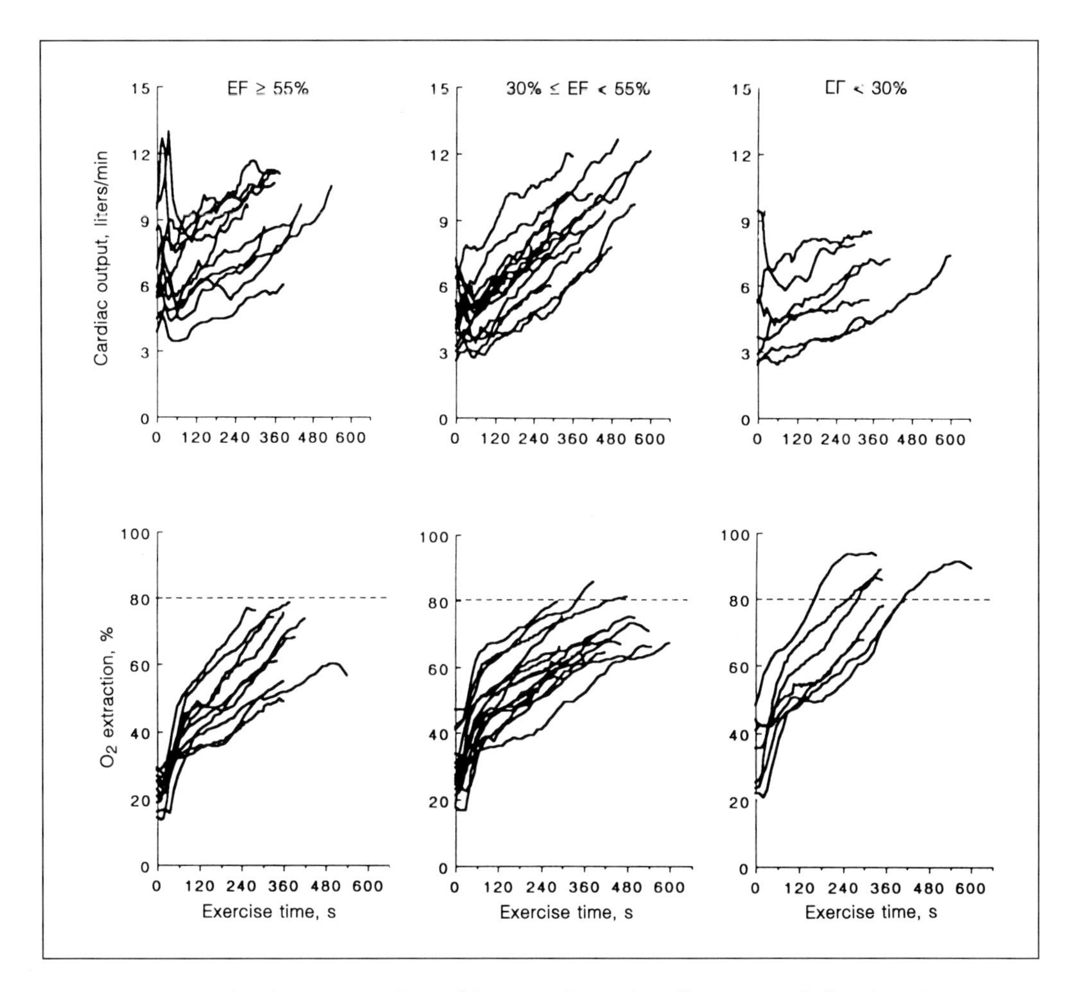

Fig. 2. Continuous recordings of O_2 extraction and cardiac output of all patients in each group: The O_2 extraction during exercise in patients with low EF was greatly increased compared with the other groups. There were no patients whose O_2 extraction exceeded 80% in the good EF group. In contrast, 5 patients out of 7 in the low EF group showed O_2 extractions over 80%.

In the leg flow study, there was no significant relationship between EF and Q1 at 30 or 60 W exercise. Figure 5 shows changes of cardiac output by the thermodilution method (CO(Td)) and Q1 during exercise in patients with good LV function and those with LV dysfunction. We defined

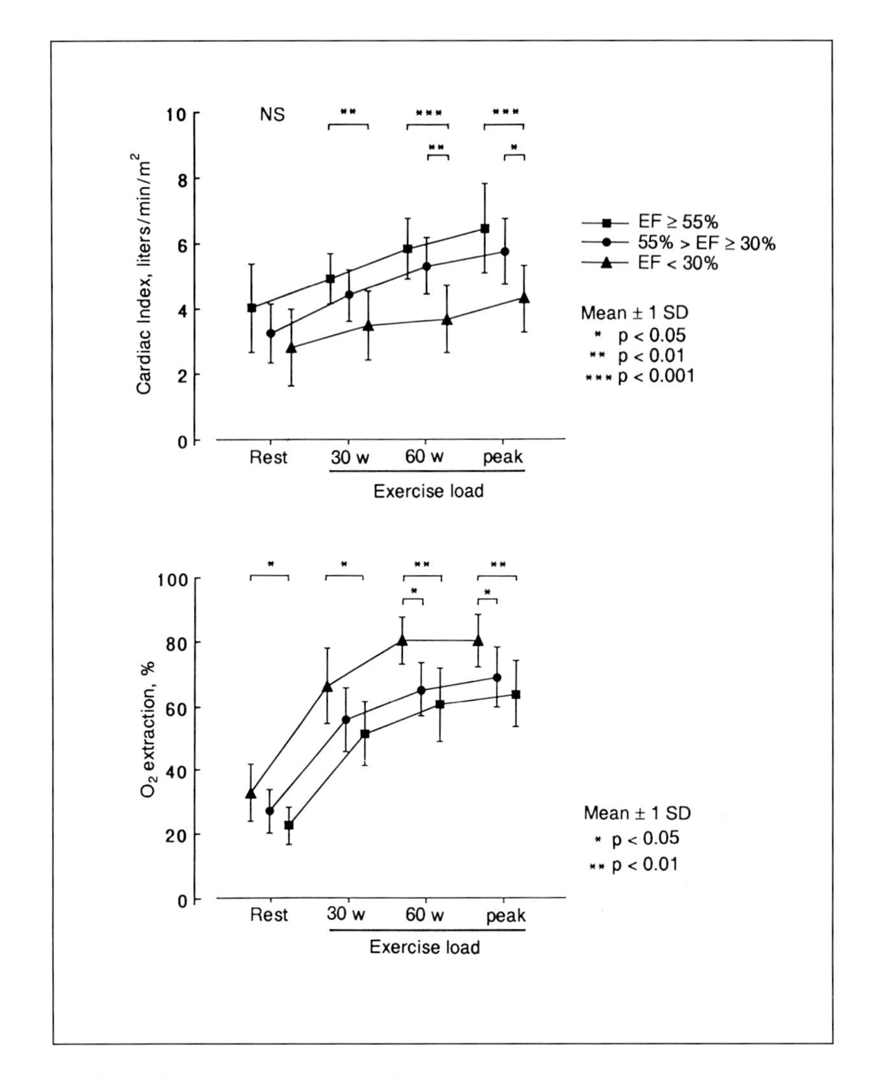

Fig. 3. Cardiac index and O_2 extraction in each group.

the difference between cardiac output and leg flow as the non-leg flow in this study. All patients cold be classified into two groups according to the patterns of the non-leg flow. One group showed increased non-leg flow (32.0 ± 9.8%) during exercise, and the other showed decreased flow (58.4 ± 5.8%), with the difference being significant.

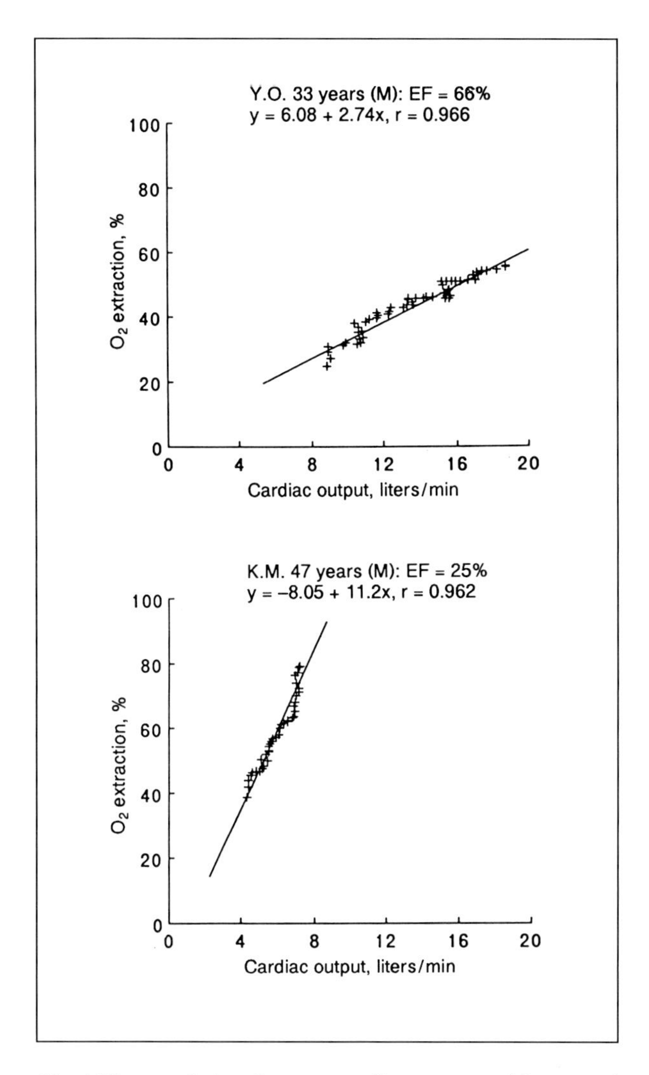

Fig. 4. The correlations between cardiac output and O_2 extraction in the subject with good LV function and with LV dysfunction.

Discussion

In recent years, even in the patients with serious LV dysfunction, such as broad MI or dilated cardiomyopathy, regular exercise has been recognized to be effective on the increase of their physical activities [7, 8], and

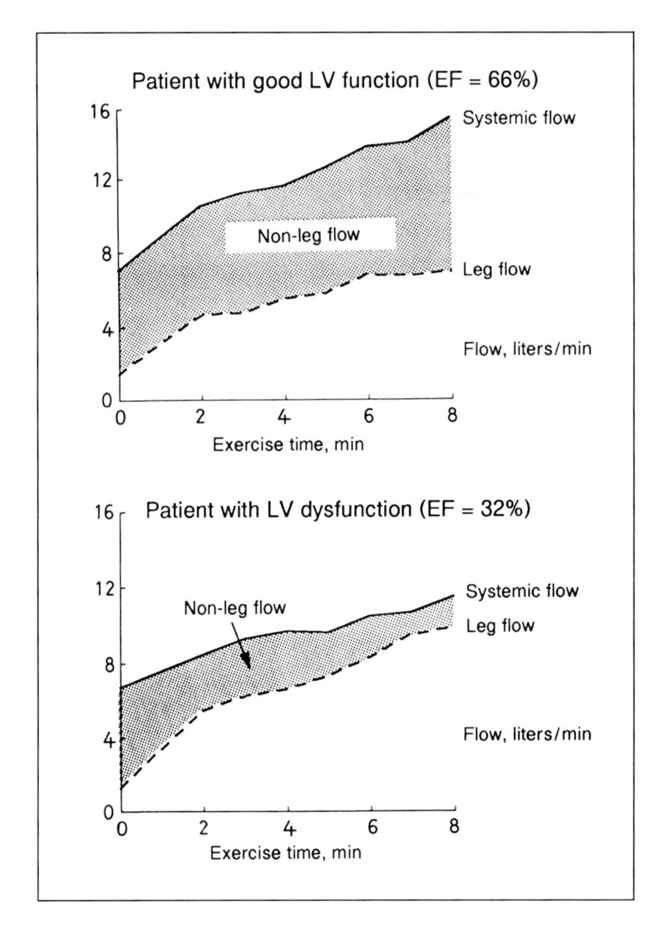

Fig. 5. The changes of CO(Td) and Q1 during exercise in patients with good LV function and those with LV dysfunction.

the measurements of the exercise capacity, including ventilatory and cardiohemodynamic responses to the exercise, have been performed in many hospitals and laboratories. In this study, we examined not only the responses of systemic cardiac output and blood flow to the exercising legs, but also systemic and peripheral O_2 utilization in patients with MI. As the purpose of this study is to clarify the compensatory mechanisms to maintain the exercise capacity of the patient with LV dysfunction, so we tried to clarify these 3 points in the patients with MI. The first point was the relationship between exercise capacities and LV functions at rest in these patients. The second point was the relation between cardiac outputs and O_2 extractions in these patients. The third point we examined was how the ratio of the blood flow to exercise muscle was different to the LV function.

First, we evaluated the relations between exercise capacity and LV function at rest. We compared the data among three groups which were thought to have obvious differences in LV function. The exercise time and mean maximal exercise load in each group were not different among the groups. In the excercising human body, to support the increased metabolic rate and the gas exchange of contracting muscles, the interaction of gas transport mechanism for O_2 delivery is very important [9]. In the patients with LV dysfunction, their exercise capacity was not necessarily reduced according to their LVEF. It has been reported that there is a poor relationship between EF and exercise endurance [2], but in our study, the patients with low EF showed high PAm during exercise. This was probably due to the elevation of the end-diastolic pressure in the LV. This suggests that patients whose EF was under 30% had, roughly speaking, poor LV functions. Exercise capacity of patients with LV dysfunction, then, should be maintained by increasing the utilization from transported O_2. The fact that the SvO_2 of the low EF group decreased markedly also supports this speculation. There should be some mechanisms to maintain the exercise capacity of the patients with low cardiac outputs. One of these mechanisms is the enhancement of O_2 utilization in the peripheral tissue of the exercised muscles. In patients with low cardiac output, the enzyme activities for O_2 utilization in mitochondria might be enhanced because of their chronic O_2 deficit during low grade exercise in the hospital which is not enough to train the muscle tissue of patients with normal cardiac output. Such a mechanism appears unlikely. Another mechanism might be the change of the distribution of the blood flow during exercise. In low EF patients, the blood flow to the exercising muscles might be maintained by reducing the flow to unexercised tissues or organs such as the kidney, liver and skin. In our study, the flow ratio to the exercising legs increased in the patients with LV dysfunction by reducing non-leg flow, and the O_2 extraction of SvO_2 increased. This is a cause of the marked increase in O_2 utilization during exercise in the exercising muscle and a preferential shift of the blood flow to the exercising muscle in the patients with LV dysfunction. For this reason, the cardiac output-O_2 extraction curves and their slopes can be used for classifying the patients with LV dysfunction.

Conclusion

The high level of O_2 extraction is a characteristic finding in patients with low cardiac output during exercise. This high O_2 extraction might be due to the high ratio of leg flow to systemic flow. The patients with LV dysfunction maintain their exercise capacity by reducing the non-leg flow.

References

1 Weber KT, Kinasewits GT, Janicki JS, Fishman AP: Oxygen utilization and ventilation during exercise in patients with chronic cardiac failure. Circulation 1982;65: 1213–1223.
2 Weber KT, Wilson JR, Janicki JS, Likoff MJ: Exercise testing in the evaluation of the patient with chronic cardiac failure. Am Rev Respir Dis 1984;129:S60–S62.
3 Fanciosa JA; Park M, Levin TB: Lack of correlation between exercise capacity and indexes of resting left ventricular performance in heart failure. Am J Cardiol 1981; 47:33–39.
4 Sorlie D, Myhre K: Determination of lower leg blood flow in man by thermodilution. Scand J Clin Invest 1977;37:117–124.
5 Sullivan MJ, Beckley PD, Hanson KM, Leier CV: In vivo validation of a thermodilution system designed to measure peripheral blood flow. Med Instrum 1985;19: 38–40.
6 Yoshioka T, Nakanishi N, Homma T, Shimouchi A, Okubo S, Kunieda T, Saito M: Cardiopulmonary hemodynamics and ventilations during supine exercise in patients with pulmonary thromboembolism; in Strano A, Novo S (eds): Advances in Vascular Pathology, vol 3. Amsterdam, Excepta Medica, 1990, pp 87–92.
7 Sullivan MJ, Higginbotham MB, Cobb FR: Exercise training in patients with severe left ventricular dysfunction. Circulation 1988;78:506–515.
8 Coats AJS, Adampolus S, Meyer TE, Conway J, Sleight P: Effects of physical training in chronic heart failure. Lancet 1990;335:63–66.
9 Wasserman K, Van Kessel AL, Burton GG: Interaction of physiological mechanisms during exercise. J Appl Physiol 1967;22:71–85.

Dr. T. Yoshioka, Department of Internal Medicine, Division of Cardiology,
National Cardiovascular Center, Fujishirodai 5–7–1, Suita, Osaka 565 (Japan)

Sato Y, Poortmans J, Hashimoto I, Oshida Y (eds): Integration of Medical and Sports Sciences. Med Sport Sci. Basel, Karger, 1992, vol 37, pp 405–415

Possible Relationship between Muscle Morphology and Capillarisation and the Risk Factor for Development of Cardiovascular Diseases

Marcin Krotkiewski

Department of Medical Rehabilitation, Sahlgren's Hospital, University of Göteborg, Sweden

Introduction

It is well known that both hypertension and non-insulin-dependent (type II, NIDDM) diabetes mellitus are frequently associated with obesity, particularly with abdominal obesity [1], and that hypertension and atherosclerosis are more prevalent among diabetic than nondiabetic individuals. Also, hyperlipidemia (high serum triglycerides, total cholesterol, low HDL cholesterol concentration) is found with increased frequency in both diabetic [2], hypertensive [3], and abdominally obese patients [4].

Thus, the substantial clustering of disturbances of carbohydrate and lipid metabolism and of blood pressure and abdominal obesity emerges as a clear phenomenon in the general population. Recently, it has been suggested that hyperinsulinemia may be a common factor accounting for the association of obesity, type II diabetes and hypertension [5].

Abdominal obesity and type II diabetes are considered the typical states of insulin resistance [6, 7] and there is convincing evidence showing that essential hypertension is most frequently also an insulin-resistant condition [3]. This lead Reaven [5] to the conclusion that insulin resistance aggregation with glucose intolerance, hypertension and dyslipidemia is to be treated as a distinct syndrome (syndrome X).

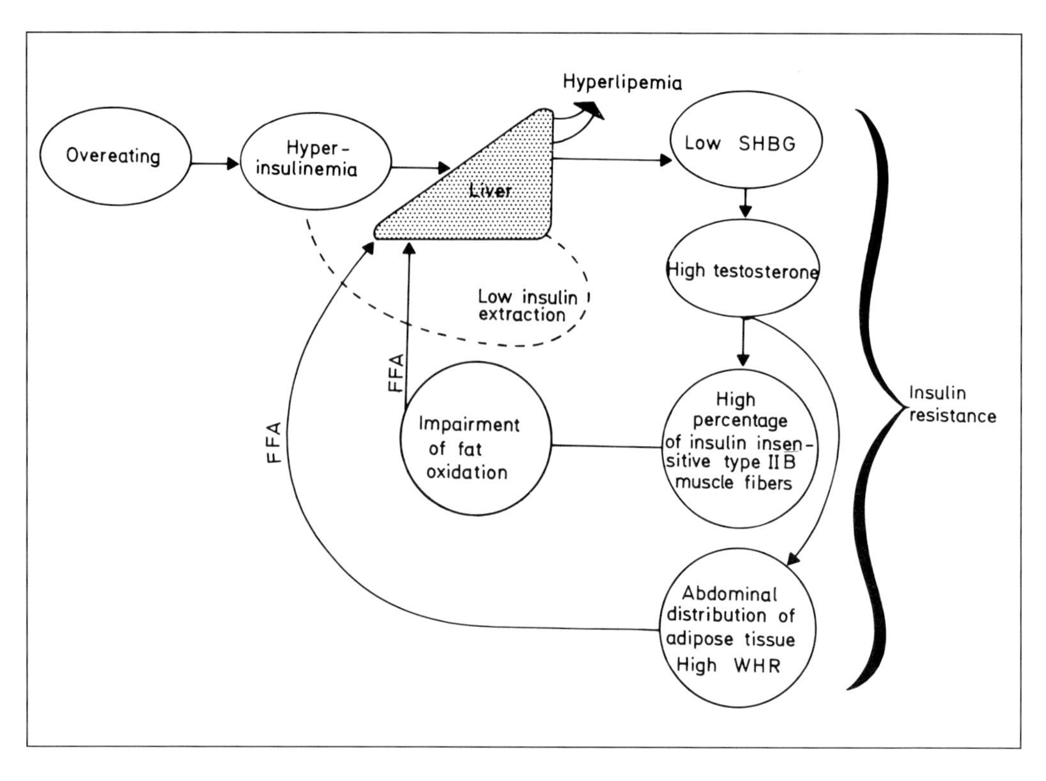

Fig. 1. Possible development of metabolic syndrome.

Diabetes, hypertension and hypercholesterolemia are generally accepted as obvious risk factors for cardiovascular disease (CVD). In Gothenburg [7], in several population studies, the abdominal accumulation of adipose tissue was found to be associated with the increased risk of early death, myocardial infarction, and stroke.

On the other hand, in some prospective epidemiologic studies [9] plasma insulin concentration has emerged as an independent predictor of coronary disease. Furthermore, Bergström et al. [10] and Haffner et al. [4] did not see an influence of body weight on the incidence of NIDDM after adjustment for insulin concentration. Several studies have also shown that an unfavorable body fat distribution is associated with insulin resistance and hyperinsulinemia [11, 12]. Providing the putative hyperinsulinemia/insulin resistance syndrome is associated with an increased risk of cardiovascular morbidity, hypertension and diabetes, the two questions arise as being fundamental for understanding the possible pathogenesis of the syndrome (syndrome X, WHR syndrome).

What is the cause of hyperinsulinemia/insulin resistance? What is the main organ responsible for the development of insulin resistance? There are three probable answers to the first question: (a) overeating; (b) decreased physical activity, and (c) genetic predisposition.

It is most probable that all three factors contribute. Overeating leads to hyperinsulinemia. Hyperinsulinemia, if chronic, leads to insulin resistance. Hyperinsulinemia itself influences the liver, suppressing among others the synthesis of SHBG (sex hormone binding globulin). Low SHBG leads to a higher concentration of free testosterone. Hyperandrogenicity leads to abdominal distribution of adipose tissue and induces the change in the distribution of muscle fibers with predominance of type IIB muscle fiber. The abdominal distribution of adipose tissue leads to overload of the portal system, the FFA liberated from the closely located abdominal fat depots. The following impairment of liver function leads to the lower extraction of insulin with aggravation of the preexisting hyperinsulinemia and hyperlipemia (fig. 1).

The changed muscle morphology contributes to the lower oxidation of fat during muscle exercise and perpetuates overweight. Furthermore, the changed muscle morphology is associated with the lower insulin sensitivity and contributes to general insulin resistance. Insulin resistance leads to the development of diabetes, hypertension and arteriosclerosis. Low physical activity contributes and accelerates all these events mainly through its effect on impairment of the nonoxidative glucose metabolism leading to local and general insulin resistance. The sequence of different events described above are summarised in figure 2.

The following findings describe and verify the likelihood of the hypothesis presented in figures 1 and 2.

Muscle Morphology

In the late 1970s when working with patients with intermittent claudication and using the catheterisation technique, we had the opportunity to note that in patients with intermittent claudication, uptake of insulin and glucose occurs in inverse proportion to the percent of type IIB muscle fibers, but correlates positively with the number of capillaries in the mus-

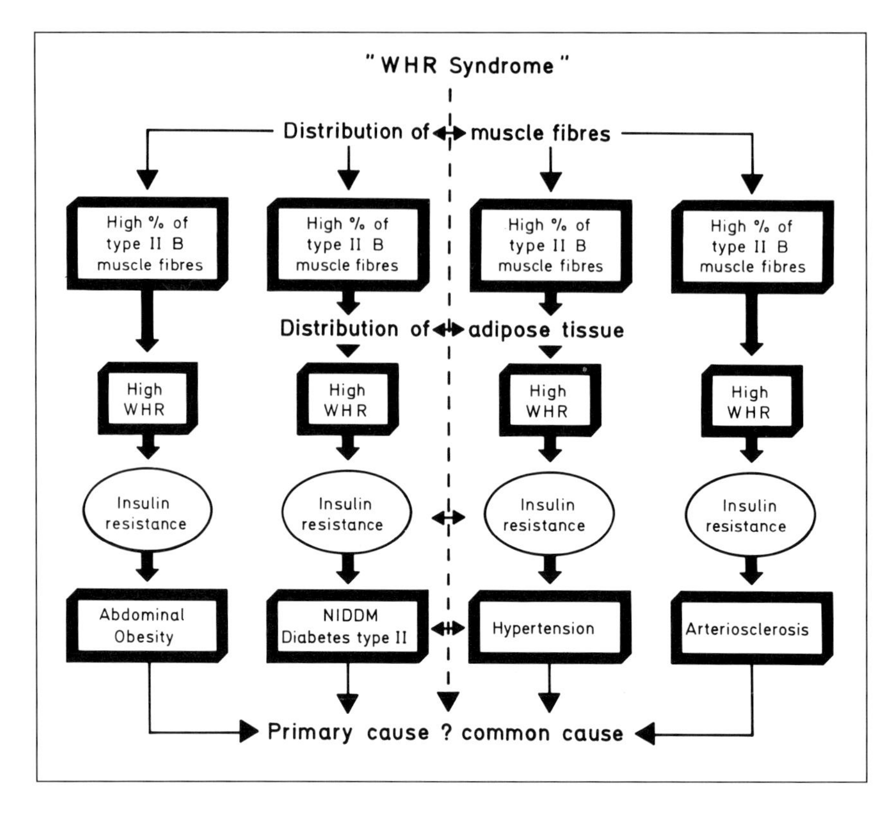

Fig. 2. WHR syndrome – metabolic syndrome.

cle tissue of the examined leg [13]. Encouraged by these observations, we started to look for muscle morphology in obesity.

As we could find, in contrast to the cross-sectional area of muscle fibers which increases in proportion to increasing body weight and body fat, the percentage to type IIB muscle fibers increases and the proportion of type I muscle fibers decreases in proportion to the abdominal distribution of adipose tissue (WHR) [14–16].

In the large population of obese women, the relative percentage of type IIB muscle fibers was found to be significantly correlated with most of the metabolic abnormalities typical of obesity, particularly hyperinsulinemia, impaired insulin extraction and insulin resistance, glucose disposal during insulin clamp, and blood pressure [14–20].

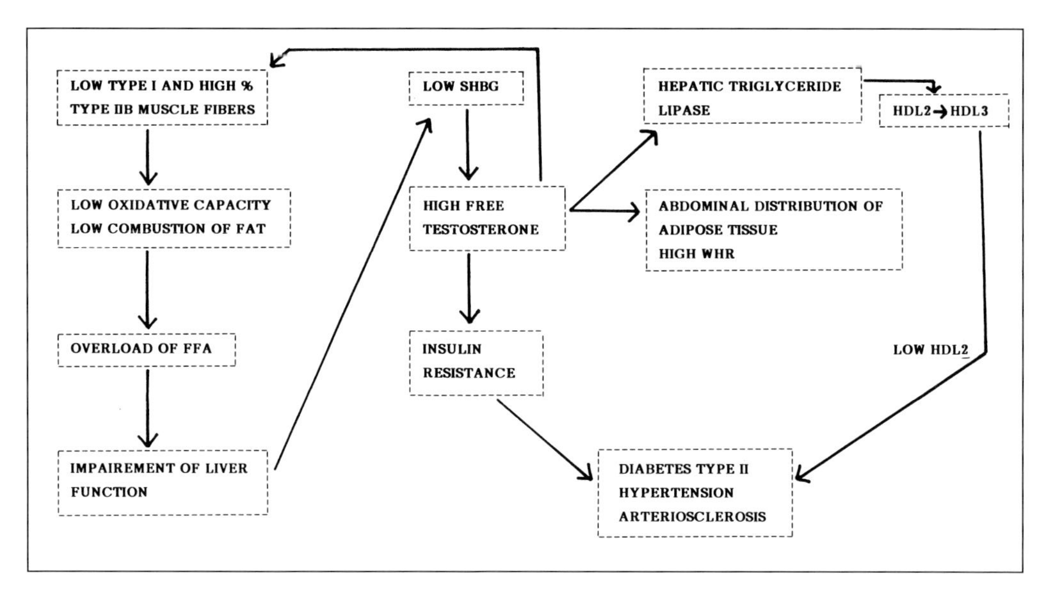

Fig. 3. Androgens, muscle morphology and the metabolic syndrome.

Furthermore, in the female obese population a higher percentage of type IIB and a lower percentage of type I was found to be significantly correlated with the concentration of free testosterone and inversely correlated with SHBG. Both the concentration of free testosterone and the percentage of type IIB muscle fibers has been found to be significantly correlated with the abdominal distribution of adipose tissue (WHR) [20] (fig. 3).

Similar observations have been made in obese patients with impaired glucose tolerance, in overweight patients with diabetes type II [16, 21–24], and by Lillioja et al. [25] in a large group of American men of varying BMI and WHR.

Recently, the percentage of type IIB muscle fibers has been found by us and the lower relative percentage of type I muscle fibers by Wade et al. [26] to be correlated with the RER. The finding has been interpreted as indicating serious impairment of the oxidation of fat during work leading to the further perpetuation of obesity [26].

Recently, examination of a large group of middle-aged men revealed several significant correlations between atherogenic factors such as Apo B, cholesterol/HDL, RER and insulin concentration (table 1).

Table 1. Correlation between percentage of type IIB muscle fibers in lateral vastus muscle and various parameters

	r	p
HDL1	-0.43	<0.01
Apo B	0.35	<0.05
Cholesterol/HDL1	0.44	<0.01
Cholesterol/HDL2	0.36	<0.05
Apo A_I/Apo B	-0.46	<0.01
Apo A_{II}/Apo B	-0.49	<0.01
RER at anaerobic threshold	0.41	<0.05
Insulin concentration	0.42	<0.01
Insulin during OGTT	0.44	<0.01

The relative percentage of type IIB fibers has been found in obese women associated with hyperlipemia and insulin resistance.
In obese men both the percentage and the relative fiber area of this type of fiber has been found to be correlated with several atherogenic factors.

Decrease of the relative percentage of type IIB muscle fibers after physical training was found [14–19, 21, 27] to be correlated with the improvement of metabolic variables, particularly the decrease in insulin concentration and the improvement of glucose disposal during insulin clamp [16, 21, 27]. Preliminary observations indicate that the administration of sulfonylurea may decrease the relative percentage of type IIB muscle fibers parallel to the improvement of insulin sensitivity.

Increased percentage of type IIB muscle fibers/decreased percent of type I muscle fibers has also been found in other conditions with increased insulin resistance such as hypertension [28], in patients with coronary heart disease [29], and Cushing's syndrome [30].

Preliminary observations also indicate an increased percent of this type of muscle fibers in PCO syndrome.

The above observations led us to the conclusion that the high percentage of type IIB muscle fibers and/or low percentage of type I muscle fibers possibly contribute in a causal manner to the clustering of risk factors observed in the states of insulin resistance and the so-called syndrome X [5], we also called this syndrome the WHR syndrome – the high WHR being present in all these conditions as well (fig. 4).

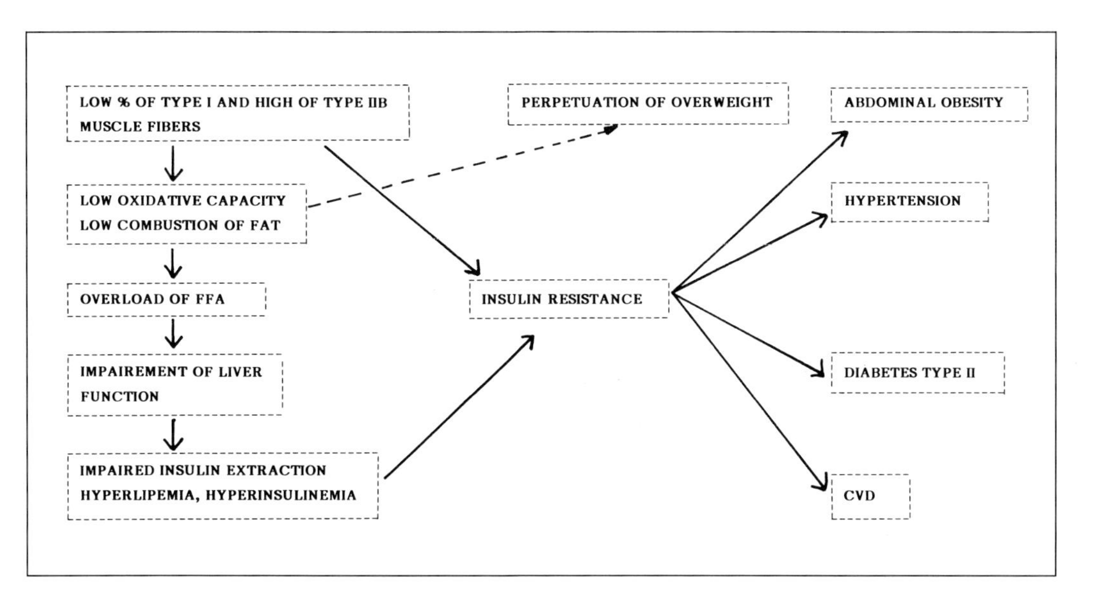

Fig. 4. Muscle morphology, fat oxidation and health hazards.

Capillary Density

Capillary density has been found to decrease with increasing body fat and to increase with physical training [17–19, 31].

Capillary density was found to be inversely correlated to plasma insulin, blood glucose concentration, and to covariate with the plasma triglyceride and muscle oxidative capacity [14–24].

According to Lithell [32], in normoglycemic middle-aged men skeletal muscle capillarization in relation to fiber size may account for about 80% of fasting insulin variation and for 70% of the variation in blood glucose at 120 min of an oral glucose tolerance test [32]. According to Lindgärde et al. [33], the weight reduction is associated with the reduction of fiber size. As the average number of capillaries around each fiber remained unchanged and the muscle fibers were found to be significantly smaller, the diffusion distance became shorter. The shortening of the diffusion distance paralleled the simultaneous improvement in metabolic variables [33]. Similar are the more recent findings of Lillioja et al. [25] who found a significant correlation between capillary density and glucose diposal during euglycemic insulin clamp [25].

Table 2. Correlation of fiber area supplied by one capillary (for type I muscle fiber)

	r	p
Body weight	0.58	<0.01
BMI	0.63	<0.001
BF	0.50	<0.01
Cholesterol/HDL1	0.52	<0.01
Cholesterol/HDL2	0.44	<0.05
Cholesterol/HDL3	0.48	<0.05
Triglycerides	0.45	<0.05
Cholesterol	0.42	<0.05
Insulin, C-peptide	0.46	<0.05

In a recent study of obese middle-aged nondiabetic men, the measure of diffusion distance, the area of muscle fiber supplied by the capillaries, was positively correlated to several atherogenic factors such as cholesterol/HDL, cholesterol/HDL2, cholesterol/HDL3, and insulin and triglycerides and cholesterol concentration.

The number of capillaries has recently been found to be surprisingly strongly decreased after the administration of testosterone to oophorectomized rats [34]. Similar trends are shown in recently obtained results in men after long-term administration of testosterone [unpubl. observations]. Table 2 shows some of the recent findings in obese men.

Respiratory Exchange Ratio

Slow-twitch type I muscle fibers are well endowed with mitochondria and oxidative enzymes and work predominantly oxidatively using fatty acids as fuel during muscle work. Fast-twitch type II, particularly type IIB, not adapted by regular exercise have fewer mitochondria, higher glycolytic capacity and use the glycolytic pathway for energy supply. The respiratory quotient (RQ) is inversely related to the proportion of fat being combusted. Regarding the differences in the substrate utilization represented by different types of muscle fibers, the different proportions of muscle fibers should be reflected by the differences in RQ (RER), especially during muscular work.

Table 3. Correlation between RER and body composition and metabolic variables

	r	p
LBM	-0.35	<0.05
BF	0.39	<0.05
HDL1	-0.40	<0.05
HDL2	-0.34	<0.05
HDL3	-0.39	<0.05
C-peptide	-0.60	<0.001
Apo A_{II}/Apo B	0.34	<0.05
Testosterone	-0.39	<0.05

Respiratory exchange ratio at the level of anaerobic threshold was found to be surprisingly high: 0.86 ± 0.03. At the level of 40% of work load corresponding ventilatory anaerobic threshold, the RER was 0.87 ± 0.04 and at 80% VAT it was 0.91 ± 0.03. The respiratory exchange ratio was not related to body weight and the waist-to-hip ratio but to the amount of body fat and negatively correlated to the lean body mass (LBM). The respiratory exchange ratio was also found to be correlated to the several atherogenic factors.

Recently, Wade et al. [26] found that the amount of body fat is significantly correlated with body fat and (in 11 men) inversely correlated to the proportion of type I muscle fibers. The authors concluded that the low combustion of fat during work perpetuates the overweight and that muscle fiber composition is an etiological factor for obesity. Similar observations have been made by our group.

The RER has been correlated to the amount of body fat, the percentage of type IIB muscle fibers, the concentration of C peptide and insulin, and inversely correlated to HDL, HDL1, HDL2, HDL3 and the ratio of Apo-A/Apo-B (table 3).

References

1 Modan M, Halkin H, Almong S, et al: Hyperinsulinemia: A link between hypertension, obesity and glucose intolerance. J Clin Invest 1985;75:809–812.

2 Nikkilä EA: High density lipoproteins in diabetes. Diabetes 1981;30(suppl):82–87.

3 Fuk MM-T, Shieh S-M, Wu DA, et al: Abnormalities of carbohydrate and lipid metabolism in patients with hypertension. Arch Intern Med 1987;147:1035–1038.

4 Haffner SM, Fong D, Hazuda HP, et al: Hyperinsulinemia, upper body adiposity and cardiovascular risk factors in non-diabetes. Metabolism 1988;37:338–345.
5 Reaven GM: Role of insulin resistance in human disease. Diabetes 1988;37:1595–1607.
6 Krotkiewski M, Lönnroth P, Mandroukas K, et al: The effects of physical training on insulin secretion and effectiveness and on glucose metabolism in obesity and type II (non-insulin dependent) diabetes mellitus. Diabetologia 1985;28:881–890.
7 Lönnroth P, DiGirolamo M, Krotkiewski M: Insulin binding and responsiveness in fat cells from patients with reduced glucose tolerance and type II diabetes. Diabetes 1983;32:748–754.
8 Larsson B, Svärdsudd K, Welin L, et al: Abdominal adipose tissue distribution, obesity and risk of cardiovascular disease and death: 13-year follow-up of participants in the study of men born in 1913. Br Med J 1984;288:1401–1404.
9 Pyörälä K: Relationship of glucose tolerance and plasma insulin to the incidence of coronary heart disease: Results from two population studies in Finland. Diabetes Care 1979;2:121–141.
10 Bergström RW, Newell-Morrish L, Leonetti DL, et al: Association of elevated fasting C-peptide level and increased intra-abdominal fat distribution with the development of NIDDM in Japanese-American men. Diabetes 1990;39:104–111.
11 Krotkiewski M, Björntorp P, Sjöström L, Smith U: Impact of obesity on metabolism in men and women: Importance of regional adipose tissue distribution. J Clin Invest 1983;72:1150–1162.
12 Fujioka S, Matsuzawa Y, Tokunaga K, Tarui S: Contribution of intraabdominal fat accumulation to the impairment of glucose and lipid metabolism in human obesity. Metabolism 1987;36:54–59.
13 Hammarsten J, Bylund-Falenius A-Ch, Holm J, Schersten T, Krotkiewski M: Capillary supply and muscle fibers types in patients with intermittent claudication: Relationship between morphology and metabolism. Eur J Clin Invest 1980;70:301–305.
14 Mandroukas K, Krotkiewski M, Hedberg M, et al: Physical training in obese women. Effects of muscle morphology, biochemistry and function. Eur J Appl Physiol 1984; 52:355–361.
15 Krotkiewski M: Physical training in the prophylaxis and treatment of obesity, hypertension and diabetes. Scand J Rehabil Med 1984;5(suppl):55–70.
16 Krotkiewski M, Björntorp P: The effects of physical training in obese women and men and in women with apple- and pear-shaped obesity; in Vague J, et al (eds): Metabolic Complications of Human Obesities. Amsterdam, Elsevier, 1985, pp 259–264.
17 Krotkiewski M, Björntorp P: Muscle tissue in obesity with different distribution of adipose tissue. Effects of physical training. Int J Obes 1986;10:331–341.
18 Krotkiewski M, Holm G: Potential importance of the muscles for the development of insulin resistance in obesity. Acta Med Scand 1988;723:95–101.
19 Seidell JC, Krotkiewski M, Björntorp P, et al: Regional distribution of muscle and fat mass in men – new insight into the risk of abdominal obesity using computed tomography. Int J Obes 1989;13:289–303.
20 Krotkiewski M, Seidell JC, Björntorp P: Glucose tolerance and hyperinsulinemia in obese women: Role of adipose tissue distribution, muscle fiber characteristics and androgens. J Intern Med 1990;228:385–392.

21 Krotkiewski M: Physical training in obesity with varying degree of glucose intolerance. J Obes Weight Regul 1985;4:179–209.
22 Mandroukas K: Physical training in obesity and diabetes; thesis, University of Gothenburg, 1982.
23 Mandroukas K, Krotkiewski M, Björntorp P, et al: The effect of long-term physical training on the relationship of muscle morphology to metabolic state and insulin sensitivity in normal and hyperglycemic obese and diabetic subjects; in Knuttgen HG, Vogel JA, Poortmans J (eds): Biochemistry of Exercise. Int Ser Sport Sciences, vol 13, 1983, pp 852–855.
24 Holm G, Krotkiewski M: Potential importance of the muscles for the development of insulin resistance in obesity. Acta Med Scand 1988;723:95–101.
25 Lillioja S, Young AA, Culter CL, et al: Skeletal muscle capillary density and fiber type are possible determinants of in vivo insulin resistance in man. J Clin Invest 1987;80:415–424.
26 Wade AJ, Maribut MM, Round JM: Muscle fiber type and aetiology of obesity. Lancet 1990;335:805–808.
27 Krotkiewski M, Holm G, Lönnroth P: Comparison of metabolic adaptations after strength and dynamic training in middle-aged men at high risk for cardiovascular morbidity. Clin Physiol 1985;5(suppl 114):4A.
28 Juhlin Dannefeldt A, Frisk-Holmberg M, Karlsson J, Tesch P: Control and peripheral circulation in relation to muscle fiber composition in normal and hypertensive men. Clin Sci 1979;56:335–340.
29 Karlsson J: Exercise capacity and muscle fiber types in effort angina. Eur Heart J 1987;8(suppl):51–57.
30 Rebuffé-Scrive M, Elfredsson J, Krotkiewski M, Björntorp P: Muscle and adipose tissue morphology and metabolism in Cushing's syndrome. J Clin Endocr Metab 1988;67:1122–1128.
31 Andersson P, Henriksson J: Capillary supply of the quadriceps femoris muscle of man: adaptive response to exercise. J Physiol 1977;270:687–691.
32 Lithell H, Lindgärde F, Hellsing K, et al: Body weight, skeletal muscle morphology and enzyme activities in relation to fasting insulin concentration and glucose tolerance in 48 year old men. Diabetes 1987;30:19–25.
33 Lindgärde F, Eriksson KF, Lithell H, Saltin B: Coupling between dietary changes, reduced body weight, muscle fiber size and improved glucose tolerance in middle aged men with impaired glucose tolerance. Acta Med Scand 1982;212:96–106.
34 Holmäng A, Svedberg J, LJennische E, Björntorp P: Effects of testosterone on muscle insulin sensitivity and morphology in female rats. Am J Physiol 1990;259:E555–E560.

Marcin Krotkiewski, MD, Department of Medical Rehabilitation,
Sahlgren's Hospital, S-413 45 Göteborg (Sweden)

Sato Y, Poortmans J, Hashimoto I, Oshida Y (eds): Integration of Medical and Sports Sciences. Med Sport Sci. Basel, Karger, 1992, vol 37, pp 416–420

Exercise, Diet and Prostaglandins[1]

Isao Hashimoto[a], *Teruichi Shimomitsu*[b], *Toshihito Katsumura*[b], *Hisao Iwane*[b]

[a]National Institute of Health and Nutrition, and
[b]Tokyo Medical College, Tokyo, Japan

Introduction

Since the discovery of prostaglandin, thromboxane, and leukotriene, much effort has been expended in studying the mechanisms of action of those compounds [1–3]. It is known that prostaglandin synthesis is influenced by disease, diet, exercise, etc. Vigorous physical exercise increases plasma and urinary prostaglandin concentrations in men [4–7] and laboratory animals [5–11]. Because the action of prostacyclin (PGI_2) and thromboxane A_2 (TXA_2) on exercise and diet has not yet been understood, this study was designed to examine the effects of acute and chronic physical exercise and diet on PGI_2, TXA_2, and prostaglandin E_2.

Methods

Acute Exercise Study

Of male participants in the triathlon (3.2 km of swimming, 162 km of bicycling, 32.2 km of running), 9 cases (body weight: 64.6 ± 0.9 kg; mean ± SEM) who volunteered to undergo our present experiment were selected as subjects for our acute exercise study. Venous blood samples were obtained 2 days before the triathlon race, right after the race and 1 day after the race. Plasma from blood of volunteers sampled in polypropylene tubes treated with 4.5 m*M* EDTA/indomethacin was immediately separated by refrigerated centrifugation. PGI_2, TXA_2 and their respective metabolites, 6-keto-$PGF_{1\alpha}$ and TXB_2 and PGE_2 were isolated with the use of an ODS column and kept at –75 °C prior to radioimmunoassay (RIA) with New England Nuclear (^3H) kits.

[1] Supported by a grant from the National Cardiovascular Center (60A-6), Osaka, Japan.

Chronic Exercise and Diet Restriction Study

Middle-aged male subjects who agreed to major items stipulated in the experimental design, participated in the experiment on diet restriction and physical training. After having undergone evaluations with medical check-ups, blood pressure monitoring, and cardiographic analysis during graded exercise testing (GXT) only 12 volunteers were accepted for the study.

Subjects were categorized into groups of diet restriction (A), diet restriction and exercise (B) and exercise alone (C), and subjected to our experiment of 12 weeks' duration. Group A (body weight, 73.3 ± 3.9 kg) was given a diet with 500 kcal energy less than the recommended dietary allowance (RDA). Group B (72.4 ± 3.8 kg) was subjected to a diet intake of minus 250 kcal in addition to 250 kcal energy expenditure by exercise. Group C (64.5 ± 4.7 kg) underwent a daily 500-kcal exercise with no restrictions on dietary intake. Groups A and B were accommodated in a hotel with a gymnasium and a 25-meter indoor swimming pool as a substitute for a metabolic ward, and meals were served with nutritional intake regulated according to the experimental diet (ordinary diet for the Japanese people, i.e. energy ratios of carbohydrate, fat and protein were about 60, 25, and 15%, respectively). As for groups B and C, exercise of 1 hour's duration with an intensity of 40–50% VO_2max was performed 5 times a week at an athletic health club according to the exercise program designated by the experimental design (physical exercise including calisthenics, stretching, bicycle ergometry, swimming, jogging, cycling, etc.). Blood analysis and determination of 6-keto-$PGF_{1\alpha}$ and TXB_2 levels were performed accordingly as described above.

Results

Triathlon race record of 9 subjects was 623 ± 23 min (mean ± SEM), including the range of 495–699 min. Body weights of subjects for 2 days before the race, right after the race, and 1 day after the race were 63.9 ± 1.9, 60.6 ± 1.6 and 62.3 ± 1.8 kg, respectively.

In the experiment on acute exercise, plasma 6-keto-$PGF_{1\alpha}$ levels increased comparably by about 58% ($p < 0.05$) and returned to the pre-exercise initial values the next morning. With regard to TXB_2, a 5% ($p < 0.01$) decrease was noted immediately after the exercise (fig. 1a). However, compared to the pre-exercise values, the PGE_2 levels were not altered. From this result, acute strenuous exercise either enhanced the synthesis or suppressed the elimination of PGI_2. In a similar manner, TXA_2 might also be explained by either suppressive effects on synthesis or enhanced elimination of TXA_2. On diet restriction/exercise in the weight loss/control experiment, mean body weight losses of 9.5 and 8.8 kg were achieved in groups A and B, respectively, though no apparent changes in body weight were noted in group C. An increase of about 15% for 6-keto-

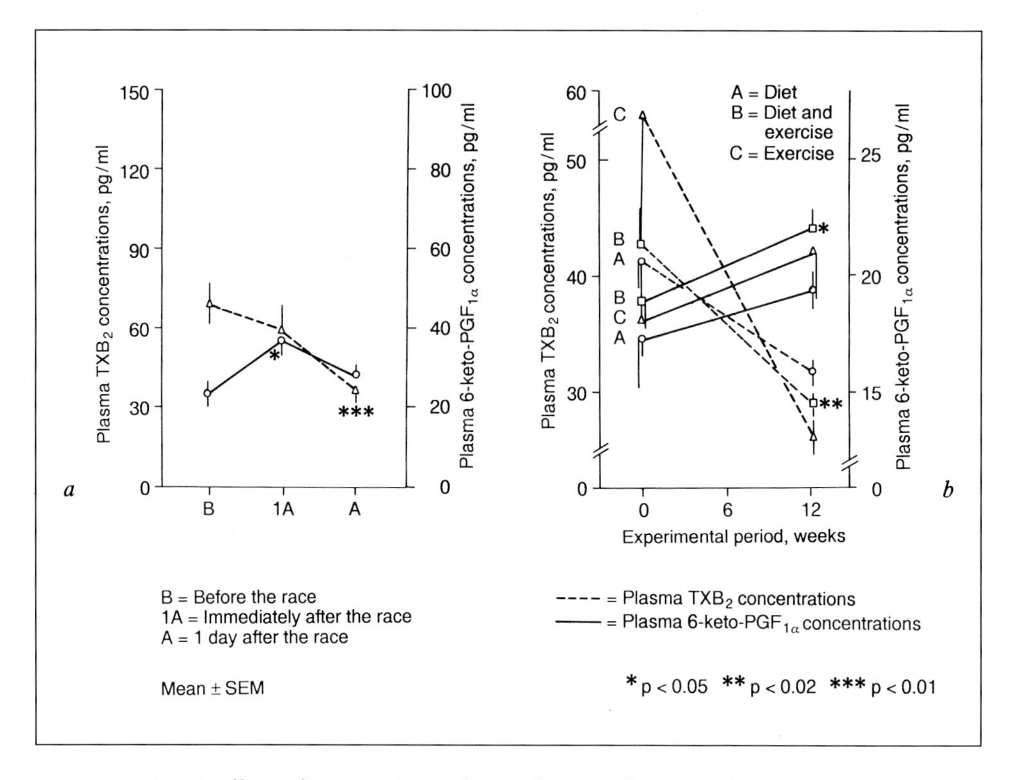

Fig. 1. Effects of acute and chronic exercise, and diet restriction on plasma 6-keto-prostaglandin $F_{1\alpha}$ (6-keto-$PGF_{1\alpha}$ and TXB_2). *a* Concentrations of 6-keto-$PGF_{1\alpha}$ and TXB_2 2 days before a triathlon race, right after the race and 1 day after the race. *b* Concentrations of 6-keto-$PGF_{1\alpha}$ and TXB_2 before and after the 12-week experiment for moderate physical training and diet restriction.

$PGF_{1\alpha}$ and a tendency to a 33% decrease for TXB_2 was achieved in all groups. However, only 6-keto-$PGF_{1\alpha}$ increased ($p < 0.05$). TXB_2 decreases ($p < 0.02$) in group B were statistically significant (fig. 1b).

Discussion

Certain organs such as lung and kidney, and vascular tissue can produce PGs, PGI_2 and TXA_2 [1, 2, 7]. In addition, there appears to be a species-dependent predisposition for producing larger quantities of one of those compounds [12]. In order to fully understand the biological roles of PGs, PGI_2 and TXA_2, it is necessary to be able to monitor their production. For PGI_2 and TXA_2, their respective hydrolysis products (6-keto-$PGF_{1\alpha}$ and TXB_2) are the parameters of choice.

Because of the opposing effects of PGI_2 and TXA_2 synthesis, there is the potential for the delicate control of hemostasis and thrombosis. That is, PGI_2 is produced by the endothelial cell lining of the vascular system, and causes vasodilation, while TXA_2 is synthesized in platelets and constricts arterial smooth muscle, promotes platelet aggregation, and induces the platelet release reaction [1, 2, 7].

The acute exercise model indicated that strenuous and exhaustive exercise caused greater production or decreased clearance of PGI_2 and lesser production or increased clearance of TXA_2. Plasma PGE_2 was unchanged. For the moderate physical training and energy restriction model, exercise and diet caused greater production or lesser clearance of PGI_2 and lesser production or greater clearance of TXA_2. These data illustrate that the delicate control of hemostasis and arterial thrombosis may shift towards less platelet aggregation and more vasodilation because of the opposing effects of changes in circulating PGI_2 and TXA_2, induced by acute and chronic exercise, and energy restriction.

References

1 Moncada S, Gryglewski R, Bunting S, Vane JR: An enzyme isolated from arteries transforms prostaglandin endoperoxides to an unstable substance that inhibits platelet aggregation. Nature 1976;263:663–665.

2 Samuelsson B, Goldyne M, Granstrom E, Hamberg M, Hammarstrom S, Malmsten C: Prostaglandins and thromboxanes. Ann Rev Biochem 1978;47:995–1029.

3 Marcus A: The role of lipids in platelet function: With particular reference to arachidonic acid pathway. J Lipid Res 1978;19:793–826.

4 Greaves MW, MacDonald-Gibson WJ, MacDonald-Gibson RG: The effect of venous occlusion starvation and exercise on prostaglandin activity in whole human blood. Life Sci 1972;1:919–924.

5 Hashimoto I: Exercise, prostaglandin E, and plasma renin activity; PhD Diss, University of Toledo, 1977.

6 Hashimoto I, Lamb DR: Immunoreactive plasma prostaglandins in men. Jpn J Phys Fit Sports Med 1980;29:1–4.

7 Nowak J, Wennmalm A: Effect of exercise on human arterial and regional venous plasma concentrations of prostaglandin E. Prostaglandins Med 1978;1:489–497.

8 Hashimoto I, Higuchi M, Yamakawa K, Suzuki S: Effect of exercise on spontaneously hypertensive rats. Med Sci Sports 1981;13:138.
9 Hashimoto I, Lamb DR: Indomethacin and exercise-induced plasma renin activity. IRCS Med Sci [Biochem] 1981;9:286–287.
10 Hashimoto I, Higuchi M, Yamakawa K: Effects of exhaustive exercise on prostaglandin metabolism in SHR rat kidney. Biochem Exerc 1983;13:657–661.
11 Zambraski EJ, Dunn MJ: Renal prostaglandin E_2 and F_2 secretion and in exercising conscious dogs. Prostaglandins Med 1980;4:311–324.
12 Hamberg M: On the formation on thromboxane B_2 and 12L-hydroxy-5,8,10,14-eicosatertraenoic acid (12 ho-20:4) in tissues from the guinea pig. Biochim Biophys Acta 1976;431:651–654.

Dr. Isao Hashimoto, Exercise Physiology Laboratory, National Institute of Health and Nutrition, 1–23–1 Toyama, Shinjuku-ku, Tokyo 162 (Japan)

Sato Y, Poortmans J, Hashimoto I, Oshida Y (eds): Integration of Medical and Sports Sciences. Med Sport Sci. Basel, Karger, 1992, vol 37, pp 421–429

A New Method for the Evaluation of Muscle Aerobic Capacity in Relation to Physical Activity Measured by Near-Infrared Spectroscopy

Takafumi Hamaoka[a], *Chris Albani*[b], *Britton Chance*[b], *Hisao Iwane*[a]

[a]Department of Preventive Medicine and Public Health, Tokyo Medical College, Tokyo, Japan; [b]Department of Biochemistry and Biophysics, University of Pennsylvania, Philadelphia, Pa., USA

In normal people and trained athletes, the training effect of primary importance is the capability to perform identical submaximal exercise with less effort. In addition to that effect, reduced myocardial oxygen consumption is also important at submaximal exercise levels in patients with coronary artery disease. Many researchers have demonstrated that peripheral factors play a favorable role in these effects [1–4]. An improvement in peripheral factors in these cases can be attributed to an increase in oxygen transport to muscle cells. Therefore, these studies indicate that one of the major objectives in physical fitness should be the improvement of oxygen transport to muscles.

However, such a method for evaluating oxygen transport to muscles has never been available.

Instead of measuring oxygen transport to muscles, two techniques allow us to estimate blood flow to muscles: a thermodilution catheter method [5, 6], and the venous plethysmography method. The catheter technique is invasive and does not distinguish between flow to working muscle, flow to nonworking muscle or flow to nonmuscular tissue. The plethysmography technique does not distinguish between muscular and nonmuscular limb flow and cannot be used to assess leg flow during upright exercise. The xenon-133 technique has also been used to measure muscle flow, but is imprecise [7]. In addition to the disadvantages of these techniques, such techniques do not monitor the oxygen transport to muscles. Another approach to assess peripheral adaptation to exercise is the muscle biopsy; morphological changes in muscles induced by exercise training, i.e. increase in capillary density, increase in the number and size of mitochondria, and so on, can be detected. However, this method is also invasive.

Our study focused on near-infrared spectroscopy (NRS), which could solve the problem of measuring muscle oxygen delivery in relation to muscle oxygen demand. The NRS unit has provided the capability for continuous, noninvasive monitoring of changes in tissue oxygen level. This original idea of Millikan [8] was further developed by Jobsis-Van der Vlieet et al. [9] and most recently by Chance et al. [10, 11].

Our purpose was to determine whether NRS can detect the level of peripheral adaptations to exercise noninvasively. To assess a level of peripheral adaptations to exercise, we have measured the half-time of oxy-hemoglobin/myoglobin (oxy-Hb/Mb) recovery ($T_{1/2}$), which indicates a repayment of the energy deficits following exercise.

Methods

Near-Infrared Spectroscopy

The NRS unit (Mini RunMan™, NIM) used for this study consisted of a probe (4.0 × 7.0 × 2.0 cm and 85 g) and a computerized control segment (14.5 × 9.0 × 4.0 cm and 240 g, fig. 1). The probe provides sensory input for the unit; it contains two broad-spectrum tungsten lights for tissue illumination and two wavelength-sensitive detectors, with the distance between the light and the detector being 3.0 cm. The frequency of the light flash could be altered by the computer software (0.9 Hz in this study). These detectors were designed to monitor the reflected light (signals) from tissue at two specific wavelengths (760 and 850 nm, 20 nm half-width, peak accuracy ± 5 nm). The light from the probe permeates the skin and enters muscle tissue. It is then either absorbed or scattered within tissue and part of the scattered light returns back through the skin to the detectors. The only significant absorption compounds are Hb and Mb. As Hb/Mb are oxygenated, the absorbance at 760 nm decreases, while the absorbance at 850 nm increases. Oxy-Hb/Mb concentration is calculated as the difference between the 760- and 850-nm signals.

The actual path of the light in tissue has been evaluated both by direct measurements [12, 13] and by Monte Carlo simulation [14]. These show that the pattern of the light path detected from input to output follows a banana-shaped figure in which the penetration depth into the tissue is approximately equal to half the distance between the light and the detector. Thus, the depth of the penetration with the probe in this study is 1–2 cm. The NRS is sensitive to changes in muscle oxygenation both at the level of the capillaries and venules and at the intracellular sites of oxygen uptake (Mb, mitochondria) [15].

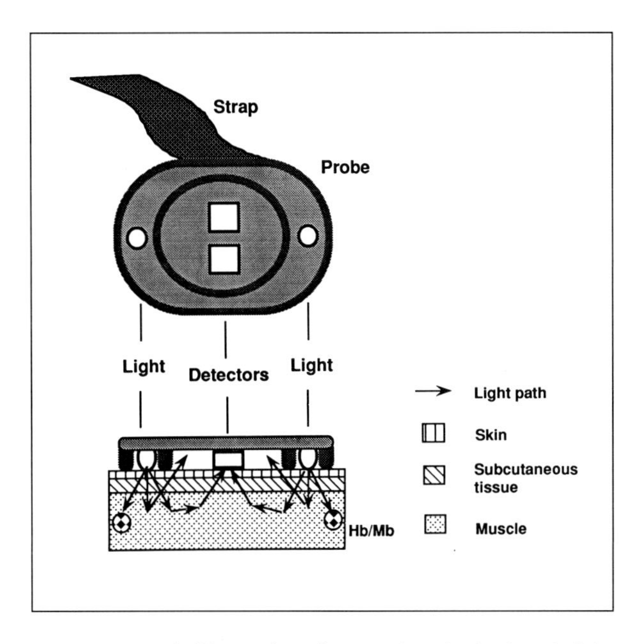

Fig. 1. Schematic illustration of the NRS probe (top) and of the optical path from the light to the detector (bottom).

An experimental model system was used to determine the use of the 760- and 850-nm wavelength. The model included a scattering medium with properties similar to muscle tissue (intralipid, 0.5–1.0%) and blood at a concentration of 10–100 μ*M*. Yeast (0.63 mg/ml) was used to deoxygenate the oxy-Hb. Adding oxygen to the yeast makes Hb convert to deoxy-Hb and vice versa repeatedly, and the addition of blood represents the in vivo increase of blood volume. According to the results of the model, optical density calculated from the subtraction of the two wavelengths is linear with the change in oxy-Hb concentration.

Reproducibility

Subjects. To determine the variation of $T_{1/2}$ measured with NRS, 5 male subjects (26–32 years of age) were tested.

Exercise Testing. The right vastus lateralis muscle was identified during an isometric contraction with keeping the knee straight. The NRS probe was placed on the flexed muscle 11–13 cm proximal from the center of the patella using Verclo strap (fig. 2). Plastic wrap was inserted between the probe and leg to prevent detectors from getting moist due to perspiration. An adequate tension in the Velcro strap was used to maintain the probe in the same position and to avoid any local ischemia on the exercising muscle. The subjects were required to stop exercise immediately following 2 min of bike exercise and to keep

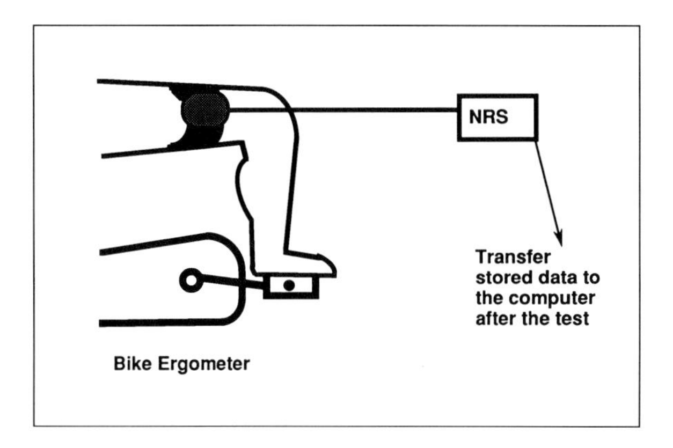

Fig. 2. Exercise test on the bike ergometer with NRS.

legs in the same position as at rest with an assistant giving support to hold the knee and thigh in a steady position. The bike used in this study was the Combi Aerobike.

The kinetics of oxy-Hb/Mb recovery were monitored for 2 min, which was enough time for the oxy-Hb/Mb concentration to recover to the resting level following the 2 min of bike exercise at 150 W. The subjects were tested 3–10 times, once a day, on different days.

Triathletes versus Sedentary Subjects

Subjects. Fifteen Japanese triathletes (29.5 ± 3.7 years, 15 males) were tested 2 days before the 1990 Hawaii Ironman race. All triathletes had trained for 15–30 h a week, for at least 3 years, and completed the race within 8–15 h (3.9 km swim, 180.2 km bicycle and 42.2 km marathon). Ten Japanese sedentary subjects (28.8 ± 3.4 years, 10 males) were tested. Sedentary subjects were untrained and performed exercise less than twice a week for durations of no longer than an hour within the last 3 years.

Exercise Testing. The workload was initially determined so that the relative systemic oxygen uptake is equivalent and relative energy deficit would be small in the two groups. The workload for triathletes was 150 W, which is equivalent to 53 ± 5% of VO_2max. The workload for sedentary subjects was 100 W, which is equivalent to 52 ± 6% of VO_2max.

The exercise procedure was the same as the test of reproducibility except for the workload of sedentary subjects. The kinetics of oxy-Hb/Mb recovery were monitored for 2 min following 2 min of bike exercise at 150 W for triathletes and 100 W for sedentary subjects.

Blood samples were taken to determine Hb concentrations, which may affect the $T_{1/2}$, in triathletes and sedentary subjects prior to the exercise test. Hb concentrations were 14.2 ± 1.2 mg/dl in triathletes and 14.5 ± 1.5 mg/dl in sedentary subjects (no statistical difference in Hb concentrations between triathletes and sedentary subjects by unpaired t test).

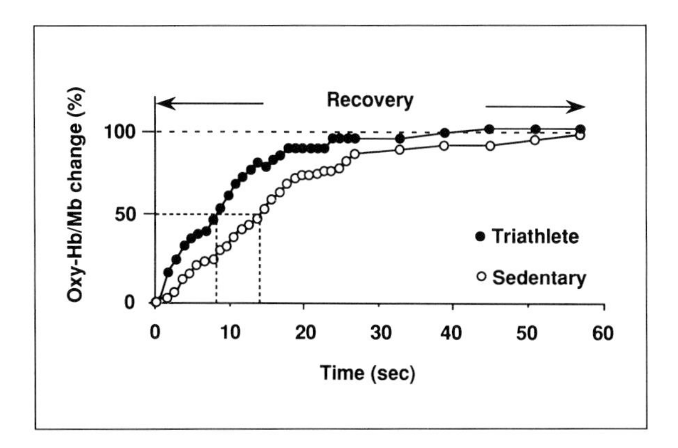

Fig. 3. Typical kinetics of oxy-Hb/Mb recovery. The measurements were made in a sedentary subject and a triathlete following 100 and 150 W of 2 min bike exercise, respectively. The half-time of oxy-Hb/Mb recovery for the sedentary subject and the triathlete was 14.8 and 8.8 s, respectively.

All the measurements were performed under controlled temperature at 22–24 °C, with a humidity condition of 50–60%. Informed consent was obtained from all of the subjects. Unpaired t test was used to determine $T_{1/2}$ differences between triathletes and sedentary subjects.

Results

Reproducibility

In 5 subjects, the value of $T_{1/2}$ ranged from 13.2 to 17.9 s. The number of tests executed varied from 3 to 10 times. One subject had the quickest recovery (13.2 s) of all the subjects. SD (in %; SD/mean × 100) ranged from 7.7 to 15.9%, which were considered reasonably reproducible.

Triathletes versus Sedentary Subjects

The typical kinetics of oxy-Hb/Mb recovery of the vastus lateralis in a triathlete and a sedentary subject are shown in figure 3. The $T_{1/2}$ for triathletes was 9.1 ± 2.0 s (ranging from 6.6 to 11 s), while that for sedentary subjects was 14.7 ± 3.2 s (ranging from 13 to 20 s). The $T_{1/2}$ for triathletes was faster than that for sedentary subjects (fig. 4; $p < 0.01$).

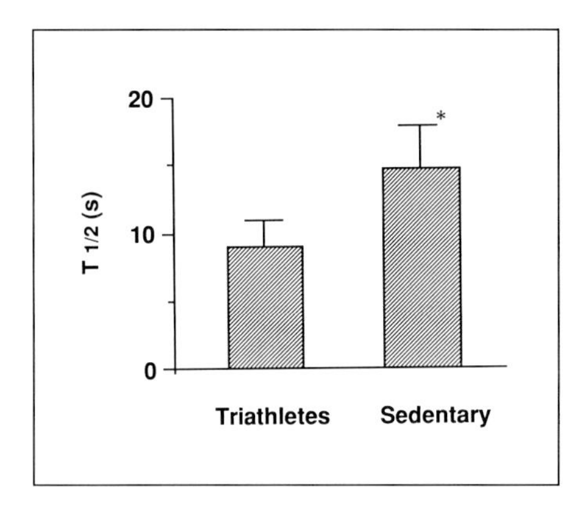

Fig. 4. Half-time of oxy-Hb/Mb recovery in triathletes and sedentary subjects. The $T_{1/2}$ for triathletes is 9.1 ± 2.0 s (mean ± SD; n = 15) and 14.7 ± 3.2 s for sedentary subjects (n = 10). * $p < 0.01$ compared to triathletes.

Discussion

We have used the half-time of Hb/Mb recovery ($T_{1/2}$) in relation to oxygen delivery and demand. The advantage of this index is that no absolute measurement is needed. The half-time has been used to evaluate recovery kinetics by many researchers in different fields [16, 17].

In the reproducibility study, one subject, a former triathlete, showed the quickest recovery of all the subjects. The subject had been training only 2–4 h per week for the last 3 years. It is interesting that the $T_{1/2}$ for the subject remained between the value of $T_{1/2}$ for triathletes and sedentary subjects. This suggests that $T_{1/2}$ may indicate the level of physical fitness, and $T_{1/2}$ could be shortened by appropriate exercise training or longer detraining.

The NRS has distinguished between triathletes and sedentary subjects with different levels of muscle adaptation to exercise. This result could be attributed to higher capillary density and larger numbers and size of mitochondria, higher oxidative enzyme activity and/or an increase in Mb content, detectable by muscle biopsy. An increase in capillary density reduces the distance between the capillary and the muscle cell, which facilitates oxygen transport to muscle cells [18]. Larger numbers and size of mitochondria also reduce the transient time of oxygen from Hb to mitochondria. Higher oxidative enzyme activity may facilitate mitochondrial respiration. An increase in Mb content will enhance the diffusion of oxygen [19]. Brodal et al. [20] showed that the mean number of capillaries per muscle fiber was greater in a trained versus an untrained group by 32%, and was almost of the same order of magnitude as the difference in VO_2max between the two groups. Henriksson and Reitman [21] showed that the activity of succinate dehydrogenase and cytochrome oxidase from the biopsy of the vastus lateralis had increased by 32 and 35%, respectively, after 8–10 weeks endurance training. The VO_2max increased by 19%. The $T_{1/2}$ in the triathletes was 34% faster than that in sedentary subjects. This magnitude of difference seems to be comparable to the capillary density differences between trained and untrained observed by Broadal et al. [20] and to oxidative enzyme changes after training studied by Henriksson and Reitman [21].

Our preliminary study observed the relationship between $T_{1/2}$ and VO_2max (ml/kg/min) with correlation coefficients of −0.75 in 7 normal subjects when every subject exercised at the same workload of 150 W (70–80% of VO_2max). When exercise was performed at higher workloads which produce higher energy deficits, the $T_{1/2}$ could reflect systemic maximal oxygen uptake. A plausible explanation is that a subject with greater muscle aerobic capacity could recover faster because the energy deficit produced during exercise for the subject is relatively smaller. The subject with greater muscle aerobic capacity accompanied by enhanced cardiac function has less bioenergetic stress induced by exercise.

Absolute concentration changes of Hb/Mb cannot be measured because of nonquantifiable biophysical quantities such as optical pathlength in accordance with the Beer-Lambert law. The kinetics of oxy-Hb/Mb during exercise can be assessed after the limit of absolute concentration change is determined; when the cuff tourniquet is used following exercise at pressures exceeding the arterial values for calibration, changes in Hb/Mb are expressed relative to overall change in the signal noted after cuff inflation. The signal after cuff inflation was considered to represent physiological maximal Hb and Mb deoxygenation.

Wilson et al. [22] have used the cuff tourniquet as a physiological calibration in the study of patients with heart failure measured by NRS. The parameter they used is relative Hb/Mb oxygenation change to the overall change in the signal noted after cuff inflation. The results showed that relative Hb/Mb oxygenation in the patients decreased more than in the normal subjects throughout exercise. The NRS might be used to monitor the training effects on muscles in patients as well as in healthy people.

In conclusion, NRS could be utilized to assess the bioenergetic stress induced by exercise and to detect muscle adaptations. The advantages of the measurements are: (1) it is noninvasive; (2) it is compact, cheap and easy to operate; (3) it requires short measurement time; (4) it requires no maximal exercise, and (5) repeated measurements are allowed. The NRS could be useful to monitor the level of physical fitness in sedentary people and athletes as well as in patients.

References

1 Clausen JP, Jensen JT: Effects of training on the distribution of cardiac output in patients with coronary artery disease. Circulation 1970;42:611–624.
2 Varnauskas E, Bjorntorp P, Fahlen M, Prerovsky I, Stenberg J: Effects of physical training on exercise blood flow and enzymatic activity in skeletal muscle. Cardiovasc Res 1970;4:418–422.
3 Tabkin BS, Hanson JS, Levy AM: Effects of physical training on the cardiovascular and respiratory response to graded upright exercise in distance runners. Br Heart J 1965;27:205.
4 Detry JMR, Rousseau M, Vandenbroucke G, Kusumi F, Brasseur LA, Bruce RA: Increased arteriovenous oxygen difference after physical training in coronary heart disease. Circulation 1971;44:109–118.
5 Pavek K: Measurement of cardiac output by thermodilution with constant rate injection of indicator. Circ Res 1964;15:311.
6 Forrester JS: Thermodilution cardiac output determination with a single flow-directed catheter. Am Heart J 1972;83:306.
7 Cerretelli P, Marconi C, Pendergast D, Meyer M, Heisler N, Piiper J: Blood flow in exercising muscles by xenon clearance and by microsphere trapping. J Appl Physiol 1984;56:24–30.
8 Millikan GA: Experiments on muscle hemoglobin in vivo: The instantaneous measurement of muscle metabolism. Proc R Soc 1937;123:218–241.
9 Jobsis-VanderVlieet FF, Pinantadosi CA, Silvia AL, Lucas SK, Keizer JH: Near infrared monitoring of cerebral oxygen sufficiency. Neurol Res 1988;10:7–17.
10 Chance B, Nioka S, Kent J, McCully K, Fountain M, Greenfeld R, Holtom G: Time resolved spectroscopy of hemoglobin and myoglobin in resting and ischemic muscle. Anal Biochem 1988;174:698–707.
11 Chance B, Conard H: Acid-linked function of intermediates in oxidative phosphorylation. J Biol Chem 1959;234:1568–1570.
12 Chance B, Maris M, Sorge J, Zhang MZ: A phase modulation system for dual wavelength difference spectroscopy of hemoglobin deoxygenation in tissue. Proc Soc Photo-Opt Instrum Eng 1990;1204:481–491.
13 Delpy DT, Cope M, van der Zee PP, Arridge S, Wrary S, Wyatt J: Estimation of optical pathlength through tissue from direct time of flight measurement. Phys Med Biol 1988;33:1422–1442.

14 Patterson MS, Chance B, Wilson BC: Time resolved reflectance and transmittance for the noninvasive measurement of tissue optical properties. J Appl Opt 1989;28: 2331–2336.
15 Jobsis FF: Intracellular metabolism of oxygen. Am Rev Respir Dis 1974;110(suppl): 58–63.
16 Zatina MA, Berkowitz HD, Chance B, Maris JM, Gross GM: ^{31}P nuclear magnetic resonance spectroscopy: Noninvasive biochemical analysis of the ischemic extremity. J Vasc Surg 1986;3:411–420.
17 Chance B, Conrad H: Acid-linked function of intermediates on oxidative phosphorylation. J Biol Chem 1959;234:1568–1570.
18 Saltin B: Hemodynamic adaptations to exercise. Am J Cardiol 1985;55:42.
19 Livingston DJ, Lamar GN, Brown WD: Myoglobin diffusion in bovine heart muscle. Science 1983;220:71.
20 Brodal P, Ingier F, Hermansen L: Capillary supply of skeletal muscle fibers in untrained and endurance-trained men. Am J Physiol 1977;232:H705.
21 Henriksson J, Reitman LS: Time course of changes in human skeletal muscle succinate dehydrogenase and cytochrome oxidase activities and maximal oxygen uptake with physical activity and inactivity. Acta Physiol Scand 1977;99:91–97.
22 Wilson JR, Mancini DM, McCully K, Ferraro N, Lanoce V, Chance B: Noninvasive detection of skeletal muscle underperfusion with near-infrared spectroscopy in patients with heart failure. Circulation 1989;80:1668–1674.

Takafumi Hamaoka, MD, Department of Preventive Medicine and Public Health, Tokyo Medical College, 6-1-1, Shinjuku, Shinjuku-ku, Tokyo 160 (Japan)

Sato Y, Poortmans J, Hashimoto I, Oshida Y (eds): Integration of Medical and Sports Sciences. Med Sport Sci. Basel, Karger, 1992, vol 37, pp 430–438

Effect of Exercise on Plasma Lipoprotein Metabolism[1]

Toshitaka Tamai[a], *Mitsuru Higuchi*[b], *Koji Oida*[a], *Tsuguhiko Nakai*[a], *Susumu Miyabo*[a], *Shuhei Kobayashi*[b]

[a]The Third Department of Internal Medicine, Fukui Medical School, Fukui; [b]Division of Health Promotion, National Institute of Health and Nutrition, Tokyo, Japan

Major advances have been made in our understanding of both the pathophysiology and metabolism of plasma lipoproteins during the last 20 years. It has been reported that endurance-type exercise induces less atherogenic profiles in plasma lipids and lipoproteins. Previous studies have shown that endurance physical exercise decreases plasma (P), cholesterol (Ch), plasma triglyceride (TG), very-low-density lipoprotein (VLDL)-TG, and low-density lipoprotein (LDL)-Ch, and increases high-density lipoprotein (HDL)-Ch in young and middle-aged men [1, 2]. In the present study, we investigated the plasma lipoprotein metabolism in trained elderly subjects, effects of physical exercise on plasma apolipoprotein (Apo) B isoprotein, Apo B-48 metabolism and lipoprotein (a) (Lp(a)) metabolism, and effects of mild physical exercise on plasma lipoprotein metabolism in hyperlipoproteinemic subjects.

Subjects and Methods

Runners and sedentary controls of both sexes from 30 to 70 years of age volunteered for this study. Each person was informed of the design and risks of this project prior to obtaining a written consent. No subject had suffered any disorders that would influence

[1] The authors are grateful to Drs. Bob Spina and John O. Holloszy, St. Louis, for valuable discussion, to Ms. Naoyo Yamaguchi for excellent technical assistance, and to Ms. Miho Maeda for the preparation of the manuscript.

the lipoprotein metabolism. No subject took a specific diet or drug. All subjects completed a progressive test on a motor-driven treadmill to determine their $\dot{V}_{O_2 max}$. Total body fat composition was estimated by the method using skinfold thickness according to Nagamine and Suzuki [3].

Blood was obtained from the subjects, who had fasted for more than 14 h. Plasma lipoproteins were fractionated by sequential ultracentrifugation at densities of 1.006, 1.019, 1.063, and 1.125 g/ml. The 1.006 g/ml top fraction, the 1.125 g/ml bottom fraction, and the fractions of 1.006–1.019 g/ml, 1.019–1.063 g/ml, and 1.063–1.125 g/ml were designated VLDL, HDL_3, intermediate density lipoprotein (IDL), LDL, and HDL_2, respectively. Only in result (IV) were plasma lipoproteins fractionated by ultracentrifugation at a density of 1.006 g/ml for 22 h at 105,000 *g*. The infranatant solution was applied to heparin-Mn^{2+} precipitation method according to the method of Lipid Research Clinics Program [4] for separation of HDL. Ch and TG were measured by enzymatic methods. Plasma concentrations of Apo A-I, A-II, B, C-II, C-III, and E were measured by a single-radial immunodiffusion method using 1% agarose plate that contained specific goat antisera [5].

Apo B isoproteins in TG-rich lipoproteins, density < 1.006 g/ml, were analyzed by slab-gel electrophoresis in 3.5% polyacrylamide gels containing 0.1% SDS. TG-rich lipoprotein fraction was dialyzed extensively against 1 m*M* EDTA and lyophilized. TG-rich lipoproteins, 100 μg protein, were dissolved in the buffer containing 2.5% SDS and 5% mercaptoethanol and boiled for 1 min at 100 °C. Then, samples were applied to the gels and electrophoresed. Gels were stained with 0.2% Coomassie brilliant blue. The Apo B bands were identified by immunoblotting using anti-Apo B polyclonal antiserum. Apo B-100 and Apo B-48 were identified from their molecular weights which were calculated by comparison with the electrophoretic mobilities of standard proteins. After destaining, the relative quantity of Apo B-48 and Apo B-100 bands was analyzed by densitometry at 565 nm. The Apo B-48 ratio which is the percentage of Apo B-48 in total Apo B was calculated [6].

Cut-off levels of P-Ch and P-TG for elimination of hyperlipoproteinemia were 220 and 150 mg/dl, respectively, which were recommended by the Japanese Atherosclerosis Society.

Results and Discussion

Plasma Lipoprotein and Apolipoprotein Concentrations

Table 1 represents the profiles of male volunteers studied and concentrations of lipoproteins and apolipoproteins. There were no significant differences in age, height, and body mass index between runners and controls of old and young groups. There was a significant difference only in body weight between the old groups. Although a significance was not observed in body mass index, there were significant differences in total body fat estimated from the skinfold measurements between old runner and old control, and between young runner and young control. Significant differ-

Table 1. Profile of volunteers and plasma concentrations of lipid, lipoprotein and apolipoprotein

	Old runners (n = 12)	Old controls (n = 12)	Young runners (n = 16)	Young controls (n = 15)
Age	62.5±0.7	64.0±0.9	34.7±0.7	34.0±0.7
Body mass index, kg/m^2	21.1±0.4	22.6±0.7	21.1±0.4	22.1±0.6
Estimated body fat, %	12.0±1.9*	15.6±3.9	11.6±1.6*	15.5±2.5
$\dot{V}_{O_2\,max}$, ml/min/kg	50.4±1.2*	30.6±1.0	63.8±0.5*	49.6±1.8
Plasma-Ch, mg/dl	207.7±8.3	211.2±9.4	191.3±6.2	194.1±7.2
Plasma-TG, mg/dl	76.2±6.5	106.2±14.6	61.3±3.1	88.0±8.0
VLDL-Ch, mg/dl	4.6±1.1*	10.9±2.7	4.1±0.6	6.9±1.2
VLDL-TG, mg/dl	23.0±4.8	47.8±13.1	16.8±2.0*	33.0±6.0
LDL-Ch, mg/dl	134.7±8.1	149.0±8.8	122.4±5.3	130.2±6.4
HDL-Ch, mg/dl	68.4±4.7**	51.3±3.1	61.4±2.2	54.8±2.4
Apo A-I, mg/dl	156.1±4.7*	143.7±3.6	148.5±4.1	146.5±4.1
Apo A-II, mg/dl	34.8±1.2*	31.8±0.7	36.3±0.9	34.7±1.0
Apo B, mg/dl	77.2±4.2	79.6±5.0	73.8±1.8	79.1±2.5
Apo C-II, mg/dl	3.8±0.2	4.4±0.4	4.2±0.2	4.2±0.3
Apo C-III, mg/dl	10.0±0.5	9.5±0.8	9.8±0.5	9.1±0.5
Apo E, mg/dl	3.6±0.2	4.2±0.2	4.3±0.2*	3.8±0.2
LDL-Ch/HDL-Ch	2.09±0.18**	3.04±0.27	2.05±0.14	2.44±0.14
Apo AI+AII/Apo B	2.65±0.30	2.29±0.12	2.52±0.08	2.31±0.07

Each value represents mean ± SEM.
* $p < 0.05$, ** $p < 0.01$ vs. control.

ences were also observed in $\dot{V}_{O_2\,max}$ between runners and controls of old and young groups. The old runner and young runner groups exercised 5–6 days a week for 3–5 years and ran 44±23 and 84±31 km a week, respectively.

Despite the absence of significant differences in plasma lipid concentration, old runners and young runners had decreased levels of VLDL-Ch and LDL-Ch and increased levels of HDL-Ch, Apo A-I, and Apo A-II. Ratios of LDL-Ch/HDL-Ch were significantly lower in old runners than in old controls, and young runners showed a tendency to lower ratios of LDL-Ch/HDL-Ch than young controls. Apo A-I + A-II/Apo B ratios in old and young groups showed a tendency to be higher in runners than in controls. These data indicated that physical exercise might have favorable effects on plasma lipoproteins and apolipoproteins, and atherogenic index, such as LDL/HDL.

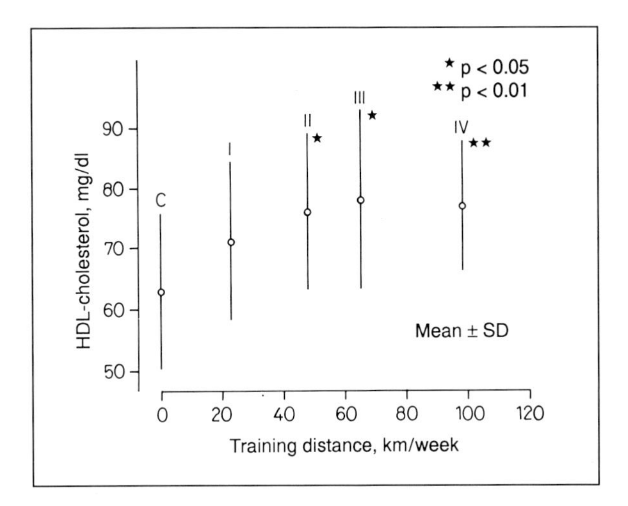

Fig. 1. Relationship between HDL-Ch concentration and running distance. C, I, II, III and IV represent untrained group (control) and grade I, II, III and IV of runner groups, respectively. * $p < 0.05$, ** $p < 0.01$ vs. control.

HDL-Ch and Running Distance

Figure 1 shows the relationship between HDL-Ch concentration and running distance. Trained subjects were divided into 4 subgroups, grades I–IV, according to running distance per week. Running distance of the grade I group was 24 ± 9 km/week. Grades II, III and IV runner groups averaged 2, 3 and 4 times of the grade I group in running distance per week, respectively. There were no significant differences in the concentrations of P-Ch, P-TG and LDL-Ch among the groups. Runner groups had significantly higher concentrations of HDL-Ch compared with the untrained groups. HDL-Ch concentrations in the untrained and the trained groups, grades I, II, III and IV were 63, 71, 76, 78 and 77 mg/dl, respectively. This study suggested that regularly performed endurance running could increase the HDL-Ch concentration concomitantly with increase of running distance up to 50–60 km/week, but further increase in HDL-Ch could not be expected by an additional increase of training distance.

Apo B-48

Apo B-48 is one of Apo B isoproteins. Chylomicron, which originates in the intestine, contains Apo B-48 while VLDL, which is synthesized in the liver, contains Apo B-100. Apo B-48 is a valuable marker of exogenous

Table 2. Lp(a) concentrations in runner and control

	Runner (n = 10)	Control (n = 21)	Significance
Lp(a), mg/l	164.7±71.5	79.1±20.3	n.s.*
P-Ch, mg/dl	132.5±5.4	168.1±6.7	$p < 0.02$
P-TG, mg/dl	66.3±9.8	97.3±9.6	$p < 0.05$
HDL-Ch, mg/dl	44.5±5.6	38.9±2.2	n.s.
LDL-Ch, mg/dl	78.4±7.6	111.6±7.6	$p < 0.01$
Apo AI, mg/dl	131.9±12.0	140.6±4.5	n.s.
Apo B, mg/dl	65.0±6.1	95.9±5.1	$p < 0.002$

Mean ± SEM.
* Nonparametric analysis.

triglyceride-rich lipoprotein metabolism. Plasma lipids, lipoproteins, apolipoproteins, and VLDL-Apo B isoproteins were studied in 19 fasted female runners (R) and 23 age-matched sedentary control women (S). Plasma concentrations of Ch, Apo A-I, and Apo E were significantly higher in runners than controls. In lipoprotein fractions, runners had lower VLDL concentrations and higher HDL concentrations mainly due to higher levels of HDL_2 than controls. Apo B-48 ratios in VLDL were significantly ($p < 0.05$) lower in runners (2.86 ± 0.39%) as compared to controls (4.6 ± 0.54%). A significantly negative correlation between Apo B-48 ratio and HDL_2-Ch concentration was observed in the runner but not in the control group. However, there were no significant correlations between Apo B-48 ratio and HDL_3-Ch concentration in both the runner and the control groups. The runners showed a tendency to a positive correlation between Apo B-48 ratio and VLDL-TG concentration, and showed a significantly negative correlation between Apo B-48 and HDL_2-Ch. These data indicated that endurance exercise accelerated exogenous lipoprotein metabolism in runners which was most likely mediated by the increase of lipoprotein lipase activities.

Lipoprotein(a)

Lp(a) was discovered by Berg [7] more than 25 years ago. He identified in human plasma Lp(a) genetically transmitted and associated with an increased risk for atherosclerotic cardiovascular disease. The liver is the main and perhaps the sole site of Lp(a) synthesis. As far as the mechanism of degradation of Lp(a) is concerned, a number of conflicting views have been reported.

Recently, Hellsten et al. [8] reported the effect of heavy physical activity on Lp(a) concentration in the longitudinal study. Sixteen physically active, nonsmoking men, aged 18–45 years, had voluntarily applied for participation in 8 days of cross-country skiing in the Swedish mountains. The very heavy physical activity, such as this cross-country skiing, significantly decreased Lp(a) concentration from 166 to 129 mg/dl.

We have measured plasma Lp(a) concentration of 10 runners and 21 healthy controls who lived in St. Louis, Mo. (table 2). Runners had significantly lower concentrations of P-Ch, P-TG, LDL-Ch and Apo B, and significantly higher concentrations of HDL-Ch than healthy controls. However, there was no significant difference in Lp(a) between runners and controls.

Extremely heavy physical activity, such as cross-country skiing, decreased Lp(a) concentration. However, there is very little evidence that regular running or exercise with mild intensity has a significant effect on Lp(a) metabolism. Further investigations are necessary to clarify the relationship between physical activity and Lp(a) concentration.

The Effects of Mild Physical Exercise on Plasma Lipoprotein Metabolism in Type IV Hyperlipoproteinemia

We studied the effect of mild exercise on plasma lipoprotein in 7 patients with hypertriglyceridemia. Patients with hyperlipidemia may be unable to perform regular running, because of complications such as coronary heart disease, hypertension, etc.

Figure 2 shows the changes of P-TG (left) and P-Ch (right) by physical exercise of more than 10,000 steps walking per day. P-TG concentration was decreased from 269 to 184 mg/dl significantly, but P-Ch did not show significant changes. Both VLDL-TG and VLDL-Ch were significantly decreased after walking exercise.

Table 3 shows significance in plasma lipoprotein and apolipoprotein changes by walking exercise. Walking exercise decreased VLDL, IDL and Apo E. However, it did not affect LDL, HDL, Apo A and Apo B.

It has been reported that physical exercise affects activities of enzymes involved in lipoprotein metabolism. Exercise increases lipoprotein lipase (LPL) [9] and lecithin:cholesterol acyltransferase (LCAT) activities [10]. LPL hydrolyzes triglyceride in both chylomicron and VLDL. As the deli-

Table 3. Significance in plasma lipoprotein and apolipoprotein changes by exercise

		Significance
T-Ch	Plasma	n.s.
	VLDL	↓ $p < 0.05$
	IDL	n.s.
	LDL	n.s.
	HDL_2	n.s.
	HDL_3	n.s.
TG	Plasma	↓ $p < 0.01$
	VLDL	↓ $p < 0.02$
	IDL	↓ n.s. ($p < 0.1$)
	LDL	n.s.
	HDL_2	n.s.
	HDL_3	↓ $p < 0.02$
PL	Plasma	n.s.
	VLDL	↓ $p < 0.05$
	IDL	↓ $p < 0.02$
	LDL	n.s.
	HDL_2	n.s.
	HDL_3	n.s.
Apolipoprotein	A I	n.s.
	A II	n.s.
	B	n.s.
	C II	n.s.
	C III	n.s.
	E	↓ $p < 0.02$

pidation of chylomicron results in the production of CM-R which are taken up by the liver, LPL accelerates the metabolism of exogenous lipoproteins containing Apo B-48. LPL also accelerates the metabolism of the endogenous lipoprotein pathway from VLDL to LDL. Increased LPL activities decrease the concentration of atherogenic lipoproteins such as LDL, IDL, VLDL and chylomicron remnants. LCAT increased by exercise stimulates metabolic process from nascent HDL to HDL_3. Exercise decreases hepatic triglyceride lipase (H-TGL) which converts HDL_2 to HDL_3 [11]. Increased activities of LPL and LCAT, and decreased activities of H-TGL result in the increment of HDL_2 concentration.

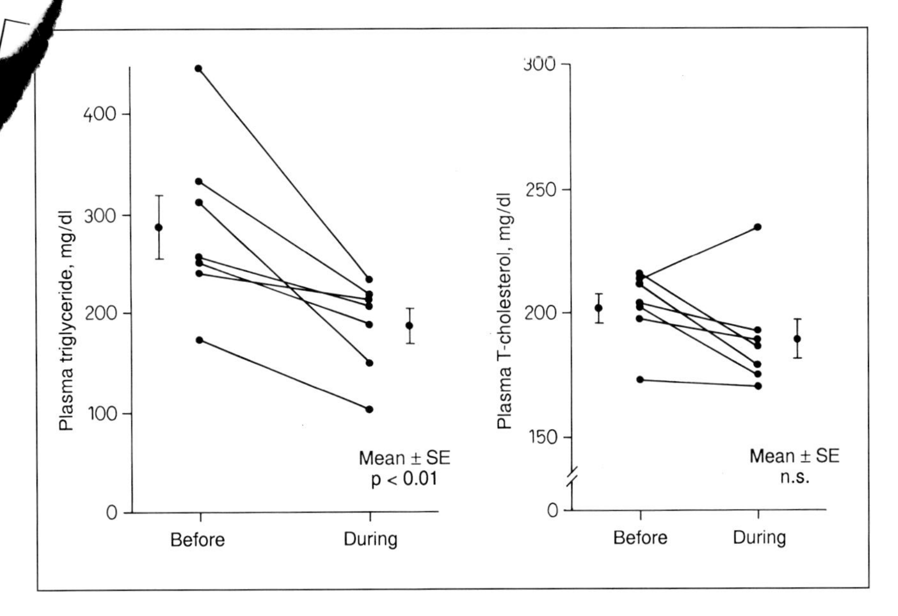

Fig. 2. Changes of plasma triglyceride and total cholesterol concentrations after walking exercise.

Conclusions

Physical exercise with mild intensity, such as walking more than 10,000 steps per day, was effective in decreasing VLDL, but not effective in decreasing LDL, and increasing HDL-Ch and Apo A-I.

On the other hand, intensive exercise, such as running more than 10–20 km per week increased HDL-Ch and Apo A-I, and decreased not only VLDL but also LDL significantly.

References

1 Kiens B, Lithell H, Vessby B: Further increases in high density lipoproteins in trained males after enhanced training. Eur J Appl Physiol 1984;52:426–430.

2 Leclerc S, Allard C, Talbot J, Gauvin R, Bouchard C: High density lipoprotein cholesterol, habitual physical activity and physical fitness. Atherosclerosis 1985;57: 43–51.

3 Nagamine S, Suzuki S: Anthropometry and body composition of young Japanese men and women. Hum Biol 1964;36:8–15.

4 Lipid Research Clinics Program: Manual of laboratory operations. Lipid and lipoprotein analysis. US Department of Health, Education and Welfare. Publ No. (NIH) 75-628, 1974.
5 Goto Y, Akanuma Y, Harano Y, et al: Determination by the SRID method of normal values of serum apolipoproteins (A-I, A-II, B, C-II, C-III, and E) in normolipidemic healthy Japanese subjects. J Clin Biochem Nutr 1986;1:73–88.
6 Takai H: Metabolism of apolipoprotein B-48 containing triglyceride-rich lipoproteins in non-insulin-dependent diabetes mellitus: The effect of glycemic control. J Jpn Diabetes Soc 1990;33:191–197.
7 Berg K: A new serum type system in man. The Lp system. Acta Pathol Microbiol 1963;59:369–382.
8 Hellsten G, Boman K, Hallmans G, Dahl'en G: Lipids and endurance physical activity. Atherosclerosis 1989;75:93–94.
9 Nikkila EA, et al: Lipoprotein lipase activity in adipose tissue and skeletal muscle of runners: Relation to serum lipoproteins. Metabolism 1987;27:1661.
10 Marniemi J, et al: Dependence of serum lipid and lecithin: Cholesterol acyltransferase levels on physical training of young men. Eur J Appl Physiol 1982;49:25.
11 Kraus RM, et al: Heparin released plasma lipase activities and lipoprotein levels in distance runners (abstract). Circulation 1979;60:73.

Toshitaka Tamai, MD, Fukui Medical School, The Third Department of Internal Medicine, Matsuoka-cho, Fukui 910–11 (Japan)

Subject Index